The Heritage of Copernicus:
Theories "Pleasing to the Mind"

The Copernican Volume of the National Academy of Sciences

The Heritage of Copernicus:
Theories "Pleasing to the Mind"

edited by Jerzy Neyman

The MIT Press
Cambridge, Massachusetts, and London, England

The preparation of the material for this volume was made possible by generous grants from the Alfred Jurzykowski Foundation, Inc., the Rockefeller Foundation, and the Alfred P. Sloan Foundation.

This book was set in Linotype Baskerville
by The Colonial Press Inc.,
and printed and bound
by The Alpine Press Inc.
in the United States of America

First MIT Press paperback edition, 1977

Library of Congress Cataloging in Publication Data

Neyman, Jerzy. 1894–
    The heritage of Copernicus: theories "pleasing to the mind."

    "The Copernican volume of the National Academy of Sciences."
    1. Science—History—Addresses, essays, lectures. 2. Copernicus, Nicolaus, 1473–1543.
I. Title.
Q126.8.N48   509   74–6415
ISBN 0–262–14021–7 (hardcover)
ISBN 0–262–64016–3 (paperback)

Contents

Preface

This book represents the implementation of a decision adopted by the Council of the National Academy of Sciences relating to the celebration of the 500th anniversary of the birth of Nicholas Copernicus. From the outset it was intended that this Copernican volume would describe a number of Copernican-type intellectual revolutions that have taken place in recent centuries. Such revolutions are characterized by the abandonment of widely held concepts and replacement by dramatically new conceptualizations that resulted in deepened understanding of natural processes.

It was the original intention of the Academy's Council that the essays would be addressed to the general educated public, including teaching staffs of colleges and high schools and college students. The combination of the intended broad coverage and the contemplated readership presented considerable difficulties and affected the choice of incidents in the history of modern science which could be discussed. As it has turned out, the material presented is quite rich and, while there is an unavoidable variation in the amount of specialized knowledge expected from the reader, we hope that the summaries preceding particular chapters are sufficient to create a broad perspective and make the whole understandable. To the extent that our goals have been approached, the help of numerous scientists is gratefully acknowledged.

The introductory chapter describes the circumstances and the essence of the initial intellectual revolution due to Copernicus himself. The subsequent six parts present in turn: (1) several revolutions in astronomy–cosmology, all in the twentieth century and one of them still in progress; (2) a number of deep revolutionary changes in the biological sciences, one of them made twenty years ago by an inspired contribution of a graduate student; (3) several revolutionary changes in the physical sciences; (4) revolutions in the thinking of mathematicians; (5) the emergence of a "pluralistic" point of view in scientific research; and (6) revolutions in technology.

The epoch when it was possible for individuals to acquire a reasonably detailed view of practically all of science is now far behind us. It was followed by a prolonged period of compartmentalization of research, with specialists in one domain becoming increasingly ignorant of developments in another. In the present epoch the pendulum appears to be moving in the opposite direction. The many problems faced by modern society depend for their solution on interdisciplinary studies and on participation of individuals with wide horizons. It is hoped that the publication of the present volume will serve not only to commemorate the achievements of Nicholas Copernicus but also to stimulate and accelerate this trend.

This volume was originally proposed by Professor Jerzy Neyman of the Department of Statistics at the University of California at Berkeley while serving as Vice Chairman of our Academy's Special Committee for the Celebration of the Copernicus Quinquecentennial. In view of his un-bounded enthusiasm, he was appointed by Committee Chairman Antoni Zygmund of the University of Chicago as head of the Editorial Board for the Copernican volume. In that capacity he was assisted by fellow Board members Melvin Calvin and Emilio Segrè of the University of California at Berkeley, Nicholas U. Mayall of the Kitt Peak National Observatory, C. R. O'Dell of Yerkes Observatory and NASA's G. C. Marshall Space Flight Center, S. M. Ulam of the University of Colorado, and Antoni Zygmund, *ex officio*. In meeting his responsibilities as principal editor, Professor Ney-man also received advice from many other academic colleagues to whom the Academy's warm appreciation is also extended. As is inevitable in such an enterprise, the hard editorial decisions and close attention to text rested with Professor Neyman, without whose remarkable vitality and inspiration this volume would never have appeared.

Harrison Brown
Foreign Secretary
National Academy of Sciences
Washington, D.C. 20418

Jerzy Neyman

1 Introduction: Nicholas Copernicus (Mikolaj Kopernik): An Intellectual Revolutionary

Introduction

According to well-established tradition, Nicholas Copernicus was the celebrated scientist who, some centuries ago, originated the idea that the earth moves around the sun, and not vice versa, and thereby laid the foundations of modern astronomy. The actual situation was much more complex.

To begin with, Copernicus was not the first to advocate the centrality of the sun. Aristarchus of Samos, eighteen centuries before Copernicus, and later Pythagoreans had the same idea. Copernicus was familiar with some of the rather vague writings of these philosophers and referred to them in his own works. Moreover, the astronomical work of Copernicus was soon eclipsed by that of Kepler, Galileo Galilei, and finally Newton. From the strictly astronomical point of view, the work of Copernicus was limited to developing numerical details of the heliocentric (sun-centered) system of planetary motions that allowed fairly accurate prediction of the future positions of the moon and of the five then-known planets. This work is published in the book *De Revolutionibus Orbium Caelestium,* which took Copernicus something like thirty years to complete.[1]

In the second century of the Christian era, similar work was done by Ptolemy based on a geocentric (earth-centered) system of planetary motions. His book is known under the original Greek title *Syntaxis Mathematica* and under the Arabic version *Almagest.* As far as the agreement with observed planetary positions is concerned, the Copernican theory has only minor advantages over the Ptolemaic.

If the achievements of Copernicus were limited to using an idea inherited from the ancients to produce a detailed theory competitive with Ptolemy's, the name of Copernicus would by now have sunk into oblivion with that of Aristarchus of Samos. The significance of Copernicus lies elsewhere. While working in astronomy, Copernicus generated a revolution in scholarly thinking in all domains, so that the day of his birth, February 19, 1473, symbolizes the birth of modern science. It is this circumstance, not the advance in the study of planetary motions, that justifies the celebrations of his quinquecentennial, not only in Poland but also in other countries, including the United States.

Statistical Laboratory, University of California, Berkeley, California 94720.

Biographical Sketch

Nicholas Copernicus (1473–1543) was the son of a well-to-do merchant, Nicholas the Elder, originally of Krakow, then the capital of Poland and the seat of one of the oldest universities. (See Chronology, Figure 1.1.) Following a business trip to Torun, the elder Copernicus moved there, married the daughter of a local patrician family, and lived there the rest of his days. He died when Nicholas, Jr., was ten years old. From then on the boy was in the care of his maternal uncle, Bishop Lucas Watzenrode, the Lord of Warmia, a tiny feudal holding of the Catholic Church, a vassal of the state of Poland.

For early education Nicholas Copernicus went to church schools, according to some sources, first in Torun and later in Wloclawek. Apparently, it was in Wloclawek that Copernicus had his first contact with astronomy, under the influence of a teacher who, according to an often quoted anecdote, was named Vodka, but "Latinized" his name into Abstemius. The sundial on the wall of the cathedral in Wloclawek is said to have been a joint effort of Vodka-Abstemius and Copernicus.

The university studies of Copernicus began in 1491 in Krakow where one of his sisters lived, the wife of a local merchant. Another reason for going to Krakow may have been that Krakow University was the alma mater of Bishop Watzenrode. The mental development of Copernicus is likely to have been affected by the news of the discovery of America (1492), which must have reached Krakow while Copernicus was there. The Krakow period of university life marks the beginning of his library. Several books on mathematics and astronomy he bought in Krakow are still extant, with some scribblings on the margins in his own hand. The next stage in Copernicus's education was in Bologna, where he went in 1496. While there to study canon law, he began to show particular interest in astronomy, under the guidance of Professor Domenico Maria da Navara. Other studies included Latin and Greek classics.

In 1500 Nicholas Copernicus went for a brief visit to Rome. Here, on Easter Sunday, a crowd of some 200,000 pilgrims received the blessing of Pope Alexander VI, the spectacular Rodrigo Borgia. The exact dates of Copernicus's visit to Rome are uncertain, but the visit lasted some time, and Copernicus must have known both the colorful ceremony of the Easter Sun-

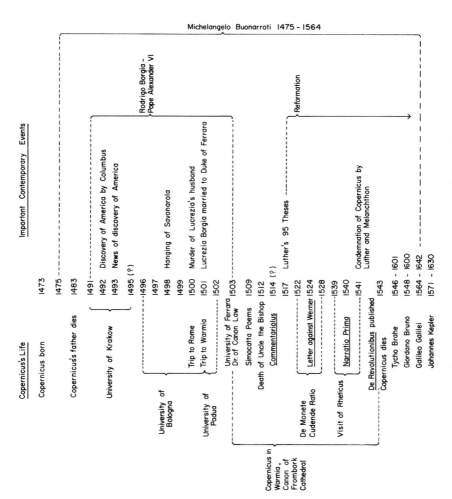

Figure 1.1. Chronology of certain events relevant to the story of Copernicus.

day and the persistent rumors of orgies and murders for which the court
of Alexander VI was famous.

After a short trip home to Warmia in 1501, to be appointed formally
canon of the Chapter of Frombork Cathedral, Copernicus returned to Italy
to continue his studies in medicine as well as law. This time he stayed in
Padua and then in Ferrara, where in 1503 he obtained the degree of Doctor of Canon Law. Soon thereafter Copernicus returned to Warmia for
good, to serve as canon of the Cathedral Chapter of Frombork. Apparently,
he never took the vows. (See Reference 2, pp. 319–320.) In modern terminology, his position was that of a high-ranking civil servant on a lifetime appointment.

In Warmia the functions of Copernicus were quite varied. Up to the death
of Bishop Watzenrode in 1512 Copernicus was his personal physician, secretary, and undoubted friend. The bishop's residence was in the castle of
Lidzbark, and predominantly Copernicus lived there too. The local political
situation was complex and the times very turbulent, with frequent outbursts
of local wars.

The area was mainly populated by Prussians, a people akin to Lithua-
nians and Latvians, now either Germanized or Polonized. In the thirteenth
century, the Order of Teutonic Knights conquered the Prussians, established its own state, and threatened Poland and Lithuania. By the end of
the fourteenth century, Lithuania and Poland united to form a single state
and in 1410 defeated the army of the Teutonic Knights. At the time of Copernicus there were along the coast of the Baltic Sea three semi-indepen-
dent vassal states of Poland: the state of Warmia, the state of the Teutonic
Knights, surrounding Warmia, and the state of Royal Prussia. Several
wealthy cities, including Gdansk and Torun, enjoyed the privileges of self-
government and some autonomy.

There was a substantial divergence of interests. The Prussians longed for
complete independence. The Teutonic Knights hoped to recover the lands
they had lost. The Bishop of Warmia did his best to keep his little state
alive and sided with the King of Poland. The cities built fortifications and
hoped for the best.

Efforts to create some modus vivendi were reflected in occasional meetings of local parliaments. These were attended by representatives of all interested sides and had functions akin to legislative. Nicholas Copernicus

participated in all this, aided his uncle to prepare letters and memoranda, accompanied him on frequent voyages, and occasionally acted as his representative. Besides these activities in diplomacy, Lucas Watzenrode and Nicholas Copernicus tried to organize a university in Warmia; they failed.

The close ties of Nicholas Copernicus with his uncle appear from a little book published in Krakow in 1509. This book, dedicated to Bishop Watzenrode, contains Latin translations by Copernicus of poems written by the Byzantine poet Theophylactus Simocatta. Poems, whether written by Copernicus or only translated by him, were of particular interest to the present writer, sufficient to make an effort to see them. They proved a disappointment.

From about 1510, the principal residence of Copernicus was Frombork, the seat of the Cathedral Chapter. Here he lived in a turret that served him as an observatory, albeit a rather modest one. Most of his astronomical observations were made here. They were not extensive. Involvement with the affairs of Warmia and with public affairs in general, continued: the administration of the lands owned by the Chapter, diplomatic contacts with the King of Poland and with the Order of Teutonic Knights, the fortification of the town Olsztyn and later its defense when the Teutonic Knights tried to capture it, and an attempt to reform the local monetary system. The latter task deserves special mention.

Reportedly, at the time of Copernicus there were in Poland seventeen political entities authorized to mint coins. As is easy to guess, all did their best to put less and less precious metal in their coins, which led to trouble in trade. Copernicus was the second of the three known discoverers of what is now called Gresham's Law. This asserts that weaker coins (with less precious metal) drive the better coins out of circulation and that the process causes inflation. The first of the three discoverers was Nicolas Oresme, Bishop of Lisieux in France, who anticipated Copernicus by more than a century. In turn, Copernicus anticipated Sir Thomas Gresham by some three decades. In a little treatise, *De Monete Cudende Ratio,* first presented at a local parliamentary meeting in 1522 and then published in 1528, Copernicus advocated the establishment of a single institution to mint coins. It seems he was only partly successful: after discussions at various meetings, the number of coin-issuing institutions was reduced from seventeen to four. The monetary chaos was diminished somewhat, yet continued.

With all these activities, Copernicus appears to have had some hobbies: painting (he is said to have painted his own portrait), reading of Latin and Greek authors, and above all, astronomy. Possibly because of the adverse climate in Warmia near the Baltic, but more likely because of the tumultuous times and his many preoccupations Copernicus did little astronomical observing. On the other hand, I am amazed by how familiar Copernicus was with the details of observations performed by other astronomers, particularly by the ancients. This familiarity shines forth from the little work, *Letter against Werner,* translated by Edward Rosen.[2] Among other things, the *Letter against Werner* deals with the timing of certain observations made more than a millennium earlier. The following passage is illustrative:

. . . and if he [Werner] will consult Ptolemy's tables for these positions, he will find them, not 149 years after Christ, but 138 years, 88 days, 5½ hours, equal 885 years after Nabonassar, 218 days, 5½ hours.

The abundance of such details caused me a degree of skepticism. However, notes by Rosen indicate that, right or wrong, the thing was not just taken from thin air but represents the result of persistent digging into and collating ancient texts. The *Letter against Werner* is dated June 4, 1524, two years after the first attempt by Copernicus (see Chronology, Figure 1.1) to convince the higher authorities of the need for monetary reform, and four years before the treatise *De Monete Cudende Ratio* appeared in print. The period must have been extremely busy and demanding for Copernicus, and his ability to produce statements of the foregoing kind indicates that at the time he must have had these and many other details fully organized in a large work. Undoubtedly, the material was assembled in connection with the book *De Revolutionibus* on which Copernicus had been working for quite a few years.

Actually, the basic heliocentric ideas underlying *De Revolutionibus* were summarized by Copernicus long before 1524 in a little treatise known as *Commentariolus.*[2] While the date of completion of *Commentariolus* is uncertain, both Rosen[2] and Rybka[3] quote evidence that it must have been before 1514. *Commentariolus* was not printed during the life of Copernicus but was in circulation in handwritten copies.

In those days, scientific journals did not exist. Their purpose was served by copies of original scholarly works that the authors prepared themselves and sent to individuals they expected to be interested. When a scholarly paper proved interesting, the first generation of its handwritten copies was followed by a second, third, and so on, each prepared by the recipients of the copies of earlier generations. The following note, reproduced from Rosen's book,[2] illustrates the situation

Tycho Brahe [the noted Danish astronomer (1546–1601)] states that ". . . a certain little treatise by Copernicus [the *Commentariolus*] concerning the hypotheses which he set up, was presented to me in handwritten form . . . by that distinguished man, ... Subsequently, I sent the treatise to certain other mathematicians . . ."

It is in such ways that, some time before 1539, the rumor of the work of Copernicus, possibly generated by his *Commentariolus*, reached Georg Joachim Rheticus, then a young professor of mathematics at the Protestant University of Wittenberg, Germany. Rheticus became fascinated by what he heard and decided to learn the details firsthand. In the spring of 1539, he arrived at Frombork and stayed there as a guest of Copernicus for about two years. The visit was not quite safe either for Rheticus or for his host. Times were rough and hazardous, and on occasion Protestants were burned alive in Gdansk and elsewhere. But the visit of Rheticus did not cause such a dramatic event. All that did happen was that for the last few years of his life Copernicus felt a kind of barrier separating him from some of his fellow canons.

After examining the manuscript of *De Revolutionibus* and discussing it with its author, Rheticus became even more enthusiastic. He joined Tiedemann Giese, a fellow canon and the closest friend of Copernicus, in trying to persuade him to publish. When Copernicus demurred, Rheticus wrote the brief summary *Narratio Prima*[2] and got it published in 1540. It was a great success and was soon republished. This must have encouraged Copernicus, and when Rheticus left Frombork in 1541 he took with him the complete *De Revolutionibus* for publication. It appeared in print in 1543, the year of its author's death. Figure 1.2, redrawn from Reference 1, represents the solar system as visualized by Copernicus, complete with times of revolutions around the sun of particular planets computed by himself. This same diagram, in its original setting, can be seen in the recent facsimile reproduction of *De Revolutionibus*.[4] Other pages of the book are quite inspiring. They show how Copernicus worked to achieve the desired level of presentation: deleting some passages, writing new versions, sketching diagrams, and so on.

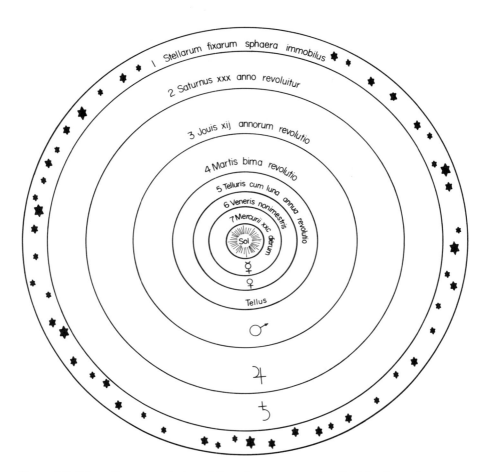

Figure 1.2. Schematic representation of the solar system as visualized by Copernicus.

Essence of the Copernican Revolution

In this essay the term Copernican revolution will designate the impressive development in all the sciences generated by the abandonment by Copernicus of a dogma backed only by tradition, in favor of an alternative idea offering better agreement with observations and more "pleasing to the mind." This is exactly what Copernicus did, as reflected in his *Commentariolus*,[2] in the testimony of Rheticus in *Narratio Prima*,[2] and above all in *De Revolutionibus*.[1] In so doing Copernicus introduced a completely novel yardstick for appraising a new theory: conformity with observations and intellectual elegance. Nowadays this yardstick is generally accepted, at least in principle, but at the time of Copernicus it was in sharp conflict with universal practice. At that time the primary criterion for judging a new piece of research was its strict conformity with the established doctrine of the ruling group whether that of the Catholic establishment or of the somewhat different doctrine of the growing Protestant establishment. Agreement with observations was indeed considered, but only halfheartedly, and whatever observations disagreed with the adopted doctrine were simply ignored. When, through the work of Kepler and Galileo Galilei a century later, the dogmas in astronomy that Copernicus abandoned eventually fell, this had a "domino effect": dogmatic thinking was abandoned in other domains of research, and modern science was born.

The degree of dogmatism at the time of Copernicus is so incredible that it needs documentation. Perhaps unexpectedly, the first known pronouncements against Copernicus's work came from two outstanding intellectual and moral leaders of the epoch, revolutionists in their own right. The following pronouncement of Martin Luther, which I reproduce from Angus Armitage,[5] is the crudest imaginable piece of dogmatism.

. . . the new astronomer who wants to prove that the Earth goes round, and not the heavens, the Sun and the Moon . . . But that is the way nowadays; whoever wants to be clever must needs produce something of his own, which is bound to be the best since he has produced it! The fool will turn the whole science of Astronomy upside down. But as the Holy Writ declares, it was the Sun and not the Earth which Joshua commanded to stand still.

Philip Melanchthon, a personal friend of Martin Luther and a theoretician of Lutheran theology, wrote in a letter to Bucardus Mithobius.[6]

... I saw the dialogue, and opposed publication. A story left to itself lapses gradually into silence; but some men think it a signal feat to honor such an absurdity like the Sarmatian astronomer who puts the Earth in motion and makes the Sun stand still. Surely, wise rulers should curb the impudence of talents. Farewell, October 16.

This letter was written in 1541, two years before the publication of *De Revolutionibus*. The news about the "impudence" of the Sarmatian astronomer must have reached Luther and Melanchthon either through the *Commentariolus* or the *Narratio Prima*.

The two quotations, particularly that from Melanchthon, illustrate a general trait of all human societies. Large or small, whether the inhabitants of a country, a religious sect, or a scholarly association, each more or less organized group creates a ruling "establishment" which erects a doctrine that justifies its power. This power is then used to enforce faith in the doctrine. The last lines of Melanchthon's letter exhibit this tendency toward repression. The same phenomenon appears in scientific associations today. A young author has little difficulty in having his research published provided it conforms with the general framework of ideas of the establishment that publishes the journal. However, if the paper contains some really new ideas reaching beyond those of the establishment, there is usually trouble over publication.

At the time of Copernicus and for some centuries before, the overwhelmingly strong ruling establishment was that of the Catholic Church, with the Holy Inquisition as the arm especially created to enforce its doctrine. This doctrine developed gradually under the influence of Greek philosophers whose teachings began to penetrate the schools in monasteries and around cathedrals since about the eleventh century of the Christian era. Among other things the Greek influences included astronomy.

Using premises that he thought self-evident, Aristotle proved that the universe must be finite. This became one of the bases of the Ptolemaic cosmology with the immovable earth in the center and with all the celestial bodies, moon, sun, the five then-known planets, and the "fixed stars" going around it in "ideal motions": circular with constant velocities. Each celestial body was supposed to be attached to its own "'sphere," a rigid transparent body. The first sphere was that of the moon. The last sphere, the all-inclusive eighth sphere, was that of fixed stars. Ptolemy's outstanding effort was to explain the motions of planets as seen against the background of the fixed stars. These motions exhibit peculiarities of several kinds, one of which will be described here as large-scale "vagaries." For a time a planet, say Mars or Jupiter, would move in a particular direction with a substantial speed, then it would slow down, stop, and reverse. These vagaries occur periodically. In order to explain them in a manner consistent with the axiomatic ideal motions, Ptolemy was led to assume, separately for each planet, the existence of a system of circles called epicycles. A planet moves ideally along its special epicycle. The center of this epicycle is not stationary but moves, again ideally, along another circle, and so on, the last circle being called deferent. In order to explain the observed motions of the planets, not only the large-scale vagaries but also the fine details, and especially in order to produce tables from which future positions of the planets could be calculated, the Ptolemaic system of epicycles had to be quite elaborate and deservedly excited the admiration of many astronomers, including Copernicus.[7] However, not all the observations of the planets conformed to the Ptolemaic system of epicycles. Among those that disagreed and were therefore ignored were the observations of the changes in apparent brightness and distances of particular planets from the earth.

Although the details of the Ptolemaic cosmology had nothing to do with theology of the Catholic Church, the centrality of the earth, created for the benefit of man the master, was an important link. Thus, imperceptibly, the whole of the Ptolemaic system came to be treated more or less on par with the basic Catholic doctrine.

In his effort to create a comprehensive scientific cosmology, Copernicus had to examine a number of details of the Ptolemaic system which he found not sufficiently "absolute," that is, not quite agreeing with the observations. Many of the discordances are rather technical and we shall concentrate on only two of them: the periodic variations in brightness of the planets, particularly of Mars, and the vagaries. The following quotation is from Rheticus's *Narratio Prima*.[8]

. . . the planets evidently have the centers of their deferents in the sun, as the center of the universe. That the ancients . . . were aware of this fact is sufficiently clear for example from Pliny's statement . . . that Venus and Mercury do not recede further from the sun than fixed, ordained limits because their paths encircle the sun; . . . . In addition to other difficulties . . . , Mars unquestionably shows a parallax sometimes greater than the Sun's, and therefore it seems impossible that the earth should occupy the center of the universe. [Parallax is a certain angle indicating the distance of a celestial body: the smaller the parallax, the greater the distance.] Although Saturn and Jupiter, as they appear to us at their morning and evening rising, readily yield the same conclusion, it is particularly and especially supported by the variability of Mars when it rises. . . . Whereas at its evening rising Mars seems to equal Jupiter in size, . . . , when it rises in the morning just before the sun . . . , it can scarcely be distinguished from stars of the second magnitude [much fainter]. Consequently at its evening rising it approaches closest to the earth, while at its morning rising it is furthest away; surely this cannot in any way occur on the theory of an epicycle. Clearly then, in order to restore the motions of Mars and the other planets, a different place must be assigned to the earth . . . my teacher [Copernicus] saw that only on this theory could all the circles in the universe be satisfactorily made to revolve uniformly and regularly about their own centers. . . .

The last sentence refers to the fact that Copernicus failed to abandon the preconception of "ideal" motions of heavenly bodies. As a result, in order to take care of the fine details of planetary motions, he had to use the device of epicycles.

The preceding quotation illustrates how Copernicus struggled for a theory that could embrace all the observations available. Now we shall deal with the more "pleasing to the mind" part of the story. This is that the abandonment of the dogma that the earth is stationary, with all the celestial bodies going around it, and the assignment to the earth of two (principal) motions explain at a single stroke a great number of observed phenomena that otherwise required specific ad hoc hypotheses: all the vagaries and the variation in brightness of the planets.

One of the motions of the earth postulated by Copernicus is the diurnal rotation around the earth's axis. This rotation, from west to east, explains the daily rising and setting of the sun and of all other celestial bodies including the fixed stars, which Copernicus assumed motionless. The second motion that Copernicus assigned to the earth was its annual revolution around the sun, which he supposed stationary. It is this second motion of the earth that will be the main subject of the following discussion. It will be seen qualitatively that this single motion explains both the large-scale vagaries of all the planets and also the periodic variation in their brightness, very pronounced for Mars and much less noticeable for Jupiter and Saturn.

To see all this, it is sufficient to examine Figure 1.3 representing sche-

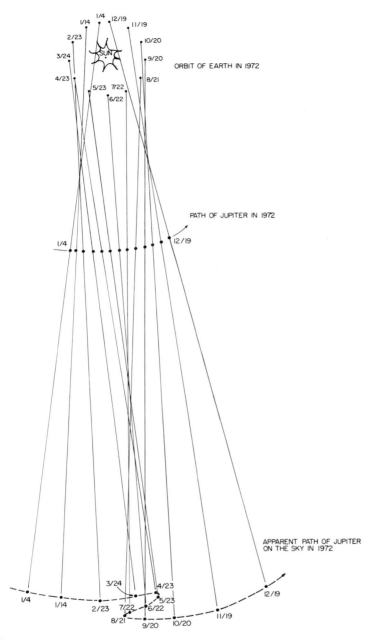

ORBIT OF EARTH IN 1972

PATH OF JUPITER IN 1972

APPARENT PATH OF JUPITER
ON THE SKY IN 1972

Figure 1.3. Schematic explanation of "vagaries" in the apparent motions of planets.

matically the *real* motions of the earth and Jupiter around the sun, as they occurred in the prequinquecentennial year 1972, and also the resulting *apparent* motion of Jupiter as seen from the earth against the background of the very distant fixed stars.

The spiked body on top of Figure 1.3 symbolizes the sun. The circular sequence of dots around the sun shows the approximate positions of the earth on the indicated dates, January 4, January 14, and so on, up to December 19, 1972. The farther sequence of similar dots indicates the corresponding positions of Jupiter on the same dates. The distance of Jupiter from the sun is about five times that of the earth. Also, the angular speed of Jupiter around its orbit is about one-twelfth that of the earth. In other words, while in one year the earth completes its trip around the sun, Jupiter covers only about one-twelfth of its entire orbit. The broad area at the bottom of Figure 1.3 symbolizes the background of the immobile fixed stars. The series of lines connecting the succession of real positions of the earth with those of Jupiter indicates the corresponding apparent positions of that planet as seen among the fixed stars. It is seen that, from about January 4 through April 23, 1972, the apparent motion of Jupiter was a rather rapid forward motion. From April 23 to May 24, Jupiter appeared motionless. Then, up to August 2, it moved in reverse, stopped, and then resumed its forward apparent motion. Here, then, the assumption that both the earth and Jupiter move around the sun, the earth in 12 months and Jupiter in (approximately) 12 years, explains the puzzling loop in the observed path of that planet.

Naturally, what was just found about Jupiter applies to the other two planets, Mars and Saturn, both more distant from the sun than the earth. The number of loops in their apparent orbits, their sizes and shapes, are determined by the distances of those planets from the sun and by their periods of revolution.

Now, what about the periodic variation in brightness of planets? As noted by Rheticus in the preceding quotation, the brightness of Mars varies impressively, but apparently not so much that of Jupiter and Saturn. Why?

Figure 1.3, combined with information that the average distance of Mars from the sun is only 50 percent greater than that of the earth, gives an easy answer. The predominant factor affecting the brightness of a planet as observed from the earth is the distance separating these bodies. As seen from

Figure 1.3, the ratio of the greatest to the smallest distance between the earth and Jupiter is about $6:4 = 3/2$. On the other hand, the same ratio for the earth and Mars is about $2.5:0.5 = 5/1$. The conclusion is that, as time goes on, the apparent brightness of both Jupiter and Mars must vary. However, for Jupiter the ratio of its maximum to its minimum brightness is comparable to that corresponding to a candle seen alternately at, say, 15 feet and 10 feet. On the other hand, for Mars this ratio compares with that for a candle viewed alternately from distances of 50 feet and 10 feet. No wonder, then, that changes in brightness of Mars leap to the eye, while those of Jupiter are much less impressive. Saturn is nearly ten times farther from the sun than is the earth. Its apparent brightness does vary periodically, but to the naked eye this variation is hardly perceptible.

The previous discussion shows that the single hypothesis of planetary revolution around the sun explains a substantial variety of the apparently disconnected observed phenomena, the large-scale vagaries of planets and the periodic variation in their brightness. This is the quality of the Copernican theory that is described by the term *intellectual elegance*. The contemplation of this theory creates the feeling that we *understand* the mechanism of the solar system.

To emphasize the difference between the mode of thinking of Copernicus and that of his contemporaries, we return for a moment to the question of whether the universe is finite or infinite. As mentioned earlier, Ptolemy accepted the proof of Aristotle that the universe must be finite. In consequence, the Ptolemaic cosmology, and that of the contemporaries of Copernicus, visualized the universe as being contained within the rigid eighth sphere of fixed stars, with no questions asked. Now, let us compare this attitude with that of Copernicus.

The question of the size of the universe is discussed by Copernicus in Chapters VI and VIII, Book 1, of *De Revolutionibus*.[4] Here are a few relevant passages in a liberal translation from Latin.

ON THE IMMENSITY OF THE SKY COMPARED TO THE DIMENSIONS OF THE EARTH. The following reasoning shows that the great size of the Earth is not comparable to the dimensions of the sky [really to the distances of fixed stars].

Here follows a reasoning and a record of observations of certain stars analogous to the modern measurements of their parallaxes. Measurements described by Copernicus indicated parallaxes indistinguishable from zero. [Instruments sufficiently accurate to measure the minute parallaxes of fixed stars were constructed in only the nineteenth century.] The summary of his discussion reads:[4]

. . . Because of this reasoning, it is sufficiently clear that, compared to the Earth, the sky is immense, having an aspect of infinity . . .

Against this Copernican approach to a natural phenomenon, we have Luther's "it was the Sun and not the Earth which Joshua commanded to stand still"!

Some other passages in *De Revolutionibus*,[4] relating to the same subject, have a humorous flavor.

But they say that beyond the eighth sphere there are no objects, no places, not even empty space into which the sky could expand; this being the case, it is surely surprising that something [the eighth sphere] could be stopped by nothing. . . . We will leave it to philosophers to argue whether the universe is finite or infinite.

When writing these lines, Nicholas Copernicus must have been smiling.

While very different from that of his contemporaries, the attitude of Copernicus toward the question whether the universe is infinite or finite is similar to the stand that a scientist in modern times would assume in an analogous situation: "So and so asserts that something or other must be finite, which is interesting. Well, let's try to measure the thing!" Thus, when scientists became aware of the phenomenon of propagation of light, many experiments were performed to measure its velocity. Eventually, unexpected results were obtained. In biology, when Mendel's work became known, ascribing the inheritable properties of living organisms to hypothetical "genes," the questions immediately arose: Can we see them? or, at least, Where are these genes located? The geneticists looked and, in due course, found first chromosomes and eventually DNA. When Einstein's theory predicted that the rays of light passing near a massive body must bend, small armies of astronomers labored to check this prediction empirically. Eventually, they got an affirmative answer.

In modern times, efforts to verify novel theories or hypotheses occur daily as current research. Most frequently, their results are negative, which indicates either that the hypothesis tested requires some modification, or, indeed, deserves abandonment, or (as in the case of Copernicus) that the observations made are not sufficiently accurate. In fact, one might say that experiments and observations intended to verify hypotheses form the backbone of modern science. In this sense, Copernicus was the first modern scientist.

Copernicus was well aware that attributing motions to the earth might cause him trouble, even though he did not see any real conflict with the Catholic creed. At the time when *De Revolutionibus* was going to press, the Holy See was occupied by Pope Paul III, a person very different from Rodrigo Borgia. Copernicus hoped that though the court of the Pope and the Holy Inquisition may have retained their earlier attitudes, at least Pope Paul III himself would see the difference between religion and morality on the one hand and the study of nature on the other. Accordingly, Copernicus prefaced the *De Revolutionibus* with a dedication to the Pope which begins as follows.

TO HIS HOLINESS PAUL III PONTIFEX MAXIMUS   PREFACE TO THE BOOKS "DE REVOLUTIONIBUS ORBIUM CAELESTIUM" BY NICOLAS COPERNICUS
I visualize, O Holy Father, that, as soon as certain people learn that in these books of mine . . . I attribute to the Earth certain motions, they will demand the condemnation both of my ideas and of myself.

In subsequent paragraphs Copernicus discusses a category of people who, because of their stupidity, are comparable to "drones among the bees." How far these pleadings affected the thinking of Pope Paul III is difficult to judge. They certainly did not convince the generations of "drones," and from 1616 to 1822 *De Revolutionibus* remained on the Index of books forbidden to Catholic readers. While Copernicus died too soon to feel the displeasure of the Catholic establishment, those who followed him had to suffer: Kepler had his troubles, Galileo had serious difficulties with the Holy Inquisition, and in 1600 Giordano Bruno was burned at the stake.

From the narrow point of view of astronomy alone, Kepler's magnificent work overshadows that of Copernicus. Kepler went further than Copernicus and abandoned the second of the Ptolemaic dogmas, the "ideal" motions of heavenly bodies. This resulted in the discovery of the famous Kepler's laws of planetary motions which threw overboard all the epicycles and paved the way for Newton. However, Kepler's own interpretation of his own work and of the achievements of Copernicus was broader in perspective:[9]

With my discoveries, I would be happy to guard the door of the House of Worship where Copernicus celebrates at the High Altar.

The process of abandoning dogmatism in science was lengthy and difficult. One example is provided by the attitude of Tycho Brahe, generally recognized as Copernicus's successor as astronomer of the epoch. Pondering the two competing systems of planetary motions, the traditional of Ptolemy and the revolutionary of Copernicus, Brahe was reluctant to adopt the latter. He produced his own system following a "via media": the earth is the immovable center of the universe, with the sun and the moon going around it. However, as in the Copernican system, the five then-known planets were assumed to move around the sun. The reader will perceive that, in effect, Tycho Brahe's system differs from the Copernican only in the frame of reference. Still, the immovability of the earth was there.

The present writer is fascinated by a question that probably cannot be resolved: Did Tycho Brahe really believe in the immovability of the earth, or was he simply taking precautions not to excite Copernicus's "drones"? For a couple of years before the death of Brahe, Kepler served as his assistant, and the following passage[10] is indicative of what may have been the case.

. . . In my opinion, humidity is present on stars . . . and therefore also living creatures. . . . And not only the unfortunate Bruno, who was roasted on glowing coals in Rome, but also the respected Brahe believed that there are inhabitants on stars.

To conclude this section on a subjective note, Copernicus seems to have been a very attractive person: a tremendous intellectual horizon, including public affairs and economics, hard work in several domains, including some military endeavors, sense of humor and straight dealings, with no falsity that I can see. For example, Copernicus appears never to have written a horoscope which, in his epoch, was the usual way for astronomers to earn their living. They were astrologists by trade, and in the eyes of their customers astronomy was only a sideline useful in their primary activity. For another thing, Copernicus's motivation for writing De Revolutionibus had nothing to do with the "publish or perish" motive of much present research. He could not and did not expect personal material advantage from his work, only condemnation and ridicule. What drove him was pure scholarly interest in the phenomena themselves and the hope of explaining them in a manner "pleasing to the mind." The original stimulus may have been in 1496–1501: a mild young man's revolt against the pomp used by respected authority to cover up its misdeeds.

The Routine of Thought and Quasi-Copernican Revolutions

Quite apart from external difficulties connected with attitudes of the existing establishments, Copernicus faced and solved a great "internal" difficulty resulting from the omnipresent psychological phenomenon of "routine of thought" or "routine of behavior." This phenomenon is well known and serves as the basis for our present systems of education, training, and upbringing, much as they may vary. However, not all the implications of this phenomenon are commonly discussed.

When confronted with the question What is two times three? we answer immediately, routinely, and without mental effort. This, of course, is the result of learning arithmetic in school, and a good thing it is too. Learning implants many similar routine reactions to all sorts of questions and situations, and usually to good advantage. However, routine behavior or thought can get hold of us and make us act and think routinely even in inappropriate cases.

An experienced driver accustomed to an automobile with conventional gearshift drives his car routinely. In so doing, he may have his thoughts far away from the conditions on the road which on occasion make him step on his brake and on his clutch, change gears, and so forth. This is all to the good, a desirable result of training. But now let us consider what happens if this same experienced driver decides to buy a car with automatic gearshift. In the few initial rides in the new car, when confronted with the need to stop, the experienced driver will not only step on the brake pedal but also will try to step on the clutch, which he well knows is not there. This is an example of the routine of behavior getting hold of us. Naturally, the driver soon recognizes that his attempt to step on the clutch is pointless, and after a few days the old routine of behavior will fade in favor of a new routine, with no harm done.

Less easily recognizable "routineness" of thought and behavior characterizes our personality traits acquired through family traditions, through upbringing in schools and through participation in particular social groups. Here again, some of the routines are intelligible and useful and fully justify the conscious effort to have them implanted and firmly established. But this is not a general rule and some of the routines are questionable. Such are, for example, our racial and religious prejudices. Nowadays it does not happen frequently that an originally happy love affair breaks down merely because one of the parties eats pork and the other does not. But similar incidents do occur from time to time, illustrating how difficult it is to notice that a supposedly high principle of a creed is a routine of thought with no other backing than a centuries-long tradition. With much suffering, we continue our efforts to step on the nonexisting clutch without any thought about checking whether the clutch is there or not.

Along with other domains of human behavior and activity, research is also affected by routines of thought and, particularly, of premises. Here again many routines are very useful by providing important economy of mental effort. But again there are exceptions. It does happen that some premises of our thought, acquired through traditional learning in schools and universities, have no other backing than tradition. And the longer the tradition of a commonly accepted premise, the more difficult it is, psychologically, to notice its routineness and to question its validity.

As described in the earlier sections, Copernicus grew up in the common belief in and admiration for the Ptolemaic system of planetary motions, sanctioned by a centuries-long tradition. Psychologically, it must have been very difficult for Copernicus to notice the routineness of this belief and to abandon it. Because of the pervasive crude dogmatism of the epoch and the "domino effect" described in the earlier sections of this essay, the original Copernican revolution affected scholarly thinking in all domains of research, not only in astronomy. However, cases when a generally accepted premise of thought is suddenly identified as a mere dogma with no backing but tradition occur even now in many domains of study. Ordinarily, after an unavoidable period of resistance on the part of some scholarly establishment, a new fruitful field of research is opened, frequently with very impressive results, and we witness an intellectual revolution. Because of the general decrease in dogmatism, this revolution is narrower than that generated by Copernicus. In fact, it may be a revolution in a subdomain of a science like astronomy or biology; a local revolution with at most a moderate "domino effect." We call it a quasi-Copernican revolution.

Not all the revolutionary developments in science or technology are initiated by the identification and rejection of a dogmatic assumption so that not all such developments are Copernican in character. Occasionally it happens that, through apparently minor contributions of many scholars, all the pieces of a complicated jigsaw puzzle are accumulated but no comprehensive overall picture is available. Then, perhaps through the genius of one or a few research workers, the jigsaw puzzle is at last completed with impressive consequences. Depending upon circumstances, this may be a revolution in science or in technology, but not a quasi-Copernican revolution: no preconception has been identified and abandoned.

Sharp distinctions are rare in the real world, and some of the intellectual revolutions we observe are only partly Copernican in character. Many of them are described in the following chapters of this volume.

Acknowledgments

I am indebted to Professor Kenneth O. May and, particularly, to Professor William Hammer for their help in locating texts written at the time of Copernicus. Hearty thanks are due to Professor Joseph Fontenrose, who helped me with medieval Latin. (For an independent recent English translation of portions from *De Revolutionibus* see Reference 11.) Finally, I am grateful to Professor Eugeniusz Rybka for a number of details relating to the post-Copernican epoch.

While preparing the essay, I used the facilities in the Statistical Laboratory provided by the research projects NIH Research Grant No. GM-10525, National Institutes of Health, Public Health Service, and US Army Research Office-Durham, Grant No. DA-ARO-D-31-124-73-631.

References

1. Nikolaus Kopernicus, *De Revolutionibus Orbium Caelestium*. R. Oldenbourg, Munich, 1949.

2. Edward Rosen, *Three Copernican Treatises*. Revised 3rd ed., including a biography of Copernicus, Octagon Books, New York, 1971.

3. Eugeniusz Rybka, *Four Hundred Years of the Copernican Heritage*. Jagiellonian University Jubilee Publication, Vol. 18, University of Krakow, 1964.

4. Nicholas Copernicus, *De Revolutionibus*. . . . Facsimile edition. Polish Academy of Sciences, Macmillan and Polish Scientific Publishers, London, Warsaw, Cracow, 1972. In the United States the book is available at Johnson Reprint Corporation, Palisades, N.Y.

5. Angus Armitage, *Sun, Stand Thou Still*. Henry Schuman, New York, 1949, p. 116.

6. *Corpus Reformatorum, Vol. IV, Philippi Melanthonis Opera*. Edited by C. G. Breitschneider, Publisher C. A. Schwetschke, 1837, p. 679.

7. Rosen, *Three Copernican Treatises*. P. 131.

8. Ibid. Pp. 136–137.

9. Max Caspar and Walther von Dyck, *Kepler in Seinen Briefen*. Vol. 1, R. Oldenbourg, Munich and Berlin, 1930, p. 304.

10. Ibid., p. 71.

11. Dorothy Stimson, *The Gradual Acceptance of the Copernican Theory of the Universe*. Peter Smith, Gloucester, Mass., 1972.

Part I

Astronomy and Cosmology

Summary

The essays in this part of the volume describe four already completed revolutions in our thinking about the universe and one revolution now in progress with an uncertain outcome. Following the Copernican revolution in the sixteenth century, the view became firmly established that the universe consists of the system of fixed stars, the Milky Way System, in which our sun occupies a more or less central position. The four successive revolutions described in this part changed this belief fundamentally. First, following the "Great Debate" of 1920, it was gradually recognized (1) that the sun is on the outskirts of the Milky Way, (2) that the Milky Way System is just one of the innumerable "Island Universes" populating the depth of space, and (3) that these Island Universes (galaxies) recede from each other with velocities proportional to distances: "the universe is expanding." (4) The fourth revolution, initiated in the 1950s and essentially completed in about 1970, is concerned with certain violent events in the realm of galaxies: it was recognized that through a mechanism not yet understood violent explosions occur in the nuclei of galaxies; these explosions result in the ejection of huge amounts of matter that can form spiral arms in the home galaxy or even an expanding cluster of progeny galaxies. These novel views reversed the earlier firmly established ideas on the evolution of galaxies. (5) The fifth revolution in the thinking about the cosmos, the revolution now in progress with no final outcome in sight, was generated by the discovery of previously only speculative celestial objects: quasars (quasi-stellar objects), pulsars, dark holes, and so on. The current thinking of the astronomers on these matters is very turbulent and involves questions, such as (a) Is it really true that all the universe was created in a single "big bang," or is there now being created continuously new matter that fills the gaps formed by the expansion of the universe, and which produces a "steady state"? (b) Do all the remarkable recently discovered phenomena obey the known laws of physics, the "terrestrial physics," or does the explanation of these phenomena demand a new physical theory, a "cosmological theory of physics"?

Bart J. Bok

2 Harlow Shapley and the Discovery of the Center of Our Galaxy

Introduction

One of the finest examples of a quasi-Copernican revolution is the story
of how Harlow Shapley, almost single-handedly, took our planetary system
with the sun and earth from their accepted central positions in our Milky
Way System and placed them in the outskirts of the Milky Way where they
belong. Until the day of publication of Shapley's epoch-making 1918 paper
"Remarks on the Arrangement of the Sidereal Universe," astronomers were
in general agreement that our sun and earth occupied a near-central position in the Milky Way System. Five years later no astronomer in his right
mind could any longer argue that this was the case. Harlow Shapley did
for the Milky Way System, our home galaxy, what Copernicus had done
400 years earlier for our solar system.

It is significant to point out, right at the start, that the Shapley revolution
did not involve difficult mathematical formulations or advanced ideas in
physics. Shapley was basically a competent and straightforward observer
who undertook the right sort of program of observation with the 60-inch
Mount Wilson reflector. The careful planning and execution of a four-year
observing program, accompanied by careful and thorough analysis of the
material, brought about the epoch-making revolution.

To set the stage, we shall first describe the beginnings of Milky Way research and show what was the state of our knowledge about the structure of
the Milky Way System at the time Shapley began his great work. We shall
then describe Shapley's researches and the resulting controversies. Our essay
concludes with a brief section about our current knowledge of the central
regions of our Milky Way System.

The Beginnings of Milky Way Research

By the middle of the eighteenth century, we find the first stirrings of interest
in the arrangement of the stars in our universe. The world would have to
wait until 1838 before Friedrich Wilhelm Bessel would measure with precision the first stellar parallax, which is a measure of the distance to a star.
But it was generally realized by the middle of the eighteenth century that
all stars are suns and that the nearest stars are at distances from us of at
least several thousand times the distance from the earth to the sun. Telescopic observation had shown that the faintly glimmering band of the Milky

Steward Observatory, University of Arizona, Tucson, Arizona 85721.

Way could be resolved into thousands upon thousands of stars. The simple fact that in the field of a 10-inch telescope we see on the average along the belt of the Milky Way five times as many stars as we see with the same instrument for an average field well away from the band of the Milky Way indicates that we are located in a flattened star system. In Germany in 1755 the great philosopher Immanuel Kant published a fine essay on "The Constitution and Mechanical Origin of the Whole Universe." Kant referred to the Milky Way as the "zodiac of stars," and he realized that our sun was located in a flattened Milky Way System. This system obviously had far greater extent in directions along the band of the Milky Way than at right angles to it. He surmised that the nebulae, such as the Andromeda Nebula, were probably island universes in their own right, and he saw the greater universe as a hierarchy of island universes. In Britain, Thomas Wright of Durham and John Mitchell were writing about the probable properties of the Milky Way System, arriving at conclusions that are in part still valid today.

It was in about 1780 that William Herschel (Plate 2.1), a German-born musician working at Bath in the southern part of England, became interested in the structure of the Milky Way. Herschel had the good sense to realize that what was needed first was basic high-quality observational material, "star gauges" he called them. We would now say "star counts." Herschel's first astronomical writings of importance date back to 1781 and 1784. At that time, he had constructed for himself a 20-foot telescope (focal length 20 feet, aperture a little under 19 inches). He had begun to gauge the heavens systematically, counting and recording, paying special attention to the nebulae and star clusters viewed with his telescope. With his relatively modest telescope he resolved many clusters into stars, and he argued that many nebulae would ultimately be resolved into stars as well. From his star gauges he attempted to "determine the length of the visual ray" for each of his fields; in other words, he tried to fix for each direction the distance to the outer rim of the Milky Way System. He assumed—not a bad first approximation—that the stars were distributed evenly within our Milky Way System and that all stars were of more or less comparable intrinsic brightness. When few stars were counted in one of his fields, then he supposed that the rim of the Milky Way System was relatively near us for that direction. If, on the other hand, he counted very many stars, then he sur-

Plate 2.1. Sir William Herschel.

Frontispiece from de Sitter's *Kosmos*. (Harvard University Press.)

mised that the distance to the outer rim would be quite large for that direction. It was in this manner that he arrived at the pictorial representation shown in Figure 2.1 which was first published in 1785. He said that our Milky Way System had the appearance of a "cloven grindstone." The cleft is shown in Figure 2.1 on the left; it results from the paucity of stars observed along the great dark Northern Rift of the Milky Way. Please note that the position of the sun and earth in Figure 2.1 is shown by a small star near the center of the grindstone. Herschel surmised that our Milky Way System contains many millions of stars. His paper suggests that 100 million might be a fair figure. This is not a bad first guess when compared to our current estimates of a population amounting to 100 billion star-suns in our home galaxy.

Herschel realized that with his 20-foot telescope he could not "fathom the profundity of the Milky Way," and hence he proceeded to construct a 40-foot telescope. His observations with the larger instrument gave him the feeling that along the band of the Milky Way he had nowhere penetrated to the outer rim of the Milky Way System. He noted that there always remains along the band of the Milky Way a faint luminous background, which seems on the verge of being resolved but which actually is not resolvable. These early doubts grew with the years. In the "Concluding Remarks" of Herschel's paper written at Slough (near Windsor) on May 10, 1817, he admits openly that he has apparently not penetrated to the outer rim of our Milky Way System. His 1817 paper contains the rather pathetic diagram we reproduce as our Figure 2.2. The small circle is supposed to contain all of the naked eye stars, and the two parallel lines indicate the position of the outer rim for directions at right angles to the band of the Milky Way. The sun is located near the center of the small circle. There is no longer any indication about the depth of the Milky Way for directions along the central band. There is also no longer any suggestion with regard to the central, or noncentral, position of our sun and earth in the Milky Way System. Before we move on from Herschel, it may be of interest to quote one of his sayings: "It is sometimes of great use in natural philosophy to doubt of things that are commonly taken for granted."

Following Sir William Herschel's death (1822), astronomers began to concentrate their efforts on measuring stellar parallaxes, which yield the distances to the stars; they investigated also the crosswise motions of the stars

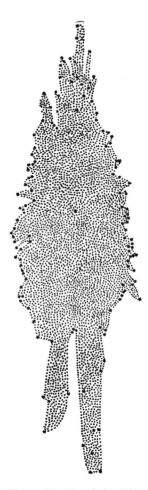

Figure 2.1. Herschel's 1785 diagram of the Milky Way System.

Herschel's cloven grindstone model of our Milky Way System. The position of the sun
is indicated by a small star near the center of the diagram, which represents the result of
analysis of Herschel's star gauges performed with his 20-foot telescope. (Copied from
M. A. Hoskin's *William Herschel and The Construction of the Heavens,* Norton, 1964.)

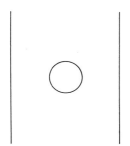

Figure 2.2. Herschel's 1817 diagram of the Milky Way System.

This diagram was reproduced in Herschel's 1817 paper. Its rather sad simplicity indicates firmly that William Herschel did not feel that he had fathomed the depths of our Milky Way System. (Copied from M. A. Hoskin's *William Herschel and The Construction of the Heavens,* Norton, 1964.)

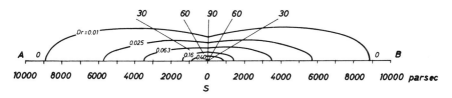

Figure 2.3. The stellar system according to Kapteyn and van Rhijn (1922).

This diagram represents the result of a careful statistical analysis of calibrated star counts, made on the assumption that there is no appreciable dimming of starlight by cosmic dust present in and near the central plane of our Milky Way System. Ten years later the assumption of complete absence of interstellar absorption was proved to have been erroneous. (A diagram first published in Contributions from the Mount Wilson Observatory No. 188.)

and undertook the first spectral investigations. Not much of great moment was achieved to improve our knowledge of the outlines of the Milky Way System, but it is worth noting that Figure 2.1, representing the cloven grindstone model of our system with the sun near its center, was far more frequently quoted than the vague and indefinite diagram reproduced as Figure 2.2. Astronomers, and their colleagues in related fields, were generally content to assume placidly that our sun and earth held a more or less central position in our Milky Way System.

It was not until the last decade of the nineteenth century that serious interest in the arrangement of the stars in our Milky Way System became active again. The two great astronomers who led their colleagues in these studies were Jacobus Cornelius Kapteyn (Plate 2.2) of the Netherlands and Hugo von Seeliger of Germany, who both established major schools for Milky Way research. Of the two, Kapteyn rightly made the greater im-

Plate 2.2. Jacobus Cornelius Kapteyn.

This is a reproduction from a photograph contained in *The Life of Kapteyn* published in 1928 by his daughter, H. Hertzsprung-Kapteyn (in Dutch, Noordhoff, Groningen).

pression on the astronomical world. He was truly a master at organizing observations of the heavens, and he was equally a master at interpretation of large bodies of data. In his works, Kapteyn took advantage of the newly developed techniques of astronomical photography, and he urged concentrated effort upon spectral classification and upon the measurement of the line-of-sight motions and crosswise motions of the stars. He organized (1904) a magnificent "Plan of Selected Areas," which still continues to produce mighty useful results. Kapteyn's approach was basically a statistical one in which the spread of intrinsic luminosities of stars was recognized as a major factor. All analyses of star counts were placed on a sound statistical basis, and the determination of star densities for the directions of the evenly distributed Selected Areas was achieved following established statistical approaches. Figure 2.3, published by Kapteyn in 1922, shortly before the time of his death, shows the Kapteyn System. Kapteyn had found that in all directions the stars thin out with increasing distance from the sun. They do this more rapidly in directions at right angles to the central plane of the Milky Way than in the plane itself, but Figure 2.3 shows that the thinning out is a universal phenomenon. Kapteyn places the sun very close to the center of our Milky Way System. In a way, Figure 2.3 by Kapteyn is Herschel's Figure 2.1 brought up to date, with, importantly, a distance scale of light-years and parsecs shown in Figure 2.3. We should explain that 1 parsec is the distance corresponding to a parallax displacement (see the next section) for a star of 1 second of arc, which is equivalent to 3.26 light-years; the light-year is the distance that a light ray travels in 1 year.

In all of his work, Kapteyn assumed that the light of the distant stars is not affected by interstellar absorption. In other words, he assumed that there are no appreciable amounts of interstellar dust present in the regions between the stars; interstellar pollution was assumed to be minimal or absent. This assumption seemed justified by the data that were available to Kapteyn when he did his work. We shall see later that the assumption of there being no interstellar pollution in our Milky Way System is decidedly wrong. Although within the vicinity of the solar system there is very little interstellar matter for directions at right angles to the plane of the Milky Way, there is much cosmic dust within a thousand parsecs in the central plane; the dimming produced by interstellar dust in the plane of the Milky Way is now known to be considerable. Figure 2.3 shows results that are still more or less valid today for directions at right angles to the central plane of the Milky Way. However, the apparent thinning out of the stars with increasing distance from the sun in the central plane of the Milky Way is now understood to be an apparent phenomenon, and no overall thinning out to 1000 parsecs is observed if proper corrections are made for interstellar absorption.

Please note that Figure 2.3 was published in 1922, four years *after* Shapley had propounded his views on the structure of the Milky Way System. Kapteyn refused to believe in the Shapley system, in which the sun holds a noncentral position in our Milky Way System, and he stood by Figure 2.3 until the time of his death (1922). By 1924, the year in which the present author entered on the study of astronomy at Leiden University, there was no longer any astronomer who believed that Kapteyn's system, shown in Figure 2.3, represents a true outline of the overall structure of our Milky Way System.

Scales of Distance

Through his star gauges, Sir William Herschel had laid the foundations for the exploration of our Milky Way System and of the universe of galaxies beyond. But he realized, probably even more so than his contemporaries and immediate successors, that a thorough exploration of the depths of the universe would become possible only if methods were developed to measure with fair precision the distance to the stars. The astronomers of the nineteenth century concentrated their efforts upon establishing such distance scales. In this section we shall concern ourselves briefly with the methods that were developed for the measurement of stellar distances during the century that followed Herschel's death.

Foremost in the researches on stellar distances are the measurements of the first trigonometric parallaxes of nearby stars made during the years 1837 to 1839. Friedrich Wilhelm Bessel measured with precision the parallax of the faint star No. 61 in the constellation of Cygnus, 61 Cygni; F. G. W. Struve measured the parallax for the bright star Vega; and P. Henderson measured the parallax of the southern star Alpha Centauri, the nearest bright double star and one of the Pointers for the Southern Cross.

How do we obtain information on the distances of the stars? The straightforward and basic technique involves the measurements of the star's annual displacement as the star is viewed from opposite ends of the earth's orbit, half a year apart. A star's *parallax* is defined as the angle under which we would view from that star the radius of the earth's orbit around the sun. From the earth at one extreme point in its annual orbit around the sun, we would observe the star as shifted to the left relative to a background of far more distant stars. Half a year later we would be in the opposite extreme position and find the star shifted to the right by the same amount. The total shift would amount to twice the star's parallax (see Figure 2.4). There are some complications, recognized fully already by Bessel, which are caused by the crosswise motions of the stars, the so-called *proper motions,* but these can be averaged out if the observations are made over several years. The technique of measuring distances by parallaxes is generally referred to as

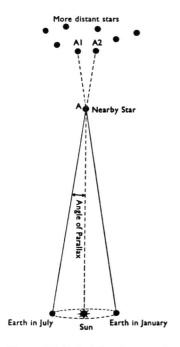

Figure 2.4. Principle of measuring a trigonometric parallax.

(Copied from M. A. Hoskin's *William Herschel and The Construction of the Heavens,* Norton, 1964.)

the technique of measuring trigonometric parallaxes, since we use as a basis for our work the principles of trigonometry. It is really a modification of the techniques used by a surveyor who wants to measure the distance to a church steeple on the other side of a river. The star with the largest trigonometric parallax is our old friend Alpha Centauri, which has a parallax close to 3/4 of a second of arc. Its distance in light-years equals 4.3 light-years. As we noted already, astronomers often make use of a unit other than the light-year for measuring distances. They say that a star with a (par)allax of 1 (sec)ond of arc (which does not exist) would have a distance of 1 parsec. Alpha Centauri has a distance of 1¼ parsecs. To convert from one unit to another, it might be helpful to know that 1 parsec equals 3.26 light-years.

In his studies of 61 Cygni, Bessel measured the position of that star with reference to two faint neighboring stars that were presumably sufficiently far from the sun and earth not to show a measurable annual parallax displacement. His value for the trigonometric parallax of 61 Cygni was $p = 0.3136$ second of arc. He trusted his result to a precision (mean error) of ± 0.0202 second of arc. Bessel's value compares remarkably well with the best available current determination, which gives the parallax of 61 Cygni as being equal to $p = 0.292$ second of arc. Bessel communicated his result with obvious pride in a letter to Sir John Herschel, the son of Sir William, who in 1838 was considered to be Great Britain's top astronomer. The choice of the three stars for the first measurement of trigonometric parallax was a fortunate one because of their wide range in intrinsic brightness. Vega has an intrinsic brightness in visual light equivalent to about fifty times that of the sun, Alpha Centauri is intrinsically just a bit brighter than our sun, and 61 Cygni has an intrinsic brightness less than 1/10 that of the sun.

During the remainder of the nineteenth century, astronomers attempted with mixed success to measure trigonometric parallaxes for many stars. The real breakthrough came with the advent of astronomical photography toward the end of the nineteenth century. The established modern technique for the photographic measurement of trigonometric parallaxes was developed by Frank Schlesinger in 1904–1905. Progress in the field has been slow, because the great majority of the stars have trigonometric parallaxes too small for precision measurement. We possess today parallaxes of reasonable reliability for only about 7000 stars, and additional parallaxes are presently being produced at an annual rate of only 60 to 80. With the best of care, the astronomer cannot now measure the parallax of a star with a mean error of less than $\pm$ 0.004 second of arc, and larger errors are not uncommon.

What is the value of the trigonometric method of parallax measurements for the study of our Milky Way System? By considering the sample of stars that are relatively nearby, say within 50 parsecs of the sun and earth, we can use the measured parallaxes to obtain intrinsic brightnesses for a considerable variety of stars. From a study of the stars within 5 parsecs of the sun, we find that Sirius, the famous Dog Star, the brightest star in the local sample, has an intrinsic visual luminosity equal to twenty-three times that of our sun, whereas the very faint third component of the Alpha Centauri system, generally known by the name of Proxima Centauri, has an intrinsic brightness of only 5/100,000 that of our sun.

We digress a moment to define intrinsic brightness in another manner. It is obvious that the intrinsic brightness of a star can be precisely determined once its apparent brightness and trigonometric parallax are known. Apparent brightnesses are expressed in a scale of *magnitudes.* Altair or Spica are used often as anchor points for first magnitude stars. The faintest stars visible with the naked eye are assigned magnitude +6. The scale of magnitudes depends on ratios of brightnesses. Each successive step of one magnitude amounts to a factor close to 2.5 for the brightness ratio. A star of apparent magnitude +1 radiates, as viewed by us, at the rate of $2\frac{1}{2}$ times that of a second magnitude star. And it produces $(2.5)^2$, about 6.3 times as much light as a third magnitude star. It takes $(2.5)^5$ stars of apparent magnitude 6 to produce as much light as we receive from a first magnitude star; $(2.5)^5$ equals just about 100. For convenience we establish our precision magnitude scales so that a difference of 5 magnitudes in apparent brightness amounts to a factor of 100 in the light received. The precise value of the ratio of brightness corresponding to one magnitude is 2.512, since $(2.512)^5 = 100$.

By international agreement, astronomers define the *absolute magnitude* M of a star as the apparent magnitude which that particular star would have if it were at a distance from us equal to 10 parsecs; at this distance the parallax $p = 0.1$ second of arc. It is not difficult to derive the basic formula relating absolute magnitude M, apparent magnitude m, and parallax p. The resulting formula is

$$M = m + 5 + 5 \log p.$$

On this scale our sun has an absolute magnitude $M = +4.8$, Vega has $M = +1.4$, and little Proxima Centauri has $M = +15.4$. The two components of the double star Alpha Centauri (which appears as a single star to the naked eye!) have, respectively, $M = +4.5$ and $M = +5.9$; they seem like cousins of our sun.

During the second half of the nineteenth century and the beginning of the twentieth century, other methods were developed for the measurement of stellar distances. Stellar spectroscopy and astronomical photography began to play their parts. As we noted already in the preceding section, Kapteyn made excellent use of the data that had begun to accumulate during the nineteenth century. Whereas Sir William Herschel professed no information, certainly very little indeed, about absolute magnitudes of stars, Kapteyn had available for his analysis a basic body of data on absolute magnitudes and on stellar distances from which he would derive a meaningful and reliable general luminosity function for the stars. This function (see Figure 2.5) represents the frequency function of absolute magnitudes for the region within 10 parsecs of the sun. Kapteyn could then use the star counts, which had been obtained especially through his plan of Selected Areas, and derive from these counts by statistical methods of analysis fairly precise data on the distribution in depth of the stars for any direction in the heavens. Figure 2.3 represents the final results of the efforts of Kapteyn. Please note again that the principal difference between Figure 2.1 and Figure 2.3 is that Kapteyn's figure does contain a scale of distances in light-years and in parsecs!

So far we have described only distance determinations for nearby stars. Very few of these are distances greater than 1000 parsecs from the sun. A new development took place early in the nineteenth century; it came very unexpectedly and it concerned the period-luminosity relations for variable stars of the Cepheid variety. Here is the story.

During the 1880s and 1890s, E. C. Pickering, the great early director of Harvard College Observatory, and his associates had undertaken a massive attack upon the study of the variable stars. The photographic plates gath-

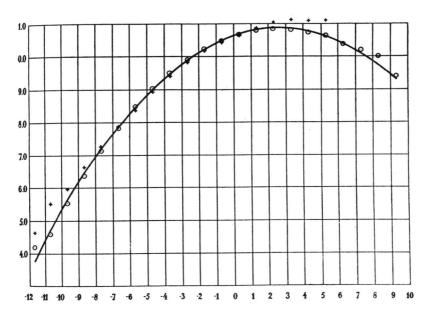

Figure 2.5 Kapteyn's general luminosity function.

The luminosity curve represents the data from stars of all spectral types taken together. Abscissae are absolute magnitudes on Kapteyn's scale in which the unit of distance is taken as 1 parsec; one would have to add 5 magnitudes to the listed abscissae to reduce them to the scale of absolute magnitudes now in use, which applies to the unit of distance of 10 parsecs. Ordinates are logarithms of numbers of stars per thousand cubic parsecs near the sun. This luminosity function was published by Kapteyn and van Rhijn in 1920. (Contributions from Mount Wilson Observatory No. 188.)

ered at Harvard Observatory's Boyden Station near Arequipa, Peru, showed that variable stars are very abundant in the two Magellanic Clouds. These are the two companion galaxies to our Milky Way System. Henrietta S. Leavitt made a special study of the abundant variable stars in the Small Magellanic Cloud (1908), and she found for these stars a close correlation between the periods of the light variation and the apparent brightnesses of the stars. Miss Leavitt concluded correctly that "since the variables are probably at nearly the same distance from the earth, their periods are apparently associated with their actual emission of light." In other words, Miss Leavitt's work suggested that for these variable stars in the Small Magellanic Cloud there exists a relation between period of light variation and absolute magnitude. This suggestion was followed up by Ejnar Hertzsprung (1913), who showed that Miss Leavitt's sample of variable stars consisted mostly of variable stars of the Delta Cephei variety. Hertzsprung (Plate 2.3) suggested that the calibration of the absolute magnitudes of the local Cepheid variables would provide the basis for a calibration of the period-luminosity relation found by Miss Leavitt in terms of absolute magnitudes. Hertzsprung was fully aware of the potential for the exploration of the depths of the universe that was provided by the Cepheid variables and their distant cousins, the shorter-period variable stars of the RR Lyrae variety. Here is where Harlow Shapley first enters the picture. In 1918, using what seems now like a very small number of local stars, Shapley set the zero point of

Plate 2.3. Ejnar Hertzsprung.

Frontispiece from A. Beer and K. Aa. Strand, Editors, *Vistas in Astronomy,* Vol. 8 (**Pergamon Press**, London and New York, 1966).

the period-luminosity relation in terms of absolute magnitudes. On the basis of his calibrations, Shapley derived a distance of 19,000 parsecs for the Small Magellanic Cloud (Plate 2.4). This value is now known to have been too small, the correct value being about 63,000 parsecs as of today. Shapley's luminosity-period curve is shown in Figure 2.6.

The important thing to note is that here we have a completely new technique, based only upon photometric (brightness) measurements and upon determinations of periods of light variation. To obtain an approximate distance for a Cepheid variable or for a shorter-period RR Lyrae-type variable star requires only the measurement of the average apparent magnitude of the star and of the period of light variation. The period-lumi-

Plate 2.4. The Small Magellanic Cloud and the globular cluster 47 Tucanae.

The globular cluster is shown in the upper left-hand corner; the greater portion of the Small Magellanic Cloud occupies the right half of the photograph. A large number of Cepheid variable stars was discovered in the Small Magellanic Cloud by Miss Henrietta S. Leavitt (1908). The illustration represents a section of a photograph made by the author with the Uppsala Schmidt telescope located at Mount Stromlo Observatory (1965). The nebulae in the Small Cloud are shown nicely because a red-sensitive emulsion was used and a red-sensitive filter was employed as well.

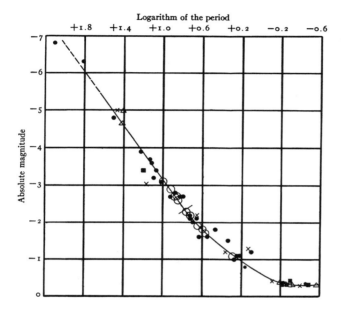

Figure 2.6. Shapley's luminosity-period curve of Cepheid variation.

The curve represents the results (1918) of Shapley's work relating absolute magnitudes (on the presently adopted distance scale using 10 parsecs as a basis for calculations) versus the logarithm of the period of the light variation of the Cepheid variable. (Contributions from Mount Wilson Solar Observatory No. 151.)

nosity relation yields an estimate of the absolute magnitude of the star, and the formula

$$M = m + 5 + 5 \log p$$

provides us with an estimated value for the parallax p, and thus yields a distance.

We note that this technique permits us to undertake the study of very faint and very distant stars. Once we discover a Cepheid variable or an RR Lyrae-type star, we can observe its light curve (Figures 2.7 and 2.8) and find the period of the light variation. All we need to do to obtain a distance estimate is to photograph the discovered variable star repeatedly with our telescope, or monitor the star's brightness variations by photoelectric techniques.

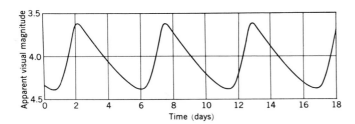

Figure 2.7. The light curve for Delta Cephei. (From B. J. Bok and P. F. Bok, *The Milky Way*, third ed., 1957, Harvard University Press.)

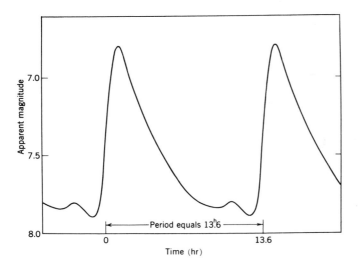

Figure 2.8. The light curve for RR Lyrae. (From B. J. Bok and P. F. Bok, *The Milky Way*, third ed., 1957, Harvard University Press.)

Whereas all of the nineteenth-century techniques were essentially limited to stars within 1000 parsecs of the sun, the period-luminosity relation permits us to go much farther, not only within our own Milky Way System but also to the galaxies that lie beyond it. The stage was set for the exploration of the distant parts of our Milky Way System.

Harlow Shapley and the Center of the Milky Way System

Harlow Shapley received the doctorate from Princeton University in 1913 on the basis of a thesis on properties of eclipsing binaries. His thesis advisor and mentor was Henry Norris Russell, the great Princeton astronomer and astrophysicist. Shapley had become interested in Cepheid variables, and he had especially followed the work of Miss Leavitt and of Hertzsprung. He received a staff appointment at Mount Wilson Observatory in Pasadena, California, and, before leaving for California, he gave much thought to the observing programs that he might undertake with the famous Mount Wilson 60-inch reflector (Plate 2.6). To gain perspective, he visited Harvard

Plate 2.5. Harlow Shapley.

From a photograph reproduced by permission of Science Service, Inc., in Washington, D.C.

Plate 2.6. The 60-inch reflector of Mount Wilson Observatory. (Hale Observatories photograph.)

College Observatory, where he was fortunate to have had some good long sessions with Solon I. Bailey, who gave him very sound advice. He suggested that Shapley should undertake the study of variable stars in globular star clusters, a project for which the 60-inch Mount Wilson reflector was eminently suited. Shapley followed Bailey's advice, and the rest is history.

Bailey had shown that some globular clusters are exceedingly rich in variable stars of the Cepheid and of the shorter-period RR Lyrae variety. The RR Lyrae stars have periods of 1 day and less. They occur not only in the globular clusters, but they are also found in considerable numbers fairly near the sun. The indications were that the RR Lyrae variables all have about the same mean absolute magnitude. Shapley reasoned that it would be possible to calibrate that mean intrinsic brightness on the basis of the variable stars in the vicinity of the sun. With the absolute magnitude M known, Shapley had only to measure the mean apparent magnitude m for these stars, and the logarithm of the parallax for the globular cluster could be obtained from the formula

$$M = m + 5 + 5 \log p.$$

The distance to the globular cluster would become known. Shapley had thus found a way for determining the distances to the globular clusters once he had finished obtaining light curves for a representative group of variable stars in the cluster. Shapley's distances to the globular clusters turned out to be very large indeed. For the bright globular cluster Messier 13, the famous cluster in Hercules, Shapley derived a distance of about 11,000 parsecs, a value so large that many astronomers doubted its correctness. All globular clusters were found to be more than 5000 parsecs distant from the sun. Obviously Shapley had developed a method for exploring our Milky Way System to far greater distances than had been dreamed of by his predecessors (or contemporaries, for that matter). The years 1914 to 1917 were spent mostly in a mass attack on the observational properties of variable stars in globular clusters (Plate 2.7). However, other properties of these clusters received due attention from Shapley as well. He derived approximate distances for 25 globular clusters.

About 1917, Shapley turned to the problem of the distribution of these globular clusters with respect to the distant parts of our Milky Way System (Plate 2.8). In Shapley's day, there were close to 100 globular clusters known.

Plate 2.7. The globular star cluster Omega Centauri.

This is one of the nearest (5200 parsecs), densest, and largest globular star clusters. The original photograph was taken by Ellis W. Miller with the Curtis-Schmidt telescope of the University of Michigan located at Cerro Tololo Interamerican Observatory in Chile. Red-sensitive emulsion and red filter were used.

It had long been recognized that the distribution of these objects over the sky was an uneven one, but, with the notable exceptions of William Herschel's son, Sir John Herschel, and (1909) of a Swedish astronomer, K. Bohlin, no one had paid much attention to the peculiar distribution shown by the globular clusters. They are somewhat concentrated toward the band of the Milky Way, but they are generally absent right along the densest parts of the band itself. If we consider the band of the Milky Way as outlining a galactic equator, then we can measure as viewed from the earth the angular distance of any observed object above or below this galactic equator and call it the galactic latitude of that object. There are very few globular clusters with galactic latitudes between $-5°$ and $+5°$, but there are many with galactic latitudes between $+5°$ and $+20°$, or with galactic latitudes between

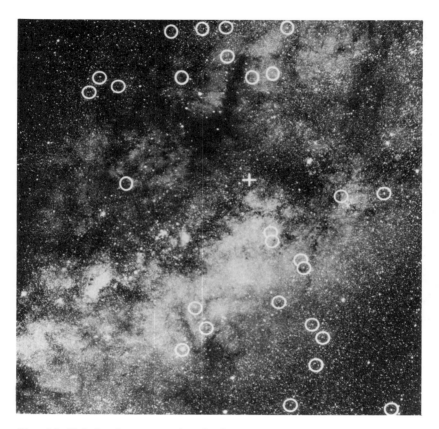

Plate 2.8. Globular clusters near the galactic center.

Positions of 27 globular star clusters have been marked on a section of a photograph of the region of the galactic center made with a 10-inch camera on Mount Wilson Observatory. The position of the galactic center is indicated by a white cross. The area of the sky covered by this photograph is about 750 square degrees, less than 2 percent of the total area of the sky. One-quarter of all known globular clusters are found in this small section. (Courtesy of the Hale Observatories.)

$-5°$ and $-20°$. Another striking fact about the distribution of the globular clusters along the Milky Way is that they seem to occur exclusively along one-half of the band of the Milky Way. There are many of them to be observed when Scorpius and Sagittarius are high in the heavens, but one should not embarrass an astronomer by requesting to be shown a globular cluster through a telescope at a time when Auriga and Orion are high in the sky. Our spring and summer Milky Way is rich in globular clusters, but they are absent from our northern fall and winter skies. One-third of all known globular clusters are found in a region of Sagittarius and Scorpius that covers only 2 percent of the entire sky.

Figures 2.9 and 2.10 illustrate Shapley's results on the space distribution of globular clusters. His three major papers on the subject were published in 1918, including the most famous one of the group entitled "Remarks on the Arrangement of the Sidereal Universe," Mount Wilson Observatory Contribution No. 157. According to Shapley, the globular clusters outline a vast system of their own, one that Shapley supposed would be concentric with our Milky Way System. Our sun is near the outer edge of this vast system at a distance that Shapley estimated to be 17,000 parsecs from the center of the system of the globular clusters. Shapley made the bold hypothesis that our Milky Way System is concentric with the system of globular clusters, and he placed the center of our Milky Way System also at a distance of 17,000 parsecs from the sun in the direction of Sagittarius and Scorpius. In one fell swoop Shapley had removed the solar system from the central position in our Milky Way System. He placed the sun and earth in the outskirts of our Milky Way System. Shapley did for the Milky Way System what Copernicus had done for the solar system. His was truly a Copernican revolution.

Looking back, it now seems perfectly natural that about 1914 someone with access to a large telescope should have undertaken the study of the Cepheid variables and of the shorter-period RR Lyrae variables in globular clusters. Shapley was the man on the spot, and he had the good sense to undertake the task of organizing the basic program, carrying out without delay the measurement and analysis, and of drawing from his material certain apparently inescapable conclusions. It is interesting to note that the work of the great often seems trivial and obvious to the freshman in college ten years later!

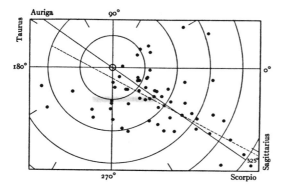

Figure 2.9. Shapley's diagram (1918) illustrating the distribution of globular clusters projected on the galactic plane.

The galactic longitudes are indicated by numbers on the outer rim of the diagram. Please note that these are galactic longitudes on the old system, now abandoned, according to which the direction of the galactic center is toward longitude 325°. The asymmetric distribution of the clusters with respect to the sun (located at the center of the cross) is clearly shown. The large circles have radii increasing by intervals of 10,000 parsecs. All of Shapley's 1918 distances were overestimated because of the neglect of interstellar absorption corrections, which were unknown in 1918. The dotted line indicates the suggested major axis of the system; the center is placed at about 17,000 parsecs from the sun. (Contributions from the Mount Wilson Solar Observatory No. 157.)

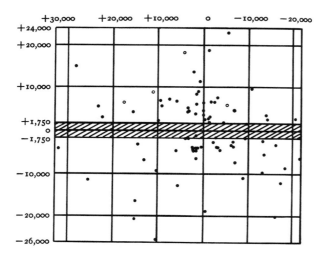

Figure 2.10. Shapley's diagram (1919) illustrating the distribution of globular clusters in a plane perpendicular to the galactic plane.

The numbers on the periphery are again distances in parsecs, now taken with reference to the center of our Milky Way System. The projection is on a plane perpendicular to the central plane of the Milky Way System passing through the sun and the Shapley galactic center. The sun, according to Shapley, should be located in the Milky Way plane (at 0 on the horizontal scale) on the right-hand side of the diagram, at about 17,000 parsecs. (Contribution from Mount Wilson Solar Observatory No. 161.)

I wish that, even today, every young astronomy graduate student could be persuaded to read Shapley's wonderful series of articles on globular clusters that resulted from his work during his years at Mount Wilson Observatory. I read them from beginning to end—and more than once—during my first year as a student at Leiden Observatory (in 1924), and they made a deep impression upon me. The series begins with some rather simple and straightforward observational papers, but gradually there emerges a wealth of relevant information on variable stars, on clusters, and on colors and apparent magnitudes of stars in globular clusters. Shapley was forced to make certain side excursions, the most significant of which was the first attempt at the calibration in terms of absolute magnitude of the period-luminosity relation for Cepheid variables. Shapley also made an important first estimate of the mean intrinsic brightnesses of the RR Lyrae variables, the variable stars that are found in terrific numbers in some globular star clusters. Shapley found their average intrinsic brightnesses to be close to one hundred times that of our sun. These RR Lyrae variables became Shapley's standard lamps in estimating distances to globular star clusters. They continue to serve as *the* basic standards in today's exploration of our galaxy and its neighbors.

Before Shapley, astronomers accepted without much questioning the basic tenet that our sun must be located near the center of the Milky Way System. Shapley changed all that. The distribution in space of his sample of globular clusters with known distances showed beyond doubt that our sun is far from the center of our galactic system. This conclusion should have been obvious for centuries from visual inspection of the Milky Way. If only more astronomers had traveled to the Southern Hemisphere to view the band of the Milky Way in all its glory, we would not have had to wait for Shapley to teach us that we are located in a spot 10,000 parsecs from the center of our home galaxy!

It is difficult to realize now how audacious and unorthodox Shapley's approach seemed to the young astronomers of the early 1920s. Those of us who were at that time interested in the structure of our Milky Way System were from the start imbued with a deep respect for advanced mathematical-statistical analysis. I, for one, had a great admiration for the intricacies of Gaussian integrals, and I truly hoped that some day I might stumble on a solution that would give the world the one and only, definitive, mathematical form for *the* density function of our galaxy!

Shapley's values of the intrinsic brightnesses and estimated distances for the RR Lyrae variable stars and globular clusters were not accepted without challenge. They became the subject for fierce debate, especially in the early 1920s. When I was still in high school in The Hague, Holland, I was an avid reader of the semipopular magazine, *Hemel en Dampkring*. There were lengthy reports about Shapley's work and very well expressed mixed praise and criticism of this work by Dutch astronomers, including the great Dutch Milky Way astronomer, J. C. Kapteyn. Kapteyn doubted the validity of Shapley's calibration of the period-luminosity relation, and he had little faith in Shapley's estimate of the distance of our sun from the newly identified center of our galaxy. Among the first-year students at Leiden the debates about Shapley's work were vigorous and concerned, but in the end few doubts were left that Shapley was basically right.

Why was it that the United States produced the young astronomer who successfully challenged the gathered wisdom of the centuries regarding the basic structure of our Milky Way System? I think that it was more than just the big telescopes that made Shapley succeed. His work was carried out very much in the spirit of the Expanding Frontier. With the Mount Wilson Observatory's 60-inch reflector as his covered wagon, Shapley set out (with Harlow Shapley in the driver's seat and his wife Martha Betz beside him) on a journey of adventure into the unknown. He did not hesitate to question established authority. One hopes that this fine spirit of true enterprise may survive in science for centuries to come. Without it Copernican revolutions will belong to the past.

Controversy and Debate
The first five years following Shapley's announcement of the discovery of the distant center of our Milky Way System (1918) were years of turbulent discussion among workers in the field. Looking back at it all after half a century, it seems in a way amazing how quickly and readily the astronomers of the world accepted Shapley's suggestion that our Milky Way System possesses a distant center. The evidence provided by the distribution over the sky of the globular clusters was just too overwhelming to be denied. With the notable exception of Kapteyn, I do not recollect anyone after 1918 questioning in earnest that the system of globular star clusters does outline the overall Milky Way System. The galactocentric hypothesis (distant cen-

ter in the direction of Sagittarius and Scorpius) won amazingly quickly over
the heliocentric hypothesis, which had held sway for so many centuries.

Some of Shapley's critics argued that our sun was perhaps not at the center of the overall Milky Way System, but they thought that the Kapteyn
System or its equivalent (see Figure 2.3) was a unit of its own. It was often
referred to as the Local System, and it was supposed that a conglomerate
of such units made up the larger Shapley Milky Way System. However, the
controversy and the debate that raged fiercely in 1920 concerned mostly the
correctness of Shapley's distance scale. Here there was indeed good cause for
questioning. Kapteyn and his successor, P. J. van Rhijn (under whom I
wrote my doctoral dissertation), attacked Shapley most directly on his assumption that the RR Lyrae variable stars had mean absolute magnitudes
of the order $M = 0$, indicating intrinsic brightnesses for these stars on the
average close to 100 times that of our sun. They based their attack upon the
fact that these stars showed large crosswise motions in the sky, large *proper
motions,* which they considered indicative of a mean absolute magnitude
4 to 5 magnitudes fainter, making these stars rather equivalent to our sun
in intrinsic brightness. We have already noted that there are many such
stars observed in globular clusters. If Kapteyn and van Rhijn had been
right, the distance to the bright Hercules globular cluster Messier 13 would
have been more nearly equal to 1100 parsecs, one-tenth of Shapley's value
of over 11,000 parsecs. Shapley answered the Kapteyn and van Rhijn criticism very effectively by pointing out that the average linear motions, as indicated by line-of-sight velocities, the *radial velocities* of the RR Lyrae stars
were at least five times as great as was the case for the average stars in the
vicinity of our sun. Shapley reasoned correctly that Kapteyn and van Rhijn's
large proper motions were not indicative of small distances for the RR Lyrae
variable stars observed in the region around our sun, but rather that they
were caused by very large linear velocities for these stars. Shapley was absolutely right; Kapteyn and van Rhijn were wrong.

There was so much discussion about Shapley's new results in the United
States that the National Academy of Sciences in Washington decided to
request Shapley to give a lecture on his views and to invite also one of his
strongest opponents, Heber D. Curtis, then of Lick Observatory in California, to present his views in lectures delivered successively at a meeting
sponsored by the Naional Academy of Sciences, arranged by George Ellery

Hale. The two had agreed to prepare their manuscripts well in advance of the meeting date, so that these manuscripts could be exchanged in advance, thus giving each a chance to revise the original text before delivery. The meeting was scheduled for April 26, 1920, and the two papers were published as Bulletin No. 11 of the National Research Council, entitled "The Scale of the Universe." (published by the National Academy of Sciences in May 1921). Please note that the topic for discussion was *not* the presence or absence of a distant galactic center. The speakers concentrated on the distance scale and related problems. N.R.C. Bulletin No. 11 is still required reading for all interested in the controversies created by Shapley's 1918 announcements.

The two lectures are generally referred to as *The Great Debate,* but in his autobiography *Through Rugged Ways to the Stars* Shapley makes it very clear that it was not really a debate, but that rather the presentations were made peacefully one after the other without further discussion; Shapley spoke first, Curtis next. Curtis vigorously attacked Shapley's new distance scale for the Milky Way System as indicated by the globular star clusters. He suggested a reduction to one-tenth of Shapley's value for the distance to the globular cluster Messier 13. He based his arguments not on the RR Lyrae variables, as had Kapteyn and van Rhijn, but he attempted to show that some of the blue stars in Messier 13 were normal late B stars and considerably fainter intrinsically than Shapley had suggested. Much of the discussion centered upon the brightest star members of the globular cluster, which Shapley supposed to be almost 1000 times as bright intrinsically as our sun, whereas Curtis gave them only a factor 10. There was much discussion about the period-luminosity relation and its validity, which Curtis questioned. Shapely argued correctly that the shape of the period-luminosity relation had been firmly established by Miss Leavitt's work on Cepheids in the Small Magellanic Cloud and that Curtis's scattered and imprecise values for the absolute magnitudes of local Cepheid variables simply reflected the poor precision of the then available basic data on proper motions and radial velocities for these stars. On this score, Shapley clearly won out over Curtis. We should, however, point out that Shapley's distances were generally somewhat overestimated. The reason for this lies in the fact that Shapley—as everyone else at that time—was not aware of the fact that there is considerable dimming of starlight by interstellar solid particles gathered mostly in

cosmic dust clouds located in and close to the central plane of our Milky
Way System.

Shapley would clearly have come out the undisputed victor if the presentations had limited themselves to the relation between globular clusters
and our Milky Way System. But the two protagonists went far beyond our
home galaxy. The concluding seven and a half pages of Curtis's presentation
deal not with our own Milky Way System but rather with "The Spirals as
External Galaxies." During his years at Lick Observatory, Curtis had been
an ardent photographer of the objects that were then called spiral nebulae.
He had come to the conclusion, which was absolutely correct, that these
nebulae are in reality galaxies very much like our own Milky Way System.
He referred to them—in the spirit of an Immanuel Kant—as "Island Universes." Shapley saw them at that time as truly nebulous objects, most of
which lay slightly beyond the limits of our Milky Way System. Shapley
looked upon them as minor affairs, in no way comparable to the majestic
Milky Way System that his researches had revealed. Why did Shapley take
this point of view? The answer is quite simple. There was a staff member
at Mount Wilson Observatory, a Holland-born astronomer named Adriaan
van Maanen, who had made studies of these spiral nebulae with the 60-inch
and 100-inch reflectors on Mount Wilson. Van Maanen had apparently detected fair-sized crosswise motions, proper motions, for some of the knots in
the spiral nebulae, and he had interpreted these motions as indicating
streamings along the spiral arms. Van Maanen's measurements had naturally
made a deep impression on the astronomical profession, for he was recognized as a competent astronomer. However, his measurements were later repeated by others, who proved van Maanen to have been wrong, but these
results were not available in 1920 either to Shapley or to Curtis. Curtis had
the intuition and vision to reject van Maanen's results. He based his conclusions not only on his own photographic work at Lick Observatory, but he
also pointed out that the spectra and radial velocities of these spirals spoke
against their being minor systems, perhaps in the words of Shapley, "little
clouds of gas traveling along with the Milky Way System." The problem
was finally resolved in 1924, when Edwin Hubble reported his study of variable stars in the Andromeda Nebula, Messier 31. His studies proved conclusively that Messier 31, a fine spiral nebula, is at a distance estimated to be
close to 1 million light-years from our sun, and that it is a giant galaxy, a

full-fledged Milky Way System, comparable to (or perhaps even larger than) our own Milky Way System, our home galaxy.

Shapley came out on top with regard to the distance scale of our Milky Way System and the system of globular star clusters, but Curtis was right with regard to the Island Universe hypothesis. The impact of the Shapley and Curtis presentations upon astronomical thought was terrific. I remember how excited I became about THE DEBATE when I read about it (in my last year in high school) in the Dutch semipopular magazine, *Hemel en Dampkring*. I wrote a letter to the public library in my home city of The Hague urging them to purchase from the National Academy of Sciences in Washington a copy of National Research Council Bulletin No. 11. I told them that no self-respecting public library would be considered complete without a copy of the Bulletin on its shelves, and I rather hope that today it can still be found there or in some of the archives! Anyhow, they purchased a copy, and I read it carefully before I entered Leiden University.

The Galactic Center, 1920 to 1973

The years following the Shapley-Curtis debate were momentous ones for the development of astronomy. We have already noted that the universe of galaxies came into bloom with Hubble's 1924 discovery of Cepheid variable stars in the Andromeda galaxy, Messier 31. What role did Shapley's identification of the galactic center play during the years following 1920?

By 1925 Shapley's outline of the galaxy, with its distant galactic center, had become generally accepted. In 1926–1927, there followed another momentous discovery relating to our Milky Way System (to which we shall from here on refer affectionately as "our galaxy"). Bertil Lindblad in Sweden and Jan Hendrik Oort in Holland had made the spectacular discovery of galactic rotation. Their work had shown that our galaxy derives its highly flattened shape from a general rotation around Shapley's galactic center. Lindblad, who was the first to suggest seriously the existence of such a rotation, proved that the characteristics in the motions of the stars near the sun, especially the phenomenon of star streaming, can be understood if we assume that the sun and all the stars near it move in roughly circular orbits around the galactic center. He suggested that the primary gravitational force is supplied by that center. Oort pointed out that, as in our solar system, the outer parts of our galaxy would then move at slower rotation rates than the inner parts, and he found from studies of radial velocities of stars within 1500 parsecs of the sun that this was indeed the case. The Oort-Lindblad theory proved once and for all that the central regions of our galaxy are very massive indeed, so much so that they, rather than local conditions relatively near the sun, control the motions of all stars in our galaxy. According to present-day estimates the total mass of our galaxy is close to 100 billion (1×10^{11}) solar masses, with probably close to 10 percent of the total concentrated within a relatively small galactic center. The best available value—the one officially adopted by the International Astronomical Union as the "correct" one!—for the distance from the sun to that galactic center is 10,000 parsecs; Figure 2.11 shows the up-dated version of Figure 2.10. Ac-

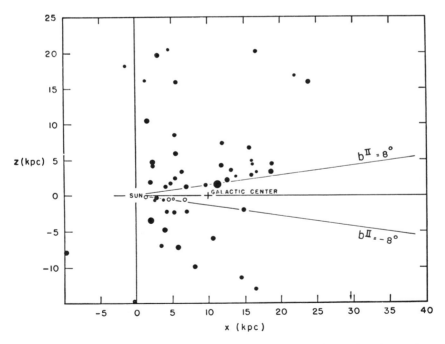

Figure 2.11. H. C. Arp's diagram (1964) for the distribution of globular clusters in a plane perpendicular to the galactic plane.

Figure 2.11 represents an up-dated version of Figure 2.10. The distances are measured in kiloparsecs, 1 kiloparsec is 1000 parsecs. (From a chapter in A. Blaauw and M. Schmidt, *Stars and Stellar Systems*, Vol. 5, 1964, University of Chicago Press.)

cording to the same IAU vote, the best available value for the circular velocity of the sun and the stars near us around the galactic center amounts to
250 kilometers per second. If these values are correct, then our sun completes one galactic revolution around the center in 250 million years.

We had a rather curious situation between 1927 and 1930. Galactic rotation seemed to have been established, and it was taken as proof that our
galaxy is a homogeneous gravitational unit with the inner parts moving
faster around the center than the outer ones. However, no one had yet disproved the validity of the Kapteyn System as a Local System, the way it is
shown in our Figure 2.3. How could such a system possibly exist and hold
together in a galaxy with differential rotation? This question was answered
in 1930, when Robert J. Trumpler of Lick Observatory proved conclusively
that all stars in or near the galactic central plane are subject to dimming
of their light by interstellar absorption. Trumpler showed that, on the average, this absorption amounts to a dimming by two magnitudes over a distance of 3000 parsecs, a value that still holds pretty well as an average today.
The neglect of interstellar absorption in Kapteyn's work had yielded the
result of Figure 2.3. When the proper corrections for this general absorption
are made, then most of the apparent concentration of stars near the sun
disappears.

At present the study of the central regions of our galaxy is increasingly
assuming greater importance. The phenomena taking place in the central
regions of our galaxy are not especially spectacular when compared to the
explosive events now studied in the nuclei of many galaxies beyond ours.
And yet, the study of the central regions of our galaxy has taught us much
that is new and unexpected about our own Milky Way System. In the concluding paragraphs of this essay I shall briefly review some aspects of today's
studies of the nucleus of our own galaxy. We must distinguish here from the
start clearly between the *nuclear region* of our galaxy, which is considered
to extend up to 5000 parsecs (half the distance to the sun) from the center and the very heart of this nuclear region, which is the center itself. By
optical means we can do much work now on the nuclear region, but the center itself has probably been seen (if one wants to call it that) only in the
very far infrared. Current estimates indicate that in visual light 25 magnitudes of obscuration are produced between us and the center itself.

The Great Star Cloud in Sagittarius marks the direction toward the galactic center. In addition to the globular clusters, RR Lyrae variable stars, novae, planetary nebulae, many other objects show marked concentration toward the center. Walter Baade demonstrated in the 1940s that the varieties of stars that are mostly found in the central region of our galaxy are very different from those for the regions near our sun. He called the varieties near us *Population I*, and those in the central regions *Population II*. General indications are that there is a far greater preference for older stars in the central region of our galaxy than we find in the parts where the solar system happens to be located. The indications are that most of the stars in the central region of our galaxy were formed earlier than the stars in the regions near the sun. It is not unlikely that star formation may have almost ceased in the central regions, while apparently it is still quite active in the outer parts, the sections where we find ourselves.

The very birth of radio astronomy was related to the center of our galaxy. Karl G. Jansky, in the early 1930s, made the discovery that radio noise was reaching us from beyond the earth and that it was strongest when the Great Star Cloud of Sagittarius was highest in the heavens over Holmdell, New Jersey! The galactic center announced its presence through its disturbing radio noise. Radio maps of the central region of our galaxy became more and more detailed, and in 1951 J. H. Piddington and H. C. Minnet in Australia located a discrete radio source in Sagittarius, which obviously marks the direction to the galactic center. This central source, generally referred to as *Sagittarius A*, emits radiation of the synchrotron variety, which can be produced only by electrons accelerated in very strong magnetic fields. Radio astronomy has made further important discoveries. There is neutral atomic hydrogen in the central regions, and it has a state of motion that is apparently a highly disturbed one. There is a rapidly rotating inner disk, and there are some vast expanding spirallike features, which are apparently being shot out of the central regions. Some of the neutral atomic hydrogen escapes from the nucleus and seems to rain back on to our galaxy after having moved first far away from the central plane. The strongest radio sources in the central region of our galaxy have proved to be happy hunting grounds for the radio astronomers interested in discovering new varieties of interstellar molecules.

In a way, the results from infrared studies have been the most spectacular. There is a strong infrared source precisely at the position occupied by the

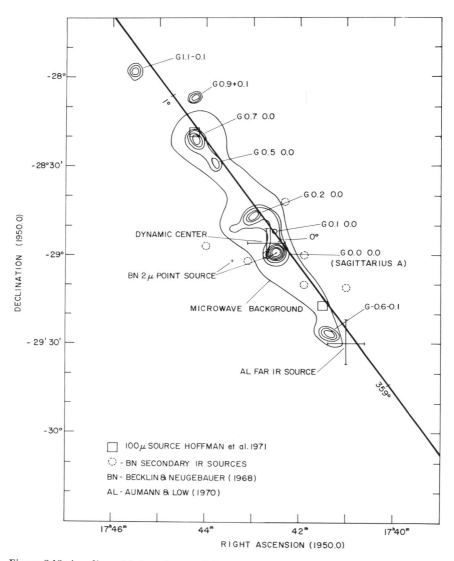

Figure 2.12. A radio and infrared map of the central region of our galaxy.

The map was prepared by Richard Capps of Kitt Peak National Observatory. It shows the positions of the more important radio sources observed in directions not very different from the direction to the center of our galaxy; these sources are indicated by their galactic longitudes (G) and latitudes (with a + or − sign) according to a survey by Downes, Maxwell, and Meeks of the distribution of radio sources at the wavelength 3.75 centimeters. The elongated contour is the result of infrared observations at 100-micron wavelength by Hoffmann, Frederick, and Emery. Positions of two secondary sources are marked by squares. Broken circles give the positions of infrared sources observed at 2.2-micron wavelength by Becklin and Neugebauer. Other important sources detected by Aumann and Low are also shown. (Courtesy of Kitt Peak National Observatory.)

radio source Sagittarius A. Its diameter has been estimated as equal to only 15 parsecs, and at the very heart of this infrared source is a pointlike source with a diameter of perhaps only 1 parsec. This central source has terrific intensity at longer infrared wavelengths than the discovery wavelength of about 2 microns. In the far infrared, between 20 and 100 microns, it radiates with an intensity equivalent to perhaps 10 million suns. We show in Figure 2.12 a combined radio and infrared map of the central region of our galaxy. There is currently much speculation about the true nature of these central sources, some of which may originate as a by-product of dust-embedded nebulae.

Radio and the infrared are not the only techniques through which we begin to resolve the central regions of our galaxy. High-energy gamma radiation has now been detected as coming from the band of the Milky Way, and a broad maximum is shown for the direction of the galactic center. Gravitational waves apparently emanate from the center. Shapley's galactic center is here to stay, and it obviously bears watching!

Selected References

W. de Sitter, *Kosmos*. Harvard University Press, Cambridge, Mass., 1932. De Sitter's Lowell Lectures contain the story of the development of Milky Way research between 1750 and 1931.

A. Berry, *A Short History of Astronomy*. Dover Publications, New York, 1961. An excellent account of eighteenth- and nineteenth-century developments in astronomy.

Michael A. Hoskin, *William Herschel and the Construction of the Heavens*. W. W. Norton and Company, New York, 1964. The original papers by William Herschel, with astrophysical notes by D. W. Dewhirst.

Harlow Shapley and Helen E. Howarth, *A Source Book in Astronomy*. McGraw-Hill Book Co., New York, 1929. A collection of historical papers published before 1900.

Harlow Shapley, *Source Book in Astronomy, 1900–1950*. Harvard University Press, Cambridge, Mass., 1960. A fair selection of the more important astronomical papers published during the first half of the twentieth century.

O. Struve and V. Zebergs, *Astronomy of the Twentieth Century*. Macmillan, New York and London, 1962. An excellent narrative with good references.

Harlow Shapley and Heber D. Curtis, "The Scale of the Universe." Bulletin of the National Research Council, No. 11, Washington, D.C., May 1921. The texts of the two famous papers.

H. Shapley, *Through Rugged Ways to the Stars*. Charles Scribners, New York, 1969. Shapley's autobiography.

H. Shapley, *Studies Based on Colors and Magnitudes in Stellar Clusters*. Mount Wilson Observatory Contribution Nos. 151, 152, 153, 154, 155, 156, 157, 160, 161, reprinted from Vols. *48, 49,* and *50* of the *Astrophysical Journal*, 1918 and 1919. Papers 6 to 14 in the Shapley series are the original papers relating to the discovery of the center of our galaxy.

P. van de Kamp, H. S. Hogg, C. Schalen, B. J. Bok, and H. Hoagland, *Five Articles Honoring Harlow Shapley on His Eightieth Birthday. Publications of the Astronomical Society of the Pacific,* Vol. 77, pp. 325, 336, 409, 416, and 442, 1965.

B. J. Bok and P. F. Bok, *The Milky Way*. Fourth ed. Harvard University Press, Cambridge, Mass., 1974. The story of our galaxy brought up to date.

Donald W. Goldsmith

3 Edwin Hubble and the Universe outside Our Galaxy

Introduction

The final stages of the Copernican revolution occurred almost fifty years ago, with the discovery that our own Milky Way is just one galaxy among countless other galaxies, as our sun is one among a multitude of stars. Far from being the center of the universe, the earth ranks only as a small follower of an average star in the outer reaches of a common sort of galaxy. As recently as 1920, well within living memory, we could not be sure whether the Milky Way galaxy contains all the objects visible to astronomers, or whether some of these objects, the "spiral nebulae," lie at immense distances *outside* our home galaxy of stars.

In the seventeenth century, the work of Copernicus resulted in the establishment of the fact that our earth is not the unmovable center of the universe. Three hundred years later, humanity discovered the general arrangement of the universe on the largest scale of distances. Before 1920, we knew only that multitudes of stars surround us as far as we can see. Ten years later, we knew that most stars are clumped into galaxies, and that the galaxies are all moving away from one another. To one man, Edwin Hubble (1889–1953), we owe much of the credit for understanding the place that our galaxy occupies in the cosmos, and for recognizing that the entire universe is actually expanding. In the space of ten years (1920–1929), Hubble made four great classifications that revealed the large-scale arrangement of the universe. First, he established beyond any reasonable doubt that spiral nebulae are in fact other galaxies similar to our own Milky Way galaxy. Second, in one of science's great moments of synthesis, Hubble realized from observations of other galaxies' distances and motions that *all* galaxies are moving away from one another. Third, Hubble found that the number of galaxies of different apparent brightnesses behaves in a manner consistent with an even distribution of galaxies in space, averaging over the greatest distances. Finally, Hubble's patient cataloguing of galaxies showed that almost all galaxies can be typed as "spiral" or "elliptical," types similar in their masses and sizes but quite different in appearance. Though Edwin Hubble was fortunate to have the use of the largest telescopes then in existence, and to receive valuable collateral effort from the work of V. M. Slipher, Milton Humason, Nicholas Mayall, and several other astronomers, the achievements listed here testify to Hubble's brilliant combination of patience and intuition.

Department of Earth and Space Sciences, S.U.N.Y., Stony Brook, New York 11790.

The Milky Way as One among Other Galaxies

In the preceding chapter, Bart Bok has described Harlow Shapley's study of the spatial distribution of globular star clusters, which determined the scale of our own galaxy and the sun's location within it. From J. C. Kapteyn's investigations of the distribution of stars in space, it was clear that our galaxy forms a flat, dishlike system; Shapley proved that the sun lies near the system's edge. But a key question remained unanswered: Does our Milky Way galaxy comprise almost all of the visible universe, or does it contain only a tiny fraction of the luminous material that astronomers photograph with their great telescopes? By 1920, this question of the Milky Way's importance in the universe demanded an answer, just as four centuries earlier Copernicus strove to determine the importance of the *earth's* position in space. Copernicus's revolutionary ideas produced a century of argument, but the question of the nature of the Milky Way caused only a decade of confusion.

By the time that the First World War began, astronomers had photographed many fuzzy-looking "nebulae" that were clearly more complex in shape than stars. The brightest and best-known of these nebulae appeared on a famous list made in the 1780s by a French amateur astronomer named Charles Messier. These nebulae, often called "Messier objects," still carry numbers like M 31 and M 42 from Messier's catalogue. Today Messier's list appears to be a rag-bag collection of disparate objects, for better telescopes and astrophysical research have revealed that all sorts of star clusters, gas clouds, and galaxies form the different fuzzy patches that Messier noted in his search for comets. Some of the Messier objects turned out to be globular star clusters, like those that Shapley used to establish the configuration of our galaxy. Others are clouds of glowing hot gas, heated by groups of extremely bright stars, like the nebula that forms the middle "star" in the sword of Orion (M 42). Still others are "planetary nebulae," spherical shells of hot gas surrounding a single, ultra-intense star. But the greatest and most significant of Messier's nebulae are the *galaxies,* such as M 31 in Andromeda (Figure 3.1), M 33 in Triangulum (Figure 3.2), and M 51, the Whirlpool Nebula in Canes Venatici (Figure 3.3). Most of these spectacular objects have an elaborate spiral structure, with their spiral arms confined to a planar system, so that a spiral galaxy seen edge-on appears thin indeed (Figure 3.4). In 1919, when Hubble began his astronomical research (he had

Figure 3.1. The great spiral nebula in Andromeda (M 31). (Courtesy of the Hale Observatories.)

Figure 3.2. Messier 33, a nearby spiral nebula in Triangulum.

Foreground stars appear as points of light; the brightest of these have diffraction spikes
produced by the telescope optics. (Courtesy of the Hale Observatories.)

Figure 3.3. The "Whirlpool" spiral nebula (M 51) in Canes Venatici. (Courtesy of the Hale Observatories.)

Figure 3.4. A spiral nebula (number 4565 in the New General Catalogue) seen almost edge-on. Note the obscuring dust lane in the central plane of the nebula. (Courtesy of the Hale Observatories.)

practiced law briefly before he decided to return to graduate school and
had enlisted in the army just after receiving his doctorate), he decided to
concentrate on problems involving nebulae. By this time, several thousand
"spiral nebulae" had been catalogued, and the phrase was used as a general
expression that included both the spiral objects and the more spherical (in
actuality, ellipsoidal) systems that we now call elliptical galaxies (Figure
3.5). Where were these spiral nebulae located in space? Were they at distances of hundreds of light-years? Thousands? Or farther still, so far that
they must lie outside the Milky Way System and form "Island Universes"
in their own right?

The second half of the "Great Debate" between Shapley and Heber Curtis in 1920 concerned precisely this point, and, as Dr. Bok pointed out, for
once Shapley was on the wrong side. From today's viewpoint, it seems hard
to believe that Shapley failed to recognize that spiral nebulae are galaxies in
their own right. Spectroscopic observations dating from 1899 on had shown
that, when the light from spiral nebulae was divided into its component
colors, the relative amounts of the light emitted with various colors were
close to the proportions found for the light from ordinary stars. The conclusion that spiral nebulae are complex agglomerates of many stars followed
rather naturally. Also, spiral nebulae were found in all regions of the heavens *except* for a narrow band around the sky, which Hubble later called the
"zone of avoidance." This phenomenon could be explained if our galaxy
were a flattened system that contains many dust particles spread diffusely
through its equatorial plane. Such dust absorbs light quite efficiently, as
seen in Figure 3.4. In addition, in 1917 a Swedish astronomer named Kurt
Lundmark considered observations of novae in the Andromeda Nebula.
Such novae, "new" stars that show a sudden outburst of light, had been well
studied in the Milky Way. The distances to some of them could be measured by trigonometric methods, and thus the true brightness of a nova could
be found from the measurement of its apparent brightness and its distance.
Since the nova outbursts in the Andromeda Nebula showed a time behavior
similar to that of the familiar, brighter novae, one could assume that the
Andromeda novae had about the same *true* brightness as the familiar novae,
and that their lesser *apparent* brightness must indicate a great distance to
the Andromeda Nebula. All of these novae showed a rapid rise in the stars'
apparent luminosities, followed by a slow fading away and no subsequent

Figure 3.5. A small elliptical galaxy (NGC 205).

This galaxy is a satellite of the Andromeda spiral (see Figure 3.1). (Courtesy of the Hale Observatories.)

outburst. Lundmark's comparison of the novae in the Andromeda Nebula with the novae that were definitely part of the Milky Way suggested that the nebula was about half a million light-years away from us. This result agreed with estimates made independently by Curtis. Even the largest diameter derived for the Milky Way galaxy, the 300,000 light-years obtained by Shapley, was less than this distance, which would have placed the Andromeda Nebula definitely outside our own star system.

For us today this seems easy to understand; in 1920 it was not. There were two main reasons for this hesitation.

First, the Milky Way appeared to be far larger than any of the spiral nebulae. If the Andromeda Nebula were distant by half a million light-years, as Curtis and Lundmark suggested, then its angular size would imply a diameter of 30,000 light-years, just one-tenth of the diameter calculated by Shapley for the Milky Way. And yet the Andromeda Nebula ranks among the most structurally developed of the visible spirals. Could our galaxy truly be by far the largest of all? The Copernican revolution had come full turn. In Copernicus's time astronomers hesitated to adopt a cosmology in which our earth's location in space was *not* central and predominant. Now in the twentieth century astronomers hesitated to assign a special, preeminent role to our location in space. This conflict was not resolved until the early 1950s, when a reevaluation of the distance to the Andromeda galaxy tripled the original value, giving a distance of 1½ million light-years. By then, Robert Trumpler had found that within our galaxy interstellar dust absorbs significant amounts of light and thus reduces the apparent brightness of globular clusters, so that the Milky Way galaxy spans only 100,000 light-years, rather than Shapley's 300,000. The two corrections in the distance scales imply that our galaxy and the Andromeda galaxy have almost the same size, but this was unknown fifty years ago.

Aside from the feeling that our galaxy could hardly be king of the universe, there was a more potent argument against the possibility of spiral nebulae lying at great distances from us. A Dutch astronomer, Adriaan van Maanen, took a series of photographs of spiral nebulae with the great 60-inch and 100-inch telescopes at the Mount Wilson Observatory. When van Maanen compared his photographs with similar photographs made a few years earlier, his measurements revealed transverse motions of individual bright knots in the nebulae. In Figure 3.6, which shows a plate from van

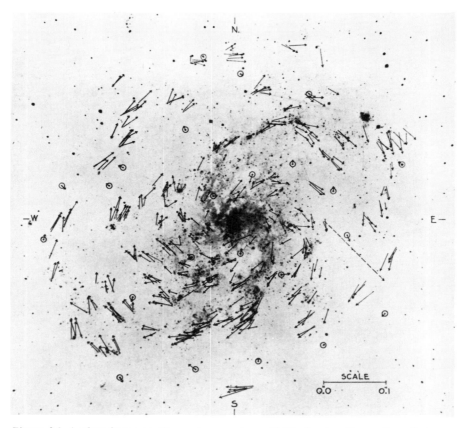

Figure 3.6. A plate from van Maanen's research on M 33, showing the motions that van Maanen claimed to have detected.

The length of each arrow shows the distance that the bright knots would appear to move (with reference to the stars in circles) during 2500 years. (Courtesy of the Hale Observatories.)

Maanen's work on the spiral nebula M 33 in Triangulum, each arrow indicates the apparent direction of motion and the length of the arrow gives the angular motion that would occur in 2500 years. Van Maanen felt that the reference stars (in circles) were far enough away to have insignificant yearly motions across the sky. In reality the nebula, which lies hundreds of times farther away than the foreground "reference" stars, has an angular transverse velocity many times less than the stars. It is hard to look at a photograph like Figure 3.6 and not be convinced that the entire galaxy is rotating, if we believe the comparison of two photographs taken twelve years apart to be free from error. The paper that included Figure 3.6 was the *tenth* in a series which van Maanen wrote during the late 1910s and early 1920s. He found rotational motions in all the famous spirals: M 31, M 33, M 51, M 81, M 101. The rate of motion that he found was about one hundred-thousandth of a revolution per year—not very large, but fatal to the idea of spiral nebulae being outside our galaxy! For if a nebula did indeed rotate once in a hundred thousand years, then a distance of even 5000 light-years would mean that the nebula's outer regions must move at the enormous speed of 300 miles per second. A distance of 500,000 light-years would imply a speed of 30,000 miles per second, one-sixth the speed of light. This was too fantastic for belief: no object could be thought to keep from flying apart when spinning about at 30,000 miles per second!

Van Maanen was a capable and stubborn man, and even in 1935, after checks of his measurements had failed to show any rotation of galaxies, and other observations had removed all doubt that spiral nebulae are outside our galaxy, he would only state that such results "make it desirable to view the motions with reserve." In 1920, when Shapley supported the contention that spiral nebulae are within our own galaxy, van Maanen had no such reserve, and Shapley believed van Maanen's measurements to be accurate. In actual fact, some form of systematic error, perhaps the shrinkage of the emulsion on the older photographs, must have affected van Maanen's comparison of the plates exposed at two different times, for no angular movement of rotating galaxies has yet been measured. It is an ironic fact that spiral galaxies *do* rotate, but their rotation cycle takes hundreds of millions of years, rather than the hundred thousand that van Maanen claimed to have found. This rotation produces an effect too small to be measured directly within a time interval of ten or a hundred years.

When in 1920 Curtis and Shapley debated the distances to spiral nebulae, the weight of hard evidence fell on the incorrect side, Shapley's. To believe that spiral nebulae lie outside our own Milky Way, it was necessary to conclude that van Maanen's measurements were totally incorrect, and (apparently) that our galaxy is ten times larger than prominent spirals like the Andromeda Nebula (M 31) and the nebula in Triangulum (M 33). On Curtis's side were the observations of novae in spiral nebulae, though one could assert that a different type of novae must exist in these nebulae. In fact, a "supernova," thousands of times brighter than an ordinary nova, had appeared in the Andromeda Nebula in 1885, and one could assume that the supernova was an "ordinary" nova and the Andromeda Nebula not far away. We now know that supernovae form a separate class of exploding stars, far brighter and much rarer than novae. On the average, supernovae occur once every 50 or 100 years in a spiral galaxy (see Figure 3.7). (The last observed supernova in our galaxy occurred in 1604, so we are due for another soon.)

Curtis did have a powerful argument for great distances to spiral nebulae that proceeded by deduction. Look at the spirals, he said, and we find some large and some small. Because all spiral nebulae have about the same structural appearance, is it not reasonable to assume that the spirals which seem smaller are farther away? But if this be so, then a spiral nebula with an apparent diameter one hundred times smaller than that of the Andromeda spiral should be a hundred times more distant, so if the Andromeda Nebula were 10,000 light-years away (like a nearby globular cluster), the smaller spiral would be 1 million light-years from us. Then the farther spirals would definitely lie outside the Milky Way System and be the most distant objects known.

Curtis's line of reasoning was cogent but inconclusive. Astronomers remained confused over the distance to spiral nebulae until Hubble made a discovery that resolved the dispute.

In the early 1920s, Hubble began a systematic observing program to find variable stars in spiral nebulae. By taking long-exposure photographs with the 100-inch reflector, then the world's largest telescope, Hubble recorded the brightest stars in the outer regions of spirals on many different occasions (Figure 3.8). In this way, he planned to detect both the novae (with one-time outbursts) and the periodic variables (with recurring cycles of brightness

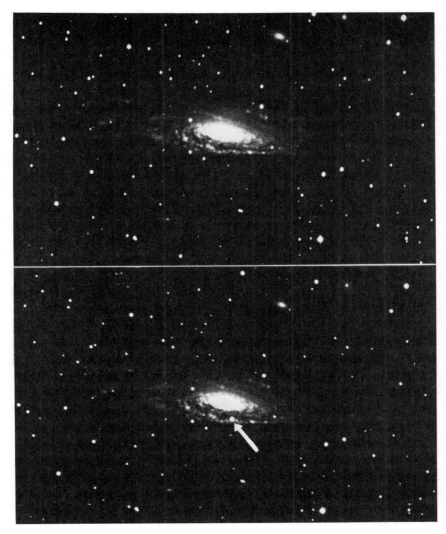

Figure 3.7. A supernova (indicated by the arrow) that appeared in the galaxy NGC 7331 in the year 1959. (Courtesy of Lick Observatory.)

Figure 3.8. An outer region of the Andromeda galaxy, photographed with the 100-inch telescope. Some of the stars in this region are Cepheid variables whose variability was first detected by Hubble. (Courtesy of the Hale Observatories.)

variation). Hubble's first good photographs of the Andromeda Nebula showed two novae and another faint new star that also seemed to be a nova. But older plates of the nebula showed the previous appearance of a new star at precisely the same location. By the end of 1923, Hubble had established that he had found a "Cepheid" variable star with a light cycle of about one month. Some fifteen years of study had revealed a great deal about Cepheids. The marked relationship between their *periods* of variation and their apparent brightness, discovered for stars in the Small Magellanic Cloud by Henrietta Swan Leavitt (see the previous chapter), allowed the calculation of a Cepheid's true brightness from a measurement of the light cycle, once Shapley established the distances to Cepheids in our own galaxy. Luckily, Cepheid variables have enormous true brightnesses and thus are visible even at great distances. Thousands of Cepheid variables had been studied in the Magellanic Clouds (now known to be satellites of the Milky Way), and their characteristic changes in brightness were familiar to astronomers. Hubble's discovery of a Cepheid variable in the Andromeda Nebula could hardly be dismissed. The star's cycle of light variation corresponded to a Cepheid with a true brightness about 7000 times the true brightness of our sun, or of the next nearest star, Alpha Centauri. Since the Cepheid's apparent brightness was 15 million times less than the apparent brightness of Alpha Centauri, one could conclude that the great distance to the Cepheid was the cause of its faint appearance. The distance to the Cepheid variable had to be about 200,000 times the distance to Alpha Centauri ($4\frac{1}{2}$ light-years) in order for the Cepheid to have its observed brightness. This gave a distance of 900,000 light-years to the star in the Andromeda Nebula, and established the nebula as a true *extragalactic* system beyond our Milky Way.

Once Hubble had detected the first extragalactic Cepheid, he increased his efforts. Within a year, he had found eleven more Cepheid variables in the Andromeda spiral, and in 1929 he published an exhaustive study of that galaxy which listed 40 Cepheids and 86 novae, and confirmed his original distance estimate. In addition, Hubble observed other galaxies, and by 1926 he found 35 variable stars in M 33 and 11 in an irregular galaxy, NGC 6822. All of these stars showed the light variations typical of Cepheids, and they allowed Hubble to find that the galaxies that contained them had distances of 1 to 2 million light-years. This confirmed the conclusion that spiral nebulae are separate galaxies, although the Milky Way did seem to be the largest star system. Eventually a correction to the true brightness of the Cepheids raised the distance to the Andromeda galaxy to $1\frac{1}{2}$ million light-years from Hubble's 0.9 million, implying a larger diameter for this spiral and other similar galaxies.

Hubble's discovery of Cepheid variables in spiral nebulae marked a turning point in our view of the distribution of stars in space. Following his work, almost all astronomers accepted the idea that most stars are clumped into galaxies, the "Island Universes" suggested by Curtis, with relatively little (if any) matter strewn among the galaxies. In addition, Hubble's discovery of Cepheids in spiral galaxies furnished a way to find the distances to the nearest galaxies, which showed Cepheid variables bright enough to be photographed as individual stars. The time was ripe for Hubble to use his measurements of the distances to galaxies to make his greatest discovery, that the entire cosmos is expanding.

The Motions of Galaxies and the Expanding Universe
As the 1920s sped by, filled with exciting events like Coolidge's presidency, Lindberg's flight to Paris, Babe Ruth's great home run seasons, and the glory days of Prohibition, Edwin Hubble determined the distances to the brightest spiral galaxies by observing Cepheid variables within them. He then sought to compare the galaxies' distances with their *motions* toward us or away from us. Because of the "Doppler effect," observations of the light from distant galaxies can determine the galaxies' speeds along the line of sight. To understand the Doppler effect, we may consider light rays as a series of undulating wave motions, with crests and troughs succeeding each other as the light moves through space. The distance between two successive wave crests (or two successive troughs) is the *wavelength* of the light. When light waves enter our eyes and impinge on our retinas, each particular numerical value of the wavelength corresponds to a particular color. Blue light has a wavelength so small that 50,000 waves of blue light can fit into one inch. Red light has a longer wavelength, and other colors have intermediate wavelengths. What we call "white light" is a mixture of light rays with many different wavelengths and colors.

Light rays can be divided into their component colors, called the *spectrum* of light, by passing the light through a prism, or by reflecting it from a surface with thousands of parallel grooves spaced regularly a few times 10^{-5} inch apart. Astronomers use this effect to photograph the wavelength components of the light from celestial bodies like stars and galaxies. These celestial spectra tell us much about the objects' composition and motions. First of all, each kind of atom tends to interact only with certain wavelengths of light, either in emitting light rays when the atoms are heated, or in absorbing light when the atoms are located in front of an intense source. For example, salt crystals (sodium chloride) thrown into a flame produce yellow light, which is characteristic of the way that sodium atoms emit radiant energy as light rays. The same shade of yellow comes from high-intensity sodium vapor lamps now being installed on many city streets. Furthermore, sodium atoms placed in front of a source of white light will *absorb* only the yellow light with the same wavelength that is characteristic of sodium atoms' emission. Similarly, calcium atoms absorb and emit a definite wavelength of blue light. From years of analyzing stars' spectra, astronomers have determined the chemical composition of hundreds of thousands of stars in the Milky Way galaxy, simply by comparing the wavelengths that are absorbed in the outer, cooler regions of the star as the "white light" from the stars' fiery interiors makes its way outward into space. In this way, more than seventy different kinds of atoms (elements) have been found in the outer layers of our own well-observed sun.

The spectra of galaxies reveal that their light resembles the light from billions of stars, because the various wavelengths that are absorbed correspond to the wavelengths characteristic of stars in our own galaxy. The same familiar wavelengths tend to be absent when the galaxies' light is spread into colors for detailed analysis. But subtle shifts do appear, and can be explained as the Doppler effect.

C. J. Doppler was a nineteenth-century Austrian physicist who considered what happens to wave patterns if the source of the wave moves with respect to the observer of the wave. Doppler realized that if the source approaches the observer or the observer approaches the source, then more wave crests will reach the observer each second than would be the case for no relative motion. The relative approach of source and observer just allows the observer to meet wave crests (or troughs) more quickly. The observer will see a reduction in the interval between wave crests and will interpret this as a reduction in the wavelength. Conversely, if the source of waves recedes from the observer or the observer from the source, then fewer wave crests will arrive per second. The wavelength will appear to increase over the case of no relative motion. When an observer measures the light from a distant object, the motion of the source of light waves will affect the wavelength observed: relative motion of the source toward the observer always decreases the observed wavelength, and relative motion away always increases the measured wavelength. This "Doppler shift" in the wavelength reflects the relative motion between the wave source and the observer and does not arise from anything intrinsic to the wave source. If we hear sound waves from an approaching ambulance, the siren's pitch will be higher (the waves have shorter wavelength) than when the ambulance recedes, so longer wavelengths give a lower pitch. The Doppler effect operates in an analogous manner for light waves, and the "Doppler shift" is the difference between the wavelength that we *do* observe and the wavelength that we *would* observe if there were no motion of the source relative to the observer. For light waves with Doppler shifts much smaller than the original wavelength, the amount of the Doppler shift of the wavelength divided by the original wavelength is directly proportional to the relative velocity between the source and the observer.

We can use the proportionality for the Doppler shift as a way to find the relative velocity, provided that we know the original wavelength and the amount of the Doppler shift. Luckily, experience has taught astronomers how to recognize certain characteristic patterns in the spectra of stars and galaxies. If astronomers find a spectrum that has several wavelengths absorbed in a familiar pattern but all the wavelengths are shifted by the same relative amount, then we can conclude that the source is in motion relative to ourselves. The Doppler formula allows us to find the amount of this relative velocity. So it goes for galaxies: their speeds toward us or away from us can be measured from the Doppler shift of the absorption "lines" in their spectra. Figure 3.9 shows a photograph (upper panel) of a distant cluster of galaxies, together with the spectrum (lower panel) of the light from these galaxies. The bright lines on either side of the galaxies' spectrum are comparison standards produced at the telescope when the spectrum was photographed. The spectrum itself shows the familiar (to astronomers) "H and K" absorption lines of calcium, but the lines are shifted toward the red (that is, toward longer wavelengths) with a displacement that indicates that

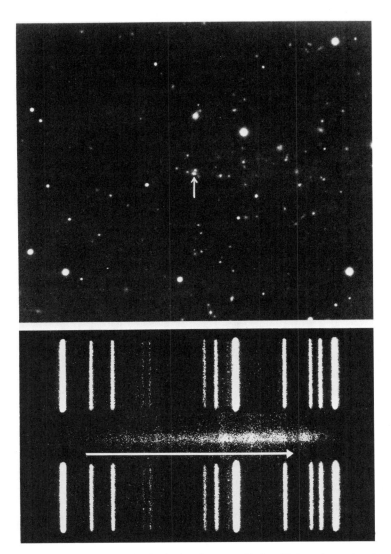

Figure 3.9. A cluster of galaxies in the constellation Hydra, shown by the arrow in the upper photograph.

The lower panel shows the spectrum of the light from the cluster (between two sets of laboratory comparison lines). The H and K lines arising from absorption by calcium appear weakly, but *shifted* toward the red by an amount that indicates a recession velocity of 38,000 miles per second. (Courtesy of the Hale Observatories.)

the galaxies are all receding from us with a relative velocity of 38,000 miles per second, more than one-fifth the speed of light!

At the Lowell Observatory in Arizona, V. M. Slipher had first measured the radial velocity of the Andromeda galaxy in 1912. By 1925, Slipher had obtained the radial velocities of 45 galaxies by determining the amount of their Doppler shifts. Most of these spectra showed displacements toward the red: in *all* directions, galaxies were moving away from us, or we from them. The galaxies' range of velocities went from 190 miles per second toward us up to 1125 miles per second away from us, with an average of 375 miles per second away from us. No one really knew what to make of Slipher's results until Hubble compared the list of radial velocities with the list of galaxies' distances that he had carefully derived during the 1920s. There were two dozen galaxies on both lists, most of them great spiral systems like M 31, M 33, M 51, M 64, M 83, M 101—the heart of Messier's catalogue, compiled 150 years before. From his comparison of distances and Doppler shifts, Hubble found the great synthesis of modern cosmology. The size of the Doppler shift increased with the distance to the galaxy! Only the nearest galaxies showed some "blue shift," indicative of relative motion toward us; all the other galaxies' spectra had "red shifts," and the amount of the Doppler shift was larger for more distant galaxies. Figure 3.10 shows Hubble's original diagram of the galaxies' velocities (from their Doppler shifts) as a function of their distance from us. The points scatter a fair amount, and Hubble realized that this scattering arises from the galaxies' individual random motions. But Hubble recognized among these random motions an underlying relationship, and concluded that as a general rule the galaxies' radial velocities *increase in direct proportion to their distances*. Thus he drew the solid line in Figure 3.10 to represent the best fit of a law of direct proportionality to the scattered points, and published his results in the *Proceedings of the National Academy of Sciences*.

One might think at first that Hubble's observations, embodied in Figure 3.10, showed only that galaxies are receding from our own Milky Way. But astronomers and cosmologists were quick to realize that Hubble's discovery implies that *the entire universe is expanding*. To see why this is so, we should consider the important currents of cosmological thought. First of all, the Copernican revolution of 1543 had an impact that reverberated through

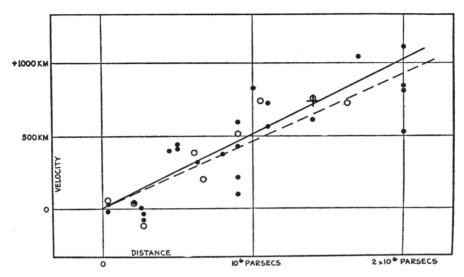

Figure 3.10. Hubble's first velocity-distance diagram, showing the best fit (solid line) to a linear dependence of a galaxy's velocity on its distance. One parsec is 3.26 light-years. (National Academy of Sciences.)

four hundred years. Once the earth was no longer thought to be the center of the universe, there was no way to fit the universe with *any* center, for if our earth were not that special, what could take its place? Copernicus and Kepler already knew that the *sun* does not occupy the center of the planets' orbits; even when Copernicus assumed these orbits to be circular, he had to move the sun off-center. Shapley discovered that our sun lies far from the center of our galaxy, and Hubble found that our galaxy is one among many others. At first glance it might appear that Hubble's discovery of the velocity-distance relationship reestablished our own galaxy as the center of the universe, since he concluded that all but the nearest other galaxies are moving away from us. But the effects of the Copernican revolution went deep, and astronomers were quick to see that such a conclusion is naive. In fact, if our galaxy's role in the universe is average—the Copernican postulate extended—then *all* galaxies are moving away from all other galaxies. Consider the galaxies to be a group of spectators at a college graduation, resting outdoors in chairs. If the spacing between the chairs somehow began to increase both along the rows and in the distance between the rows, then *every* spectator would see *all* the others in motion away from him or her at a speed that is directly proportional to the spectators' distances. This moving-chair model gives a good representation of the motion of the galaxies in the universe, with two major points of difference. First, galaxies move in three-dimensional space and not on the two-dimensional surface of a college courtyard. Second, the group of spectators does have a set of boundaries and this can have a center. However, the universe has no boundaries and no center: *all* the galaxies can move away from all the other galaxies, everywhere. Strange as it sounds, the entire three-dimensional universe could be finite in volume and still have no boundaries, something like the two-dimensional surface of an expanding balloon. A discussion of this possibility lies outside the scope of this article. With Copernicus's dethronement of the earth, people grew accustomed to assuming that our position in space represents an *average* position. And if we are average, and galaxies move away from us with speeds proportional to their distances, then observers on other galaxies should see the same thing, like spectators at a universal graduation with moving chairs.

Thus astronomers and physicists were ready to believe that Hubble, by detecting a "local" pattern of motions of the galaxies within a few million light-years, had indeed discovered a universal phenomenon. Any other explanation must assign a special role to the Milky Way's location, so that galaxies might recede from *us* but coalesce somewhere else beyond the bounds of visibility. Such special hypotheses had become anathema to scientists, largely as a result of the Copernican revolution.

In addition to the Copernican refusal to center the universe on ourselves, the general theory of relativity formulated by Albert Einstein dealt with the problems of other galaxies' motions. Einstein's equations, first published in 1915, describe the relationship between the presence of matter (and energy) and the way that space reacts to this presence as the matter produces a gravitational field of influence. Einstein and others set out to find solutions of the relativity equations that would provide a mathematical expression for how all of space behaves. Work by a Dutchman, Willem de Sitter, by a Russian, Alexander Friedmann, and by a Belgian priest, Georges Lemaître, showed that all reasonable solutions of Einstein's equations predicted an expansion of the distance between the "test particles" used to idealize galaxies in space. The expanding universe leapt out of the theory of relativity! Originally, Einstein himself was skeptical of these solutions. During the 1920s, with no observational evidence for an expanding universe, he introduced an extra term into the equations which were postulated to govern the behavior of space and matter. This extra term appeared to allow for a solution in which the universe neither expanded nor contracted. However, Hubble's observations showed the new term to be totally unnecessary. Years later, Einstein told George Gamow that introducing this term "was the biggest blunder he had made in his entire life."

Hubble and the other astronomers who studied galaxies continued to make observations amid the debates of the relativity theorists. The last paragraph of Hubble's original paper states that "the outstanding feature . . . is the possibility that the velocity-distance relation may represent the de Sitter effect [that is, the expansion of space]." But as the rest of the paragraph shows, Hubble was not too clear about what exactly this effect was, and in fact the theorists had done an imperfect job of explaining to themselves and to others how to solve Einstein's equations. Certainly Hubble did not pursue his investigations because de Sitter had predicted that space is expanding.

Hubble's research aimed at finding out how the universe *does* behave, not at a mathematical model of its behavior. This was natural for him, since he had access to what was then the world's largest telescope for extragalactic studies. With the capable assistance of Milton Humason, Hubble was able to use the instrument wisely and well, and thus he became the first to find the universal expansion.

Outside the astronomical community, the world was slow to realize what had been found. In those days, science and scientists were not thought to have much importance for the rest of society. Describing Einstein's forthcoming visit to Caltech in 1930, a newspaper as conservative as the *New York Times* could write that "strange things are going forward at the Institute of Technology, where Dr. Millikan makes the atoms and the molecules lie down, roll over, and say 'Uncle.' " Throughout 1930, newspapers would occasionally print a notice of the measurement of a new "largest" velocity for a galaxy, without any indication of the general expansion of the universe that these measurements had been shown to imply. Finally, on Christmas Eve of 1930, the *New York Times*'s London correspondent sent a report of a talk given on the British radio by Sir James Jeans, the famous cosmologist. Next to first-page stories such as "Bank rate cut to 2%, lowest in world, as spur to business," a short notice quoted Jeans as saying that "we are probably living in the midst of a great explosion of the universe . . . if [the measured velocities] are real, the whole universe must be expanding."

Within the smaller world of science, there arose a greater effort to understand how Hubble's results meshed with the equations of Einstein's relativity theory. Famous physicists like Jeans, Sir Arthur Stanley Eddington, and Caltech's Richard Tolman tried to explain the relativity equations and to derive the theoretical description of the expanding universe in an exact way. Astronomers achieved sudden recognition. Einstein came to visit Hubble at Mount Wilson in December 1930; Jeans followed in April 1931. More newspaper stories trumpeted the motions of the cosmos. By coincidence, the discovery in 1930 of Pluto, the ninth planet, also focused attention on the heavens. Good news was scarce in 1931, and finding that the universe is expanding could be regarded as some progress in understanding how things work.

Hubble dealt easily enough with the flow of visitors and continued his work. After 1929 he had the valuable assistance of Milton Humason, who began work at Mount Wilson in 1909 as a mule driver, graduated to janitor, learned how to help make observations, and eventually became the astronomer most experienced in obtaining spectra to measure the radial velocities of distant galaxies. Profiting from Humason's patient work, Hubble turned his attention toward faraway clusters of galaxies, groups with hundreds or thousands of galaxies, plainly close to one another but tremendously distant from us. Figure 3.11 shows a galaxy cluster in the constellation Hercules. Many of these galaxies are spiral galaxies, but some are "ellipticals." Hubble's years of observations led him to classify galaxies as spiral, elliptical, or (more rarely) "irregular." The Magellanic Clouds, satellites of our own Milky Way, are examples of the last category. (See Plate 2.4 of Dr. Bok's article, Chapter 2.) Hubble's observations showed that the number of galaxies with an apparent brightness greater than any given value behaved in the way one would expect if, on a scale of great distances, the galaxies are distributed evenly through space. This means that matter appears to be homogeneously strewn throughout the visible universe, aside from "local" events like its clumping into galaxies and of galaxies into clusters.

Figure 3.11. A cluster of galaxies in Hercules, showing spirals, ellipticals (for example, at the lower right), and irregulars. (Courtesy of the Hale Observatories.)

Hubble thought it reasonable to assume that clusters of galaxies are similar to one another, as individual galaxies are. He tried to compare the very brightest stars in the largest galaxies of the Virgo cluster with the brightest stars in our own galaxy. This method seemed valid, but many years later (in 1958) Allan Sandage concluded that these brightest "stars" are in fact hot gas clouds surrounding a group of tremendously bright stars, like the Orion Nebula (M 42). On the average, the brightest glowing clouds are five or six times more luminous than the brightest individual stars, so the distance to galaxy clusters that Hubble derived must be increased by a factor between 2 and 3 to include this correction. The new understanding of Cepheid variables in the early 1950s increased the estimated distances to nearby spirals like the Andromeda galaxy by two or three times, so the distance scale for clusters of galaxies has increased almost ten times since the pioneering work by Hubble and Humason.

Hubble's method worked well enough to show the basic motions of the universe. He extended the idea of comparing the apparent brightness of two objects thought to have the same true brightness. The brightest members of the Virgo cluster of galaxies turned out to have about the same true brightness as the Andromeda galaxy. Once this was known, Hubble and Humason could compare a prominent member of a distant galaxy cluster with a prominent member of the Virgo cluster. If the distant galaxy were, for example, one hundred times fainter in apparent brightness, it was reasonable to conclude that it lies ten times farther away, since apparent brightness decreases with the square of the distance. Using this approach, Hubble extended his velocity-distance diagram to clusters of galaxies distant by $\frac{1}{2}$ billion light-years (now revised to 5 billion light-years). Soon he was able to guess the red shift that Humason was likely to find before Humason obtained the spectrum of the cluster galaxies! Figure 3.12, produced by Hubble and Humason in 1931, shows how far this method reached in two years. The original velocity-distance diagram now appears far down in the left-hand corner, and the direct relationship between the distance to a galaxy or a galaxy cluster and its speed of recession emerges much more clearly.

Figure 3.13 gives a modern version of the sort of measurements made by Hubble and Humason. The appearance, distance, and spectra for each of five representative cluster members shows the increase of the spectral red

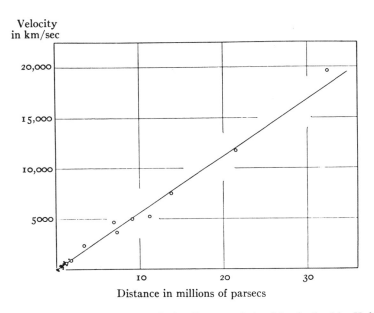

Figure 3.12. The improved velocity-distance relationship obtained by Hubble and Humason in 1931, after they had extended the observations to distant galaxy clusters. (*Astrophysical Journal.*)

shift with the increase in the galaxies' distances. All of these galaxies obey what we now call Hubble's law: for distant objects, the amount of the red shift equals a constant times its distance. The constant number, called *H*, the "Hubble constant," gives the algebraic formula $v = H \times d$. Like the other "laws" of physics, Hubble's law summarizes observational results in a coherent way. Since Hubble's law has been established as generally valid from observational evidence, astronomers can use it to find the distances to far-distant objects with large radial velocities.

The Triumph of the Expanding Universe
In contrast to the reception of Copernicus's thesis of the earth's motion around the sun, which generated a great resistance among the astronomers and theologians of his time, Hubble's great discovery gained rapid acceptance. His observations indeed startled the scientific world, but in a very

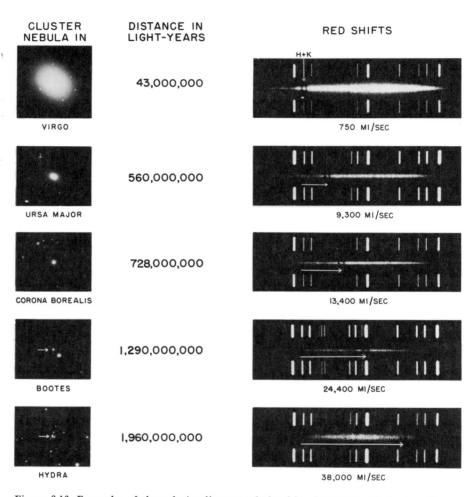

CLUSTER NEBULA IN	DISTANCE IN LIGHT-YEARS	RED SHIFTS
VIRGO	43,000,000	750 MI/SEC
URSA MAJOR	560,000,000	9,300 MI/SEC
CORONA BOREALIS	728,000,000	13,400 MI/SEC
BOOTES	1,290,000,000	24,400 MI/SEC
HYDRA	1,960,000,000	38,000 MI/SEC

Figure 3.13. Examples of the velocity-distance relationship, drawn at a time when the Hubble constant was calculated to be 19 miles per second per million light-years.

For each galaxy, the spectrum shows a red shift in the wavelength of the H and K calcium absorption lines, and the red shifts increase in proportion to the galaxies' distances. (Courtesy of the Hale Observatories.)

short time astronomers and physicists realized that his measurements fitted neatly into the theoretical framework provided by the solutions to Einstein's equations. With the assumption that our position in space is average, Hubble's law implies that observers everywhere should see other objects with a velocity of recession proportional to distance. Everything made sense except for one uneasy contradiction.

If galaxy clusters are expanding away from one another, then they used to be closer together. If we run time backwards in our mind, all observers see the other galaxies approaching them, and at some point in the past they would all coalesce. In the velocity-distance law, $v = H \times d$, the Hubble constant H simply gives the fractional expansion of the universe per second. The inverse of the Hubble constant, $T = 1/H$, is called the "Hubble time." Most solutions to Einstein's equations reveal that the universal expansion must have begun at a time no longer than T years ago. Hubble's original determination of the constant H gave $H = 105$ miles per second per million light-years. That is, he found each additional step of 1 million light-years would reveal a galaxy moving about 105 miles per second more rapidly away from us. Since 1 light-year is close to 6 trillion miles, we can convert light-years to miles, cancel the miles, and see that Hubble's number was $H = 17\frac{1}{2} \times 10^{-18}$ per second, so the universe was apparently growing larger by $17\frac{1}{2}$ parts per billion billion each second. The inverse of H is $T = 5.7 \times 10^{16}$ seconds, or 1.8 billion years. However, even in 1929 the age of the earth and of the solar system had been found to be more than 3 billion years (the best value today is $4\frac{1}{2}$ billion years).

The contradiction between the Hubble time on the one hand and the geological records and calculated ages of stars on the other posed a major obstacle to the acceptance of Hubble's results. Some physicists doubted that the observed red shifts truly represent motions away from us, and suggested other possible (though unlikely) mechanisms for producing the red shifts. Hubble himself always expressed a lingering caution in interpreting his observations as definite indicators of radial velocities, and he was probably influenced in this by the low value he found for the time $T = 1/H$.

Twenty years after Hubble's discovery, Hermann Bondi, Thomas Gold, and Fred Hoyle proposed that, as the universe expands, *new matter constantly appears* at just the rate needed to keep the amount of matter per cubic light-year always the same. This "steady-state" model of the universe may seem to call for the incredible, but we should admit that if matter were created *sometime,* matter could be just as easily created today. The steady-state model predicts that, although the universe expands, it always looks the same, since no change in the rate of expansion or in the average density of matter could occur. Hence there would be no beginning or "big bang" at some finite time in the past, and no contradiction between the Hubble time T and the astronomical evidence of objects greater than Hubble's original value for T.

Though the steady-state model has been by no means disproved, our *need* for such a solution has decreased because the discrepancy in the time scales has vanished. Allan Sandage, who has inherited Hubble's place as the chief telescopic investigator of the most distant galaxies, has recently (1971) redetermined the value of the Hubble constant H. Using the corrected brightnesses of Cepheids and his own detailed studies of gas clouds and other bright objects in galaxies, Sandage has found a best value of $H = 11$ miles per second per million light-years. This increases the Hubble time by a factor of 10, to 18 billion years, and removes almost completely any contradiction between the age of the universal expansion and the ages of the earth, the sun, and the oldest stars.

Theories of how the expanding universe should behave all show that *if* we can measure the distances and radial velocities of tremendously faraway objects, then we should be able to discover whether or not the universe will ever stop expanding, and indeed whether the entire universe—all of space and everything in it—is finite or infinite. Measurements of the velocity-distance relationship for the farthest objects might reveal deviations from the straight line of Hubble's law, $v = H \times d$. Such deviations would indicate that the rate of universal expansion has changed with time. When we observe distant galaxies, we are looking backward in time, because light travels at the finite speed of 186,000 miles per second. Far-distant galaxies shine with light that is billions of years old. Unfortunately, our present observations of distant galaxies do not allow us to decide whether the expansion speed has changed significantly, for they extend over only the last few billion years. However, we can hope that future observations by Dr. Sandage and other astronomers will eventually let us know whether the expansion of the universe that Hubble found will always continue, or whether the universe will someday cease its expansion and start to contract.

In *The Love Song of J. Alfred Prufrock,* T. S. Eliot (born the year before Hubble) asked

Would it have been worth while
To have bitten off the matter with a smile
To have squeezed the universe into a ball
To roll it towards some overwhelming question . . .

Thanks to Edwin Hubble, astronomers find the possibility of the squeeze as exciting as any question they may consider.

Selected References

Edwin Hubble, *The Realm of the Nebulae*. Dover Publications, New York, 1958. This republication of Hubble's lectures at Yale in 1935 contains a fine account of his discovery of the velocity-distance relationship.

Allan Sandage, *The Hubble Atlas of Galaxies*. Carnegie Institution of Washington, 1961. Before his death in 1953, Hubble had planned an atlas to illustrate his classification criteria for galaxies. Sandage carried out this project and produced a marvelous assortment of the best photographs of different galactic types.

Charles Whitney, *The Discovery of Our Galaxy*. Alfred Knopf, New York, 1971. A clear and intriguing account of the astronomical investigation of our surroundings from ancient times to Hubble's time.

Donald Goldsmith and Donald Levy, *From the Black Hole to the Infinite Universe*. Holden-Day, San Francisco, 1974. Includes a readable account of how the universe can be finite or infinite and expand in either case.

Ronald Clark, *Einstein: The Life and Times*. World Publishing, New York, 1971. A fascinating biography of Einstein that deals mostly with nonacademic topics but also explains the scientific discoveries.

Otto Struve, "A Historic Debate About the Universe." *Sky and Telescope,* Vol. *19,* May 1960, p. 318. An account of the Shapley-Curtis debate on the distances of spiral nebulae in 1920.

Otto Struve and Velta Zebergs, *Astronomy of the 20th Century*. Macmillan Company, New York, 1962. Gives a thorough explanation of the important developments in astronomy 1900–1960.

V. C. Reddish, *The Evolution of Galaxies*. Oliver and Boyd, London, 1967. A slightly technical discussion of how galaxies are thought to live and die.

N. U. Mayall, "Milton Humason: Some Personal Notes in Recollection." *Mercury,* January/February 1973. A good account of observatory life at Mount Wilson during the time that Humason, Hubble, and Mayall worked there.

Wlodzimierz Zonn

4 Explosive Events in the Universe

Introduction
It appears to be an element of human nature to believe that violent events,
such as explosions, occur only in the variously defined immediate vicinity
of where we live. We like to think that the faraway regions are serene and
dignified. Then, a series of undeniable facts forces us to abandon the preconception, each time with reference to increasingly wider volume of space.
At the time of Copernicus (1473–1543) there was the common belief, inherited from Aristotle, that all matter around us is composed of four elements, earth and water tending to move down toward the center of the
earth, and fire and air moving in the opposite direction. These motions
were supposed to be limited to the "corruptible" sphere of "elements," below the moon. The space outside of this sphere was supposed to be "ethereal," filled with a transparent substance "ether," incorruptible, unchangeable, and majestic.

The purpose of the present essay is to sketch two incidents in the history
of human thought in which the traditional preconceptions of incorruptible
unchangeability and calm and "dignity" had to be abandoned for increasingly larger volumes of space. The first quasi-Copernican revolution of this
kind seems to have been generated by the Danish astronomer Tycho Brahe
(1546–1601), the immediate successor of Copernicus in his status of the astronomer of the epoch. In addition to the invention, construction, and use
of a variety of astronomical instruments which yielded observation of an
accuracy that could not be surpassed for generations, Tycho Brahe's achievements include the finding that very faint or unknown stars can suddenly
"explode" into apparently very bright or new stars ("novae"). The second
similar incident to be described is the recent recognition that, contrary to
general belief, not only stars but also galaxies, particularly nuclei of galaxies, can explode. This is how the dramatic change in attitude is described
in a report[1] of the U.S. National Academy of Sciences:

From the time of the ancient Greeks to the mid-twentieth century, the universe was
conceived as an unchanging, or at best slowly changing, cosmos of fixed stars. The first few
decades of this century replaced this view with a steadily expanding universe of galaxies
[see preceding essays, Ed.]—each galaxy a majestic, slowly rotating collection of stars
intertwined with dust and gases . . .

The last decade of exciting discovery has added to that picture a general cosmic violence,

Astronomical Observatory, University of Warsaw, Poland.

exploding galaxies and quasars [see next essay, Ed.], an almost universal presence of high-energy particles and magnetic fields, and events suggesting relativistic collapse . . .

The report containing this passage was published in April 1972. Thus, the words "last decade" are likely to refer to the 1960s.

The astronomical literature is immense, and it is very difficult to be sure of the actual origin of any particular significant idea. Nevertheless, the quotations given below do seem to indicate that the originator of the quasi-Copernican revolution connected with the explosions in the nuclei of galaxies may be the Soviet-Armenian astronomer, V. A. Ambartsumian. His basic paper hypothesizing explosions, not so much with reference to particular galaxies, but as an essential part of the mechanisms of the universe, was published in 1958. While circumstantial evidence about explosive events in particular galaxies kept accumulating for some time, the first definitive evidence was published by two groups of astronomers who worked in California. The completely direct, convincing evidence of an explosion in a galaxy, M 82, was published in 1963 by C. R. Lynds and A. R. Sandage. Slightly less direct evidence of the same phenomenon in two other galaxies was reported by E. Margaret Burbidge, Geofrey R. Burbidge, and K. H. Prendergast. Their joint papers appeared in 1959 and 1963. Since that time, references to explosive events in the realm of galaxies have become commonplace.

Supernova of 1572

One of the most striking but very rare events in the realm of stars, occurring about once every 50 to 100 years in a galaxy comparable to our own Milky Way System, is the sudden explosion of a star, previously unknown, called a *supernova*. The duration of the maximum light is rather brief. But, when the event does occur, the sight is most impressive. In fact, a supernova can be visible even at midday. The following quotation[2] is from the diary of Tycho Brahe. It recounts the story of his first notice of the supernova of 1572, of his musings about it, and of his observations and calculations. Eventually these studies led Brahe to abandon the idea that explosive events must be limited to the "corruptible" sphere of the elements.

Last year [1572] in the month of November, on the eleventh day of that month, in the evening, after sunset, when, according to my habit, I was contemplating the stars in the clear sky, I noticed that a new and unusual star, surpassing the other stars in brilliancy, was shining almost directly above my head; and since I had, almost from boyhood, known all the stars of the heavens perfectly (there is no great difficulty in attaining that knowledge) it was quite evident to me that there had never before been any star in that place in the sky, even the smallest [faintest], to say nothing of a star so conspicuously bright as this. I was so astonished at this sight that I was not ashamed to doubt the trustworthiness of my own eyes. But when I observed that others, too, on having the place pointed out to them, could see that there was really a star there, I had no further doubts. A miracle indeed, either the greatest of all that have occurred in the whole range of nature since the beginning of the world, or certainly that is to be classed with those attested by Holy Oracles, the staying of the Sun in its course in answer to the prayers of Joshua, or the darkening of the sun's face at the time of the Crucifixion. For all philosophers agree, and the facts clearly prove it to be the case, that in the ethereal region of the celestial world no change, in the way either of generation or of corruption, takes place; but that the heavens and the celestial bodies in the heavens are without increase or diminution, and that they undergo no alteration, either by number or in size or in light or in any other respect; that they always remain the same, like unto themselves in all respects, no years wearing them away.

This passage, illustrating the scholarly thinking of the epoch, is followed by Tycho Brahe's reminiscences that of all the testimonies of the ancients he read, there was only one, by Pliny ("if we are to believe Pliny") mentioning that a "new" star, different from all others previously seen, was noticed by the Greek astronomer Hipparchus. Next follow the accounts of Brahe's efforts to estimate the distance of the supernova he observed. By all rights, it should be located in the sphere of elements, below the moon, but is it?

Meticulous observations performed by Brahe showed that the new star could not be located below the moon. Furthermore, because over six months the new star did not appear to have moved from its original position among the known fixed stars, Brahe concluded that its distance from the earth must be greater than that of the then-known outer planets, Mars, Jupiter, and Saturn.

Therefore, this new star is neither in the region of the Element, below the Moon, nor among the orbits of the seven wandering stars [planets], but in the eighth sphere, among the other fixed stars. . . . Hence it follows that it is not some peculiar kind of comet or some other kind of flying meteor become visible. For none of these are generated in the heavens themselves, but they are below the Moon, in the upper region of the air, as all philosophers testify; unless one would believe with Albategnius that comets are produced, not in the air, but in the heavens. . . . But, please God, sometime, if a comet shows itself in our age, I will investigate the truth of the matter.

This prayer was granted and the achievements of Tycho Brahe include the determination that, indeed, the presumption of Albategnius was right: "comets are produced, not in the air, but in the heavens."

Over the 400 years that elapsed since the incident just described, the phenomenon of "new" stars became fairly familiar to the astronomers. They are not really new, but just faint stars that suddenly burst into brilliance. There are two broad categories of them, labeled "novae" and "supernovae." They differ greatly in brightness at maximum and in frequency of occurrence. On the average the maximum brightness of a nova is estimated at about 60,000 times that of the sun. The corresponding estimate for a supernova is 240,000,000 times the brightness of the sun. Because of the great distances and obscuration by dust clouds, the appearances of novae are not generally noted. The frequency of those noticed in the Milky Way System is about once every few years, but there may have been more. Because the supernovae occur only once in several centuries, their systematic study must be based on the observations in other galaxies. Several years ago, on the initiative of Fritz Zwicky of the California Institute of Technology, a cooperative program of several observatories was arranged aiming at the accumulation of data on supernovae. Every few weeks each cooperating observatory takes photographs of the same regions of the sky. Now and then the comparison of such successive photographs reveals that in one of the visible galaxies there appeared a supernova. This news is promptly distributed worldwide, and the new star is subjected to a variety of studies. The *Transactions of the International Astronomical Union,* Vol. XIV of 1970, indicate that up to July 1969 the Zwicky program resulted in the discovery of 254 supernovae. While there is little doubt that the explosions of novae and of supernovae result from nuclear reactions, the complete understanding of the phenomenon is not yet in sight.

Violent Events in Galaxies I: The Thinking of 1950s
As described in the preceding essays, by Bart J. Bok and by Donald W. Goldsmith, the last half century is marked by a kaleidoscopic output of new observational astronomical discoveries and also of innovative theoretical developments. A special discipline, astrophysics, after a few decades of ger-

mination, burst into blossom with its aim to connect the findings of modern physics with the observations of happenings in space. These aims were supported by the development, after the Second World War, of a new tool of observational studies, radio astronomy. A number of sources of radio emissions were found both in our own galaxy, the Milky Way System, and outside it. Some of these radio sources were identified as galaxies. A series of physical-mathematical cosmological theories sprang up, each attempting to explain not only the structure of the universe as contemporarily observed, but also its history in the remotest past and its future. Each of the several overlapping periods of a few years was marked by the predominance of a particular theory, to some neglect of its competitors. The following regrets, expressed by one of the authors, G. Lemaître (*Astronomical Journal,* Vol. 66, 1961, p. 603), illustrate the situation:

My model of the universe. . . . It is realized that this model, proposed some 30 years ago, may be considered as "old fashioned cosmology." This is not the place to discuss the strong prejudices that have arisen against it due to reverence of an authority whose influence can only be compared to that of Aristotle in older times.

On the observational side, several epoch-making compendia were published that continue to exercise a strong influence on astronomical studies all over the world. Because of frequent use, some of these publications acquired special nicknames: (i) "The Shapley-Ames Catalogue";[3] (ii) "The HMS Catalogue";[4] (iii) "The Palomar Atlas";[5] (iv) "The Hubble Atlas of Galaxies," [6] which contains 186 magnificent photographs of galaxies of various types and also a brief explanation of the contemporary views on how these types evolve; while very important for research purposes, the book is most suitable for college and school libraries and can be an ornament of a bookshelf in a private home; (v) "Reference Catalogue of de Vaucouleurs." [7]

Among the many ideas that the above and many other publications brought out was the realization that stars in the Milky Way System and also galaxies are organized in identifiable systems. The star systems range from "double" or "multiple" stars, to what are called "star clusters" or "associations," with an obvious meaning to these terms. Similarly, among the systems of galaxies we distinguish "double" or "multiple" galaxies (see Figure 4.1), and "groups" of perhaps four to a dozen objects, clearly separate and yet grouped on the photographs so closely as to suggest some kind of connection (see Figure 4.2). The largest galaxy systems, with membership running into thousands, are labeled "clusters" or "superclusters."

Figure 4.1. Wilson-Zwicky Connected Pair.

This is one of the numerous streamer-connected double systems found on survey photographs taken with the 18-inch and 48-inch Palomar Schmidt telescopes. The joining streamer and countertidal filament show plainly as extensions of the arms of the barred spiral (lower) member.

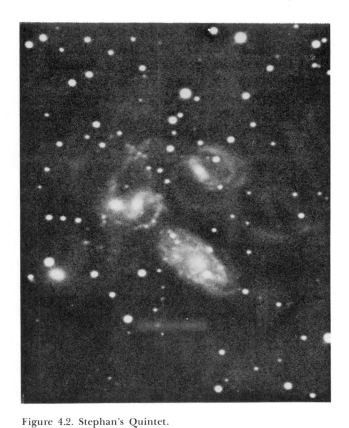

Figure 4.2. Stephan's Quintet.

Plate taken at prime focus of the 82-inch McDonald telescope. North is at the top, West at the left. Scale: 1 mm = 2".6.

Marking a shift of interest toward systems of galaxies, Fritz Zwicky and G. O. Abell used the Palomar Atlas to compile catalogues of clusters. Also, B. A. Vorontsov-Velyaminov[8] compiled a catalogue of multiple or otherwise peculiar galaxies. His catalogue has the nickname the V-V Catalogue. Finally, Halton Arp used the V-V Catalogue to produce an *Atlas of Peculiar Galaxies,* 338 of them,[9] Arp's Atlas. The questions asked about all these systems relate to their origin, to their stability, and to their future.

The thinking of the late 1950s on these questions is reflected in the proceedings[10] of the Solvay Conference held in Brussels in 1958. The views on stability of systems varied considerably. Summaries of observations made over several preceding years brought out the finding of cataclysmic events in the realm of galaxies. Among the first interpretations were "head-on" collisions of giant galaxies at relative velocities in thousands of kilometers per second.

The first paper at the conference, by Georges Lemaître, explained his theory, leading from cosmic rays to gaseous clouds to "proto-galaxies," with star formation in them, and finally, to the arrangement of galaxies in clusters. Some of the galaxies in these clusters were supposed to be endowed with large velocities from the start, eventually making them escape from the home clusters. These losses of membership of a given cluster were assumed to be gradually compensated by "captures" of new members, galaxies originally belonging to other clusters. This hypothetical mechanism, then, preserves particular clusters but allows their memberships to vary in time.

In a paper concerned with the actual large-scale distribution of galaxies in space, Jan H. Oort discussed the problem of temporal stability of systems of galaxies. Based on data in the Shapley-Ames catalogue, the sky survey by C. D. Shane and C. A. Wirtanen and others, Oort's opinions favored both stability and large-scale homogeneity of the universe.

The occurrence of violent events in the realm of galaxies was mentioned in the paper by Alfred C. B. Lovell, who summarized the contemporary radio-astronomical observations. At the time of the conference only sixteen normal galaxies had been identified as radio sources; none of them were ellipticals. In addition to these galaxies, which on the photographs of the sky appeared "normal," Lovell mentioned seven others, not "normal." All of them appeared marked by unusually strong radio emissions, and their closer study by optical means (that is, through inspection of photographs taken with large telescopes, study of spectra yielding velocities in the line of sight, and so on) indicated violent events. In four cases the conclusion was that the observed objects were pairs of colliding galaxies (see Figure 4.3). One other case was that of a "possible" collision. Then there was a case of an apparent beginning of spiral formation and a case of a blue-colored jet of luminous matter streaming from the nucleus of an elliptical galaxy, M 87 (see Figure 4.4).

Efforts have been made by Baade, Minkowski, and by several other outstanding observers to estimate the relative velocities of colliding galaxies. In three cases they were able to do so, and the estimates were 500 km/sec, 1800 km/sec and 3000 km/sec. The luminous jet from M 87 evoked little comment at the Solvay Conference, except for the reported remark by Minkowski that, in order to see a phenomenon of this kind, one has to look at the galaxy from an appropriate direction. Many such galaxies may exist in nature, and their photographs may be in the files of various observatories, but without the jets being apparent.

To complete the description of the contemporary thinking on the "realm of the nebulae" (this is the title of the very interesting book[11] by Hubble), it is appropriate to summarize briefly the ideas on the evolution of galaxies that prevailed by the end of the 1950s. More details of these ideas are to be found in the introduction to *The Hubble Atlas of Galaxies*.[6] According to these thoughts, galaxies begin their existence from accumulations of dust and gas, particularly of hydrogen. Condensations of gas and dust result in the formation of "young" stars—the O and B types, which are intrinsically very bright and blue in color. The original form of a "young" galaxy is irregular. In due course, spiral arms are formed, populated by O and B stars and laced with opaque dust lanes. Gradually, spiral arms wind themselves more and more tightly around the nucleus of the spiral galaxy. Dust and gas in the arms become exhausted by the formation of stars, and eventually the process of forming new stars terminates. In the meantime the stars formed in the earlier epoch evolve and become "old stars," reddish in color. The ultimate morphological type in this evolutionary trend is elliptical, composed of myriads of faint stars, with no dust and very little gas. Their appearance is like M 87, but without the blue jet. The inspection of the pictures of spiral galaxies in the *Hubble Atlas* shows their nuclei as amorphous bright bodies in the center, frequently very symmetric. However, as a result

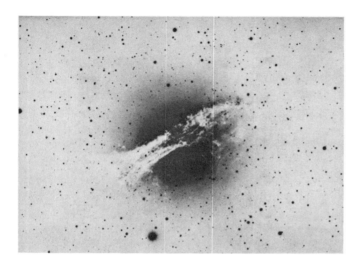

Figure 4.3. NGC 5128.
Taken with 200-inch telescope of the Hale Observatories.

Figure 4.4. M 87 jet.
Plate taken with 84-inch Kitt Peak National Observatory telescope by C. R. Lynds.

of an inspired effort, Walter Baade obtained photographs of the Andromeda Nebula (a spiral close neighbor of the Milky Way System) showing the "resolution" of its amorphous nucleus into a great accumulation of stars, somewhat similar to those in elliptical galaxies. Further study of nuclei of a number of galaxies revealed that they contain but very limited amounts of gas.

Violent Events in Galaxies II: Ideas of Ambartsumian
With unavoidable contractions and oversimplifications, the foregoing summarizes the thinking about the realm of galaxies that at the time of the Solvay Conference appeared traditional. Among the nontraditional items at this conference was the paper by V. A. Ambartsumian about what must be happening in galaxies.

The ideas that appear basic in the paper of Ambartsumian are that (i) in addition to common stars, the nuclei of galaxies must contain highly massive bodies of "prestellar" matter, also labeled "protostars"; (ii) that in these protostars explosions can occur producing hydrogen and some solid parts that move away at great velocities; (iii) that the scale of such explosions must vary, from relatively miniature events like the jet of M 87 (Figure 4.4) to cataclysmic events resulting in multiple galaxies or groups or clusters; (iv) that the spiral arms of galaxies result from explosions in their nuclei and that such phenomena can occur repeatedly in the same galaxy, occasionally resulting in several systems of spiral arms, not always in the same plane; (v) that explosions in protostellar bodies are sources both of the interstellar gas and of bright blue stars, rather than blue stars being formed out of gas.

With one exception, the evidence adduced by Ambartsumian in support of his ideas was circumstantial. The exceptional piece of evidence, that can be considered as indicating directly that explosions occur in galaxies, is the image of the jet streaming from the nucleus of M 87, one of the largest known elliptical galaxies. The jet contains condensations consistent with the assumption that they are made up of young stars, of gas, and of high-energy particles, possibly the material for formation of a small young galaxy. A few other objects similar to M 87 were found by Ambartsumian.

The strictly circumstantial evidence cited by Ambartsumian in favor of his ideas is based, predominantly, on the analysis of stability of small groups of galaxies and of clusters. For a system of galaxies to be stable, a certain harmony must exist between the velocities of particular members, on the one hand, and the forces of gravity that could keep the system together, on the other. The analysis of a number of particular cases convinced Ambartsumian that this harmony could not exist because the velocities of system members differed so much that their mutual gravitational attraction could not stop them from dispersing. In particular, Ambartsumian's analysis of the well-known Coma cluster of galaxies led him to the conclusion, contrary to that of Oort, that this cluster must be disintegrating. It is the conviction that a great many systems of galaxies must be flying apart and the subsequent quest for the possible sources of energy capable of causing such dispersals that led Ambartsumian to hypothesize both the solid (or quasi-solid) nature of the nucleus of a galaxy and the possibility of its exploding.

These ideas of Ambartsumian contrasted very much with many of those that at the time of the Solvay Conference of 1958 were traditionally accepted. For one thing, the thought of spiral arms of galaxies being generated by explosions in the nucleus was contrary to the contemporary ideas about the general direction of the evolution of galaxies (see earlier). For another thing, where the contemporary leaders Baade and Minkowski hypothesized collisions of galaxies, Ambartsumian claimed separation of parts of an original galaxy split by an explosion in the nucleus. There were many such conflicts of ideas. Explosive events in the nuclei of galaxies appeared to connect a number of seemingly diverse phenomena.

One example is the appearance of the spectacular galaxy M 51, see Figure 4.5. At the end of its spiral arms there is a companion galaxy NGC 5195, which is recognized as an elliptical. One of the possible interpretations is purely ad hoc: the elliptical companion just happened there by chance; possibly this companion is really far away from M 51, either in the foreground or in the background, but is so located in the line of sight that the photograph creates the impression of some sort of connection with the spiral arms of M 51. In principle, this is a possibility. However, there are other similar pairs in existence, and Arp's *Atlas* contains photographs of several of them. Ambartsumian's hypotheses unify all such cases: in the nuclei of galaxies gigantic explosions occur from time to time; the material ejected in such explosions can form spiral arms and companion galaxies.

In the discussion that followed the presentation of the paper by Ambartsumian, Oort expressed doubts as to whether there are "sufficiently compelling phenomena . . . to justify the adoption of such a revolutionary idea as the fission of galaxies." The hypothesis of collisions was convincing to Oort. On the other hand, the luminous jet from M 87 did seem to Oort to require some "explosive explanation." The several proponents of the steady-state cosmology [see next essay, Ed.], Bondi, Gold, Hoyle, and McCrea, discussed ties that the Ambartsumian hypothesis may have with their own theories.

The discussion initiated at the Solvay Conference of 1958 continued at a special "Conference on Stability of Systems of Galaxies," [12] arranged in 1961 at Santa Barbara, California. With the number of participants considerably larger than that at the Solvay Conference, all addressing themselves to one particular problem, the Santa Barbara Conference brought out a very substantial number of cases in which particular investigators felt that the estimated total masses of systems of galaxies would not be sufficient to hold these systems together. In order that these systems be stable, with the relative velocities of their members as large as those observed, the existence within the systems of very large quantities of dark unobservable matter would be necessary, quantities so large as to be unbelievable. Of the 16 systems discussed in a single paper by E. M. and G. R. Burbidge, 9 systems were found "probably unstable," 4 doubtful, and 3 "probably stable." In addition to papers admitting instability there were several, including the paper by G. Lemaître, defending the contrary opinion. As a summary, one might say that the Santa Barbara Conference of 1961 marked the admission of the problem of instability of systems of galaxies into the category of respectable problems for study, but nothing beyond that. In order to admit or even to discuss the possibility that instability of systems of galaxies is due to explosions in the nuclei of "mother galaxies," empirical evidence was needed that these explosions do occur. Such empirical evidence came to light two years later.

Violent Events in Galaxies III: M 82
The Hubble Atlas of Galaxies contains three photographs of the galaxy M 82 or NGC 3034 (see Figure 4.6). It is described as irregular "of second type" "which presents a mystery." The presence of much dust throughout the galaxy is emphasized. Also emphasized are certain characteristics that appear contradictory. Some features suggest that the stellar content of M 82 is like that of elliptical galaxies or centers of spirals, meaning "old stars." These features conflict with the analysis of the galaxy's spectrum.

Figure 4.5. M 51, a spiral galaxy with an elliptical companion NGC 5195 at the end of one arm.

Taken with 120-inch telescope of Lick Observatory by N. U. Mayall.

Figure 4.6. Composite photograph (by C. R. Lynds) of M 82 (NGC 3034) made by superimposing several $H\alpha$ photographs taken with the first 36-inch telescope at Kitt Peak National Observatory.

The system of filaments seen extending to large distances perpendicular to the apparent, nearly edge-on disk of the galaxy emits strongly in $H\alpha$. The filaments (hydrogen 6563Å) have systematic motions that have been interpreted as resulting from an explosion in the center of the galaxy. M 82 is a radio source of moderate strength.

Figure 4.7. Jet streaming from galaxy NGC 4676 like a "mousetail."

Plate taken at 120-inch telescope of Lick Observatory by N. U. Mayall. North is at the top, East is at the left.

With a reference to Ambartsumian who "for many years considered problems concerned with energies in the nuclei of galaxies," C. R. Lynds and A. R. Sandage performed an extensive study of M 82 (see Figure 4.6). Their paper,[13] published in 1963, contains the following summary statements.

New observations in optical frequencies of the peculiar optical and radio galaxy M 82 reveal a massive system of filaments which extends along the minor axis to the height of 3000 parsecs above and below the fundamental plane. Emission lines, typical of low excitation gaseous nebulae are present. The filaments on both sides of the plane appear to be expanding from the center along the minor axis with velocities ranging up to about 1000 km/sec. The data suggest that an explosion of matter took place from the central regions of M 82 about 1.5×10^6 years ago . . . The ion density in the filaments . . . requires the mass of the expanding material to be 5.6×10^6 M. (5,000,000 solar masses) as an upper limit. . . .

As the two authors state in the body of their paper,

The nature of the initial explosion in the central regions of M 82 is an enigma. But M 82 may be the first recognized case of a high energy explosion originating in the central regions of a galaxy, and as such may provide features required for a general explanation of many extragalactic sources.

Violent Events in Galaxies IV: Seyfert Galaxies NGC 1068 and NGC 7469

Empirical evidence of explosions in the nuclei of galaxies is contained in the two joint papers by E. Margaret Burbidge, G. R. Burbidge, and K. H. Prendergast,[14,15] one published in 1959 and the other in 1963. Compared to that in the paper by Lynds and Sandage described earlier, this evidence is a little less direct. The conclusions of Lynds and Sandage are based on measurements of velocities of matter in the filaments streaming in opposite directions from the central part of M 82. The colossal differences between these velocities in spots on the two sides of the plane of the galaxy constitute evidence of the explosion as direct as can be. The two galaxies studied by the Burbidges and Prendergast are seen from the earth almost exactly "face on." Not only is it impossible to measure anything on both sides of the main planes of these galaxies but also, whatever masses of matter there may be streaming from the visible "faces" of the two objects in the direction toward the observers, they could not be seen against the bright background of the galaxies themselves. Thus, the evidence from the study of these galaxies could be only indirect: because of such and such observations we conclude that there must have been an explosion.

The nature of this reasoning will be explained presently. In the meantime, the reader's attention is called to Figure 4.7, reproduced from a photograph taken by the Burbidges, which may be considered as a more direct empirical evidence of an explosion in a galaxy. The figure shows a phenomenon akin to that observed in M 87, see Figure 4.4: a jet streaming from a galaxy. The difference is that the galaxy M 87 is much more impressive than NGC 4676 in Figure 4.7, while the jet streaming from NGC 4676 is much more impressive than that of M 87. Incidentally, Vorontsov-Velyaminov, who found a number of objects like NGC 4676, describes these jets as "mousetails," quite appropriately. Photographs of many galaxies with "mousetails" are published in the Arp's *Atlas*.[9] These photographs are direct evidence not merely of explosions but of colossal ejections of material, presumably an initial stage in the formation of a multiple galaxy.

Returning to the evidence of explosions in NGC 1068 and NGC 7469, it is important to know that these objects belong to a rather rare category named Seyfert galaxies. They are characterized by certain special features including strong and very broad hydrogen lines in the spectra of their nuclei. The strength of these lines is evidence that the nuclei of Seyfert galaxies contain large quantities of luminous hydrogen, which is an unusual feature. The great width of these same lines is evidence that the masses of luminous hydrogen are in a violent motion. To understand this, the reader must recall the nature of the Doppler effect. This is the fact that, when the source of light of a particular wavelength approaches the observer, then the spectral line corresponding to this wavelength is shifted toward the blue; similarly, if the source of light recedes, a "red shift" is observed. Normally, the width of spectral lines depends on the instrument used. If the celestial body studied contains some luminous matter that is approaching the observer (providing a "blue shift") and also some matter of the same kind that recedes (yielding a red shift), this may result in two distinct spectral lines instead of just one line. However, if the analyzed light comes from turbulent gas, some of it approaching the observer and some receding, with varying velocities, then there is a large number of resulting spectral lines that blend together and form a single abnormally wide spectral line.

With this in mind, the unusual width of strong hydrogen lines in the spectra of nuclei of Seyfert galaxies is interpreted as evidence that these nuclei contain masses of luminous hydrogen animated by velocities, some toward us and some away. Actually, the spread of these velocities can be computed. For the two galaxies studied, NGC 1068 and NGC 7469, these computed values are tremendous, ± 1450 km/sec and ± 2500 km/sec, respectively. They alone indicate not just one explosion but many explosions. However, the question is whether these explosions result in the ejection of matter from the nuclei of galaxies. The alternative possibility is that masses of gas, once sent out of some central region, gradually slow down and then, attracted by gravity, return to the nucleus.

The subject of the two papers now discussed includes the calculation of masses of the nuclei of the two galaxies and of corresponding "escape velocities." For the two galaxies the computed escape velocities are 410 km/sec and about 400 km/sec, respectively. In both cases, the spread of velocities of hydrogen in the nucleus is much larger than the escape velocities, and the authors conclude that "we are witnessing a violent event in the evolution of the nuclear regions in which . . . matter is being ejected in large amounts. . . ."

The foregoing reasoning may perhaps be labeled the "mass-velocity" budgeting of the nuclei of Seyfert galaxies. Somewhat different kind of "budgeting," the "pure-energy" budgeting of the universe, is outlined in a paper of 1958 by G. R. Burbidge, which quotes a lecture of Ambartsumian delivered in 1955, apparently unpublished. Burbidge's attempts to balance the energy budget includes the possibility that, in the early epoch of the evolution of a galaxy, large-scale thermonuclear explosions occur, generating big motions of mass.

Concluding Remarks

As mentioned earlier, the birthday of a novel idea, and also its authorship, are rather difficult to establish. Even more difficult is the determination of the time when the novel idea is "generally accepted." Nevertheless, the incidents recounted earlier do indicate that over a decade, beginning with Ambartsumian's lectures in 1955 or 1958, a quasi-Copernican change occurred in the thinking of the astronomical community regarding the possibility of gigantic explosions in the realm of galaxies. At the Solvay Conference of 1958 Oort called this idea "revolutionary." Against this, Vol. XIV

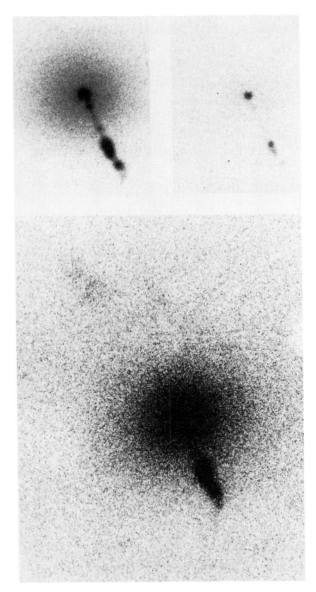

Figure 4.8. Three photographs of M 87 and its jets reduced to a common scale by H. C. Arp.

A of the *Transactions of the International Astronomical Union*, published in 1970, contains a longish enumeration of recent studies by a number of authors, all concerned with galaxies in which explosive events appear to have occurred. Among other things, this enumeration lists the discovery, by H. C. Arp, of a "counter-jet" streaming from M 87 in the direction opposite to that of the original jet, and largely hidden behind the main body of the galaxy. In 1970 this discovery was confirmed in a joint study of J. E. Felten, H. C. Arp, and C. R. Lynds (see Figure 4.8). However, the main subject of this study is not the mere occurrence of an explosion in the nucleus of M 87 but several alternative mechanisms through which this explosion could have been produced. Another quasi-Copernican revolution in attitude toward explosive events in the cosmos seems to have been completed.

Acknowledgments

The author is highly indebted to the Editorial Board for arranging the translation of the original Polish text of this essay and for adjusting it to the general style of the Copernican volume. Hearty thanks are due to Dr. C. R. Lynds and Dr. N. U. Mayall for their beautiful photographs used here as illustrations. Finally, the author is particularly grateful to Dr. H. C. Arp for preparing for the present essay a montage of three photographs of M 87 and its jets, reduced to a common scale.

References

1. *Astronomy and Astrophysics for the 1970's*, Vol. 1. Publication of the U.S. National Academy of Sciences, Washington, D.C., 1972.

2. Harlow Shapley and Helen E. Howarth, *A Source Book in Astronomy*. McGraw-Hill Book Co., New York, 1929.

3. Harlow Shapley and Adelaide Ames, "A Survey of the External Galaxies Brighter than the Thirteenth Magnitude." *Harvard Observatory Annals*, Vol. 88, 1932, pp. 44–75.

4. M. L. Humason, N. U. Mayall, and A. R. Sandage, "Redshifts and Magnitudes of Extragalactic Nebulae." *Astronomical Journal*, Vol. *61*, 1956, pp. 97–162.

5. *National Geographic Society—Palomar Observatory Sky Atlas*. This is a collection of large-scale photographs of the sky, unbound, 1954– and continuing.

6. Allan Sandage, *The Hubble Atlas of Galaxies*. Publication 618, Carnegie Institute of Washington, Washington, D.C., 1961.

7. Gerard de Vaucouleurs and Antoinette de Vaucouleurs, *Reference Catalogue of Bright Galaxies*. University of Texas Press, Austin, 1964.

8. B. A. Vorontsov-Velyaminov et al., *Morphological Catalogue of Galaxies*. Four volumes, publication of the University of Moscow, U.S.S.R., 1962–1968.

9. Halton Arp, *Atlas of Peculiar Galaxies*. California Institute of Technology, Pasadena, Calif., 1966.

10. *La Structure et l'Evolution de l'Univers*. R. Stoops, Brussels, Belgium, 1958.

11. Edwin Hubble, *The Realm of the Nebulae*. Yale University Press, New Haven, Conn., 1936.

12. *Conference on Instability of Systems of Galaxies*. Edited by J. Neyman, T. Page, and E. Scott, *Astronomical Journal*, Vol. *66*, 1961, pp. 533–636.

13. C. R. Lynds and A. R. Sandage, "Evidence of an Explosion in the Center of the Galaxy M 82." *Astrophysical Journal*, Vol. *137*, 1963, pp. 1005–1021.

14. E. Margaret Burbidge, G. R. Burbidge, and K. H. Prendergast, "Mass Distribution and Physical Conditions in the Inner Region of NGC 1068." *Astrophysical Journal*, Vol. *130*, 1959, pp. 26–37.

15. E. Margaret Burbidge, G. R. Burbidge, and K. H. Prendergast, "The Rotation and Physical Conditions in the Seyfert Galaxy NGC 7469." *Astrophysical Journal*, Vol. *137*, 1963, pp. 1022–1032.

G. Burbidge and M. Burbidge

5 Modern Riddles of Cosmology

Five hundred years after the birth of Copernicus, astrophysics appears to
the lay world, as well as to the professional astronomer and physicist, to be
probably the most exciting of the physical sciences. Why is this? The simple
answer is that man has always been full of wonder concerning the universe
around him, and his wonder has only increased as he has understood. And
while we have come a long way since medieval times in our understanding
of the solar system and the stars, the major questions that everyone at some
time in his life asks—Where did we come from? Was there a beginning?
What is the meaning of it all?—remain unanswered. The great strides that
we have taken in modern astronomy, which have enabled us to understand
many things using the known physical laws, still leave these great questions
entirely open.

In this volume, three of the major revolutions in astronomical thinking,
all in the twentieth century, are described by Bart Bok in his article entitled
"Harlow Shapley and the Discovery of the Center of Our Galaxy," by
Donald Goldsmith in his article on Edwin Hubble's discovery of the expansion of the universe, and by Wlodzimierz Zonn, who has described recent
evidence for violent explosions in the central regions of galaxies.

The first two of these developments, dating from, or gaining general acceptance, about fifty years ago, are now part of the accepted body of facts.
The third is much more recent, and belongs to the 1960s. What we shall do
in this article is to discuss the possibility that a new revolution is currently
under way in the 1970s. It is much more difficult to do this than to give an
account of what is by now firmly established. For it must be admitted at the
outset that we are going to consider some of the controversial questions, and
there is no general acceptance of some of the views that we shall describe,
nor is all of the astronomical evidence in. But the situation is worth describing, even if it turns out eventually that no radically new outlook is forthcoming.

We are going to discuss the questions: Did the universe start with a big
bang? and, closely allied with this, Must we assume that all that we can ever
deduce about the universe will be found by invoking the laws of physics as
they apply in the laboratory, or is it possible that something new in fundamental physics is to be learned from astronomy?

University of California, San Diego, Department of Physics, La Jolla, California 92037,
and Royal Greenwich Observatory, Herstmonceux Castle, Hailsham, Sussex, England.

Current Cosmological Thinking—The Bandwagon Approach

A strong belief which has persisted through the ages is that there must have been a beginning. Not all religions or mythologies require this, but it is embedded deep in the Christian and the Jewish religions, and while many have questioned the time scale of Genesis as described in the Old Testament, and its date has been continuously put back, the view that there was a beginning is implanted deep in the early training of many Western scientists. The alternative view is that the universe is eternal and unchanging, and this also has a strong aesthetic appeal to many.

The roots of the modern *scientific* view that there was a beginning started with Einstein's development of general relativity. In his own work Einstein assumed that the universe in the large is unchanging, but it was the Russian meteorologist Friedmann and independently the Belgian priest Georges Lemaître who showed that the Einstein model of the universe is unstable so that, when perturbed in the direction of expansion, it will continue to expand. While there are infinite numbers of solutions of Einstein's equations, the most interesting model is one that starts at a finite time in a state of infinite density. This then is the beginning! Incidentally, we now know that an expanding universe is also predicted in a purely Newtonian cosmological theory.

Now this was only a theory, but very soon after it was advanced, Edwin Hubble with Milton Humason produced strong evidence from studies of the distances of the galaxies and their red shifts (interpreted as Doppler shifts) that the universe is expanding. This discovery has been discussed fully by Donald Goldsmith elsewhere in this part. It was not only (in 1930) considered to be one of the most successful predictions in the history of science, but it also gave a powerful scientific impetus to the idea of a cataclysmic beginning. For the next twenty years the drive in observational cosmology, almost completely in the hands of Hubble and his colleagues, was to decide which of the myriad of general relativistic models best represent the universe. They all demand that there was a beginning, though there is one kind of model, the so-called oscillating model, which requires that the universe successively expands from a state of very high density for a finite time, the rate of expansion gradually slowing down until it halts, and then contracts at an ever increasing rate until it is compressed into a small volume of very high density. It then expands again and the cycle repeats. While such a model is mathematically possible, no physics concerning the state of matter is put in, and so there is no understanding of whether or not it is possible for the high-density contracting universe to "bounce"—in fact, it may well collapse indefinitely, in which case the oscillating universe model is not realistic.

While a number of variants on general relativity have been proposed, they only give rise to very minor effects as far as cosmology is concerned. A beginning is always required.

One of the early problems faced by the modern cosmologists was concerned with the comparison of the age of the universe with the ages of the earth, the sun, and the stars, and the ages of the chemical elements determined from measurements of radioactivity of uranium and thorium. In particular, it was found that the earth appeared to be older than the universe when its age was derived from the rate of expansion given by Hubble. Since this is an impossible result, ways out had to be found. One was a model of the universe in which it expands from a point, but then the expansion slows down and ceases for an arbitrarily long waiting period after which it expands again (the so-called Lemaître universe). A second is the idea that we live in a steady-state universe, which is infinite, with galaxies receding from each other as found by Hubble and with the resulting gaps being filled by new galaxies, formed out of matter that is continuously created. The impetus for the modern development of this theory by Hoyle, Bondi, and Gold in 1948 was the apparent age paradox. However, by now, as frequently happens when new ideas are generated, the evidence that led Hoyle and his colleagues to doubt the theory has evaporated. This is because the age of the universe derived from measures of its expansion rate has increased tenfold since Hubble derived it in the 1930s, and more than threefold since the steady-state theory was developed, while age determinations for the stars and the solar system and the uranium-thorium ages have been refined. At present there is no conflict since the age of the universe ($\lesssim 20 \times 10^9$ years), determined from its rate of expansion and the cosmological model, is greater than the age of the oldest stars ($\sim 10 \times 10^9$ years) and it is also greater than the age of the solar system ($\sim 4.6 \times 10^9$ years). In addition, the age of the universe is also greater than the age of the elements ($\sim 12 \times 10^9$ years).

Even though there is no age paradox, it was soon realized that the steady-state model is an attractive alternative to an exploding universe and it must be subjected to observational tests. It should also be mentioned in passing that the theory was, early in its history, subjected to considerable criticism and hostility of a largely nonscientific character, and vestiges of this still linger on.

Before we discuss the observations that bear on the cosmological theories, as they have accumulated in the last twenty years or so, it is necessary to discuss the second discovery which strengthened the scientific case for a beginning in a hot big bang.

About twenty-five years ago, George Gamow with his colleagues Ralph Alpher and Robert Herman considered in some detail the physics of the early universe under the assumption that it began in a highly dense state. They were largely motivated by the idea that perhaps one could form the heavier elements through nucleosynthesis of the elementary particles in the early universe. The results they obtained, which have been borne out by the most recent work that utilizes all of the modern experimental evidence of nuclear physics, showed that it is impossible to build the heavier elements in a big bang starting with only the elementary particles and radiation. Only deuterium, helium 3, helium 4, and possibly lithium 7 may be built in this way, and of these, only helium 4 has a comparatively high abundance.

Gamow and his colleagues did show that, as the universe expands, the matter and the radiation would cease to interact together effectively after a very short time, and the radiation would cool due to the expansion while retaining its blackbody[1] form, so that if we know the rate of the expansion, the effective temperature of this radiation field at present can be estimated, and they concluded that it should have a temperature of a few degrees Kelvin. Thus there was a clear prediction. If the universe began in a hot fireball, there should be a remnant blackbody radiation field present today with a temperature of a few degrees, and an energy density of perhaps a few electron volts per cubic centimeter.

In the early 1960s a group of physicists in Princeton under the leadership of Robert Dicke were reinvestigating the ideas concerned with an early universe, and they not only recalculated the temperature of this radiation field but considered the possibility (not envisaged by Gamow and his colleagues) that it might be possible to detect it. If the radiation has a temperature of a few degrees, the peak of intensity is at a wavelength near a millimeter. At the long wavelength side of the peak the intensity $I \propto \lambda^{-2}$, where

λ is the wavelength, while on the short wavelenth side the intensity falls off exponentially.

In 1965, Arno Penzias and Richard Wilson at Bell Laboratories announced that they had detected a smooth background radiation component at a wavelength of 7 cm which might be primeval fireball radiation. Soon after this the Princeton group made measurements at about 3 cm, and since then a considerable number of direct measurements have been made in the centimeter range, indicating that the curve does have a λ^{-2} form in this part of this spectrum. If the curve has a blackbody form, a single measurement defines the temperature, and this now appears to be about 2.7°K (Figure 5.1).

This remarkable discovery, if interpreted along the lines just described, is the most powerful direct evidence for an initial state for the universe. At the same time, the theory predicts (i) that the radiation must be highly isotropic,

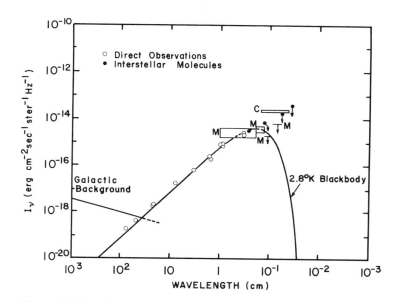

Figure 5.1. Plot of the background radiation curve—the so-called hot-fireball radiation.

The black line shows the blackbody curve for a temperature of 2.8° Kelvin. The points represent measurements of different types. The plot is taken from a review paper by P. Thaddeus in *Annual Reviews of Astronomy and Astrophysics*, Vol. 10, 1972, p. 305 and reproduced with permission from the Royal Society of London.

and (ii) that it must truly have a blackbody form. At wavelengths of a few centimeters the tests show that the radiation field is extremely smooth and isotropic, so much so that it is becoming rather embarrassing. For we would expect to see a small amount of anisotropy corresponding to the fact that the solar system is moving in an orbit about the center of our own galaxy at about 250 km/sec, and our galaxy also presumably has motion of a few hundred kilometers a second relative to the ultimate standard of rest defined by the radiation field. No anisotropy has yet been detected though we are exactly at the level at which it must appear.

To establish that the curve has a blackbody form is more difficult. This is because direct measurements at the peak of the curve and on the short wavelength side cannot be made from the surface of the earth, because of the effect of the atmosphere. Thus we must rely on indirect measurements or on direct measurements made from space. The indirect measurements use interstellar molecules as thermometers. From the spectra of such molecules as CN and CH the temperature of the radiation field in which they are bathed can be deduced, and it has been found in all cases studied that at wavelengths of a few millimeters the temperature is not greater than 2.7°K, though often only an upper limit is given, rather than a positive result. On the other hand, measurements made from rockets and balloons have given results that not only conflict with each other in some cases but also with the limits set by the interstellar molecules. This situation now may be resolving itself, since the latest observations suggest that some of the measurements made above the atmosphere are not correct so that most measurements are compatible with the blackbody curve.

If the universe had no beginning, then to explain these observations we must argue that this radiation is generated by a large number of discrete sources, galaxies, or other more peculiar objects. For the combined effect of many sources to give a blackbody curve appears remarkable and implausible to many, though Hoyle and his colleagues have shown that it is possible. There is one other interesting alternative. It has recently been proposed that, if we live in an oscillating universe which successively expands and contracts, the blackbody radiation may be starlight emitted in the previous cycle of pulsation.

The reasons for the acceptance of the big bang are clear; a feeling by many that there must have been a beginning, the existence of a beautiful theory predicting an expanding universe and a primeval radiation field, and the discovery that one and probably both of these exist.

Thus, most cosmologists believe that the direction of understanding is fairly obvious. They feel that we know in general terms how the universe started, and how it is evolving. Only the details need to be worked out, and further observations made. But is that really all?

Further Cosmological Studies

Following the lines of investigation started by Hubble and Humason, it is possible to study the history of the universe and determine its age by extending the red-shift–apparent brightness relationship to fainter and fainter galaxies—looking farther and farther into space and, therefore, back in time. The form of the curve of the red shift z plotted against apparent brightness if measured to large enough values of z allows us to decide between different big bang cosmologies or the steady-state cosmology.[2] While many people believe that the data already available rule out the steady-state theory, we believe that this question is still not settled, since the data available for the very largest red shifts are very sparse as yet.

Another major input to observational cosmology came about twenty years ago with the discovery of the powerful extragalactic radio sources. Optical identification of one of the first of these sources (Cygnus A) led to the discovery that it was a highly peculiar galaxy with quite a large red shift ($z = 0.057$). Since it is a very powerful source, it was realized that even with the radio telescopes existing in the early 1950s, objects similar to this could be detected at very much larger red shifts corresponding to much greater distances. Since very large numbers of fainter radio sources were soon detected, it was argued that they could be of the greatest value for cosmological investigation, since if they were intrinsically similar to Cygnus A they must be very distant. To use them for such investigations, several groups of radio astronomers, and particularly Martin Ryle and his associates at Cambridge University, counted the number of sources at different flux levels, since it is known that it is possible to study the large-scale structure of the universe by determining the form of the N-S relation where N is the number of sources with an observed flux greater than S. A schematic diagram of this curve is shown in Figure 5.2. For all cosmological models in which the sources are

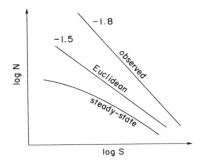

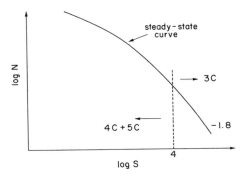

Figure 5.2. Schematic diagram of counts of radio sources.

(a) A schematic plot showing how the curve of log N against log S should look for different cosmological models and the observational curve for the brightest sources.
(b) The plot obtained for sources measured in the Cambridge 3C, 4C, and 5C surveys. The 3C survey covers the brightest sources, the 4C survey involves fainter sources, and the 5C survey covers the faintest sources so far examined. The flattening of the slope of the curve is obvious. These diagrams are taken from the Bakerian Lecture of the Royal Society given by F. Hoyle in 1968 and published in *Proceedings of the Royal Society, A,* Vol. *308,* 1968, p. 1 and are used here with permission from the Royal Society of London.

distributed uniformly in space, it is easily shown that the slope of the log N–log S plot should be -1.5 for the nearest sources (those for large S) and then become progressively flatter as one goes to fainter and fainter objects. If the exact shape of the curve could be determined, it would be possible in principle to determine which cosmological model, among the myriad of exploding models and the steady-state model, is appropriate. However, when the first surveys were carried out, it was found that the slope appeared to be significantly steeper than -1.5, and this was interpreted by Ryle and his colleagues as being due to the fact that there are more faint (distant) sources than there are sources comparatively close by. Since this means that the space density of sources was greater in the past than it is now, this was considered a powerful argument against the steady-state cosmology. Extension of the counts by many groups to fainter and fainter sources has led to the result that the curve does steadily get flatter, with the slope gradually changing from about -1.8 through -1.5 and ultimately to a value close to -0.8 for the faintest sources so far observed. This led Hoyle to argue some years ago that once one considers sources beyond the brighter ones the results can be interpreted as being due to a constant density of sources with the normal cosmological effects entering—in fact Hoyle concluded that the bulk of the data was compatible still with a steady-state interpretation.

This interpretation, that the steep slope at the bright end can equally well be interpreted as a small deficiency of strong sources as by a large excess of weak ones, appears to be borne out by a careful study made by K. Kellermann and his associates at the National Radio Astronomy Observatory. It alone casts considerable doubt on the often-quoted statement that the counts conclusively rule out the steady-state model.

The reason for the controversy in this field is that for the majority of the radio sources we really do not know either where we are looking or what we are looking at. While thousands of radio sources have been catalogued and counted to give the $N(S)$ curve, only a few hundred have been optically identified, and of these only a fraction (perhaps 500) have measured red shifts. And, as we shall describe, knowledge of the red shifts does not necessarily mean that we know the distances of these objects. It is what was discovered when the optical objects associated with some of these radio sources were investigated that first led to the glimmer of an idea that perhaps another revolution in our understanding of the universe might be coming.

Violent Phenomena in Dense Objects

It was pointed out earlier that Cygnus A, one of the first extragalactic radio sources to be optically identified and studied (by W. Baade and R. Minkowski in 1954), is peculiar. It is very different from a normal galaxy in the following sense. A normal galaxy, made up of stars, shows this in its color and its line spectrum, which is a composite of absorption lines arising in the atmospheres of the stars. Sometimes comparatively weak emission lines arising in the interstellar gas in the galaxy can also be seen. In Cygnus A, however, the spectrum consists entirely of strong, broad emission lines emitted by a hot, highly excited gas cloud, and a continuum that may be of nonthermal origin, that is, it is unlikely to be due to stars. This spectrum is somewhat reminiscent of the spectra of the nuclei of galaxies (the so-called Seyfert nuclei) which were discovered much earlier. It is now known that these nuclei are in a state of violent activity (see Chapter 4 by Wlodzimierz Zonn in this volume).

As more radio sources were investigated optically, it was found that many of them were otherwise normal giant elliptical galaxies, but a significant fraction showed much evidence in their spectra of violent activity and the ejection of matter. It was gradually realized that systems like Cygnus A might be comparatively normal galaxies in which violent explosions have taken place. The explosions are so violent that they not only produce the very large fluxes of high-energy particles which are responsible for the radio emission and probably much of the optical emission, but the light from the stars in the galaxy is overwhelmed by the light emitted due to the explosions. Accumulation of evidence, starting with the radio emission, has shown that the gigantic explosions in the nuclei of galaxies in which very large amounts of energy are released are fairly frequent. An example of what can be seen in optical wavelengths in some galaxies is shown in Figure 5.3. These discoveries have led to a revolution in our understanding of galaxies, a revolution that has changed our concept of the universe as did the Copernican revolution. Until they were made, it had been supposed that galaxies were large aggregates of stars which formed (by some mysterious process) early in the history of the universe, and since the stars evolve only very slowly, it was assumed that the galaxies also evolve very slowly in time with very gradual changes in their light and color over times of $\sim 10^{10}$ years. The fact that very large and sudden outbursts take place in times as short as 10^6 years means

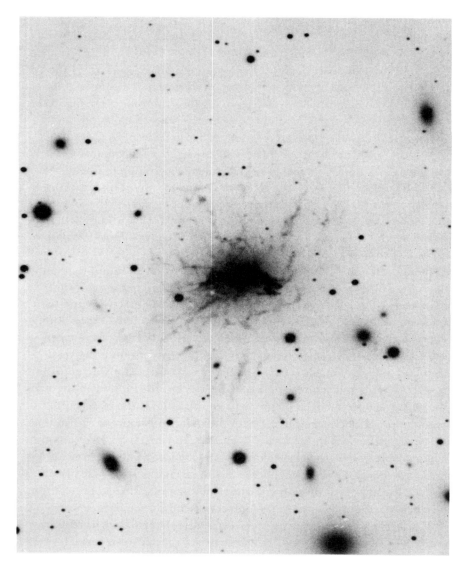

Figure 5.3. Photograph of the exploding galaxy NGC 1275 taken through a narrow filter to show the filamentary structure.

The plate was taken by Dr. C. R. Lynds with the 84-inch telescope of the Kitt Peak National Observatory. We are grateful to Dr. Lynds and KPNO for giving us permission to reproduce this picture.

that colossal amounts of energy are released by processes other than those involving normal stars.

It was soon realized that to release the amounts of energy that are often seen, perhaps 10^{60} ergs, corresponding to the conversion to energy of all of the mass contained in about ten million stars like the sun, processes very different from any dreamt of before must be operating.

While there is no general agreement as to how this energy release occurs, two approaches are being explored.

1. If we stay within the framework of known physics, we can understand the phenomenon only if we suppose that in the nucleus of a galaxy a very large mass is contained within a very small volume. Such a mass might be made up of billions of stars—at least 10^{10} within a sphere of radius no greater than about one light-year has been suggested. If such a high star density is present, the stars will interact very strongly together, they will collide at a very rapid rate and will tend to stick together to form larger stars. Some people have suggested that much of the energy will be released when the stars collide at high speed. Alternatively, it has been argued that the more massive stars formed in collisions will rapidly evolve and explode as supernovae at a rate of several per year, and that this could explain the large-scale outbursts. But there are several observational facts which suggest that these mechanisms may not be adequate.

As the dense nucleus of stars evolves further, it is easy to show that it will coagulate into one supermassive star. Now general relativity tells us that at this point such a star must gravitationally collapse, and this collapse will go on indefinitely. About ten years ago, Fred Hoyle and William Fowler suggested that it was the energy released in such a gravitational collapse which is responsible for galactic explosions. But the dilemma associated with such objects is the following. We need to suppose that the object is very small and collapsing, in order that the large amount of energy observed can be accounted for, but general relativity tells us that as a star collapses closer and closer to what is known as its Schwarzschild radius,[3] and that means closer and closer to a regime where the gravitational energy becomes a large part of the rest mass energy it becomes harder and harder to get this energy out. More recently it has been argued that if the object is rotating very fast, as it might naturally be, most of the gravitational energy will be in the rotation, and the object, called a spinar by P. Morrison, may release it in much the same way as does a pulsar.[4] While this is a very attractive proposal, it has been developed partly by analogy with the behavior of pulsars. Pulsars, however, are thought to be rotating neutron stars that are formed following a supernova explosion. Thus pulsars and spinars have completely different antecedents. Rotating massive objects may not release energy in the same way as do pulsars. In addition, they are undoubtedly unstable and, at some stage, will break up. The pieces may contribute to the energy output.

This sketch of some of the schemes proposed to explain explosions in galaxies shows that they all invoke violent energy release in very strong gravitational fields. Now the very fact that the observational discoveries lead us to this situation is very intriguing. For it has only been possible to test the general theory of relativity by experiments in the solar systems under conditions in which the gravitational field is weak. Tests of the theory in the strong gravitational field approximation are not possible except in regions such as the nuclei of galaxies. Thus the fact that the theory gives rise to difficulties when applied to these regions may simply mean that for strong gravitational fields the theory itself is not correct but needs modification. While it is very difficult to get to a point at which we can either confirm or deny this possibility, we see for the first time that it is situations such as this —in the strong gravitational field laboratories in the galaxies—that we may learn something new about physics from astronomy.

However, the observation of galactic explosions may require an even more radical explanation.

2. In an evolving universe we do not really know how condensation leading to the formation of galaxies took place. The common view at present, dating in this century from the work of Carl von Weizsacker and others, is that in the beginning there was a primordial field of turbulence already containing the density fluctuations from which the galaxies were born. Thus the galaxies have a cosmological origin; that is, we simply assume that the basic ingredients were there in the right form originally. To understand explosions, we must then assume that the density of stars in the nuclear regions steadily increases to a very high value, such that the sequence of events leading to the formation of a massive object can occur, and then that a very large amount of gravitational energy can be extracted from it. Thus, the sequence of events must be continuous contraction up to the point at which explosive phenomena occur. One of the many problems associated with this scheme is that it is not obvious that evolution to a very high density can take place in

the time available, or if it does, how explosive processes can recur in galaxies.

A possible alternative solution is to suppose that the nuclei of galaxies are places in which matter is being created, or as Sir James Jeans put it in another connection more than forty years ago,

The type of conjecture which presents itself, somewhat insistently, is that the centres of the nebulae (galaxies) are of the nature of singular points at which matter is poured into our Universe from some other, and entirely extraneous spatial dimension, so that to a denizen of our Universe, they appear as points at which matter is being continually created.

The eminent Soviet astronomer Viktor Ambartsumian has argued since the early 1950s that galaxies are being formed from matter that he calls "pre-stellar matter" which is pouring out of the nuclear region. He has also argued that whole groups of galaxies which may be unstable, and flying apart, have been generated in this way. In more recent years Hoyle and Narlikar have suggested, within the framework of the steady-state cosmology, that it is in the nuclei of galaxies that matter is being created.

The wealth of observational data showing that nuclei are exceedingly active and are pouring forth matter and radiation almost continuously, albeit at a very low level in most galaxies, can certainly be taken as evidence in favor of these radical ideas. However, at present there is no general acceptance of the correctness of these views, both for philosophical reasons, and also because there is no well-defined theory of such processes. Thus, most astronomers are still trying to understand the explosive activity in the galaxies on the assumption that one of the paths outlined in (1) is followed: that is, that nuclear explosions are manifestations of the death throes of a nucleus rather than birth. But the ideas expressed in (2) may be more nearly correct and, if this is the case, a new revolution in thinking may be under way.

The Riddle of the Red Shifts

Not only were the radio astronomers the first to show us that some galaxies are highly active and are exploding, but their investigations also led to the discovery of the quasi-stellar objects (the quasars). These are starlike objects with very large red shifts, and we now know that only a small fraction of them are strong radio emitters. The fact that they have very large red shifts means that, provided the red shifts are due to the expansion of the universe, these objects are exceedingly distant and are very much brighter intrinsically (by a factor of 50 or so) than whole galaxies of stars. They are similar in many of their optical and radio properties to the radio galaxies in which gigantic explosions are taking place. They are frequently seen to vary in light or in high-frequency radio waves in times as short as a few months. This indicates that the sizes of the emitting regions may be no larger than a few light months, and at a guess they may, in some cases, be no bigger than the solar system. And yet, on the cosmological hypothesis, energy is being radiated from this volume at a rate corresponding to the total conversion of the mass of a star equal to the sun each year.

Again, if we assume that these objects are at such great distances (indicated by their red shifts) that we are looking at them as they were billions of years ago, it has been shown by Maarten Schmidt, who identified the first red shift in a quasar, that their space density depends very sensitively on their red shift. It increases very steeply as the red shift increases until a maximum is reached close to a red shift $z \simeq 2.5$, and then it decreases sharply. This implies that the rate at which the quasars have been formed is a very strong function of epoch, and this is a strong argument against a steady-state universe in which things should remain constant and independent of time.

However, the most controversial aspect of the investigations of the quasars is the question of whether or not the red shifts are really of cosmological origin. If they are not, there are two immediate consequences:

1. Quasars are irrelevant as far as cosmological investigations are concerned because they must be comparatively close by.
2. There must be an alternative mechanism that can give rise to large red shifts.

If this last statement is correct, it seems likely that we are dealing with an entirely new phenomenon related to the questions described earlier concerned with the physics of galactic explosions, and perhaps new physics is involved.

How has the controversy developed and where do we stand at present? A major factor has clearly been that since the discoveries of the 1920s, astronomers have been very strongly conditioned to interpret all red shifts as being due to the recession of galaxies produced by the expansion of the universe. To paraphrase the present governor of the state of California in a statement he made concerning the preservation of the redwood trees, "When you've seen one red shift, you've seen them all." Another factor that has influenced many astronomers against a new interpretation has been the realization that, if the red shifts are not cosmological in origin, we probably have no explanation grounded in a theory that we are familiar with. The very idea of something entirely new inhibits many of us. The realization of this, following a discussion of related objects at the Vatican Symposium on Galactic Nuclei held in 1970, led Fred Hoyle to remark, "I don't see the logic of rejecting an observation because it's incredible!"

What tests can we make to determine the distances of the quasars and decide whether or not their red shifts are of cosmological origin, that is, due to the expansion of the universe? The following facts must be taken into account.

(a) If we plot the red shifts of the quasars against their apparent brightnesses, we do not get a smooth relationship similar to that found by Hubble and Humason for the galaxies (Figure 5.4). The discovery of the latter relation was a key step in understanding the nature of the red shifts of normal galaxies. The fact that such a plot for the quasars shows a very large scatter does not mean that the red shifts may not be cosmological in origin, but at the same time this diagram gives no support to the cosmological hypothesis. Put another way, if such a plot had been found for galaxies, we might still not know that the universe is expanding.

(b) The distribution of the red shifts of the quasars and related emission-line objects is most peculiar. There is a progressive cutoff for values greater than about 2.2 which appears to be real. If the red shifts are cosmological, this tells us something of great importance cosmologically; that is, the conditions in the universe were very different at times before the epoch corresponding to this red shift than they were after. If the red shifts are not measures of distance, this cutoff must be due, in part at least, to the physics of the sources. There are also some sharp peaks in the red-shift distribution. In particular, there is a sharp peak at $z = 1.95$ and another involving objects related to quasars at 0.061. There has been considerable argument about the reality of these peaks and related periodicities in the distribution of red shifts. At present the most detailed statistical analysis suggests the 1.95 peak is not statistically significant, but that a periodicity in the red shifts with a period of 0.031 ($\simeq 0.061/2$) is significant at the 97.5 percent confidence level. Such a result suggests rather strongly that a new type of red-shift phenome-

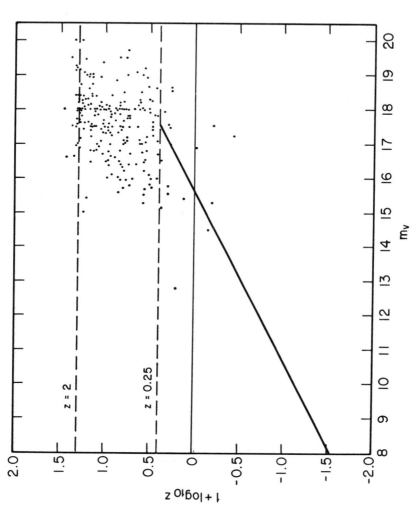

Figure 5.4. Plot of the logarithms of the red shifts against the apparent magnitudes (apparent brightness) of more than 200 quasi-stellar objects.

The line that falls largely below the points is the same relation for bright galaxies, that is, it represents the Hubble's law. The dashed lines are drawn to show that the majority of QSOs have red shifts z between about 0.25 and 2.

non is present and that the red shifts of these objects are not measures of distance. But the argument is still indirect.

(c) There are similarities between some of the radio and optical properties of the quasars, objects looking very similar to them except that they are not quite stellar, and the nuclei of Seyfert galaxies. Thus, if we plot one of these quantities against the red shifts of the objects, we find a smooth relationship. Since in such a plot the objects with the smallest red shifts—the Seyfert galaxies—undoubtedly have red shifts largely of cosmological origin, it has been argued by some that this continuity is evidence in favor of the idea that all of the red shifts have a cosmological origin. However, the form of the smooth relationship from which this conclusion is reached depends itself on the assumption that the red shifts all have the same origin. For example, if we suppose that the red shifts of the quasars are largely noncosmological, then their intrinsic properties (optical luminosities, and so on) are very similar to those of the Seyfert nuclei so that there is still continuity between those properties. Consequently, the continuity argument cannot be used as evidence for or against the cosmological red-shift hypothesis.

(d) The fact that there are rapid light variations and radio variations in quasars and in the nuclei of exploding galaxies means that the regions generating the energy are very small. In particular, for very highly luminous objects like cosmologically distant quasars, it is difficult to understand how the radiating mechanism—the radiation by high-energy electrons in magnetic fields—can work effectively. This is because the radiation density is so high in these objects that high-energy electrons find it very difficult to survive long enough to radiate their energy as optical photons. One way of avoiding this apparent difficulty is to suppose that the quasars are much closer. In this case, they are no brighter than the nuclei of comparatively nearby galaxies and the paradoxical situation that we get into in trying to understand them is largely avoided.

(e) The sizes of some very small radio sources in quasars and related objects are changing and appear to be expanding with time. If they are at cosmological distances, they appear to be expanding with velocities several times that of light. While there are ingenious ways of explaining this result, an alternative explanation is that these objects are much closer than the distances derived from their red shifts. If this is the case, the expansion speeds never apparently exceed that of light. It is interesting that the only expanding source that lies in a galaxy whose distance is certainly known is expanding with a speed of only about 20 percent that of light.

All of the arguments just given are indirect and cannot establish conclusively that the quasars do or do not have cosmological red shifts. What we really need is a direct method of measuring the distances. Now in extragalactic astronomy we still only have two methods open to us:

1. Measure the red shift of a galaxy or a quasar. *Assume* that the red shift is due to the expansion of the universe, and determine a distance from the Hubble constant. For objects that are not too distant, $D = cz/H \times 10^6$ parsecs, where H is the Hubble constant; the present value of H is 50 km/sec per 10^6 parsecs.

2. If we know the distance of object A (from its cosmological red shift) and if we can prove that object B is physically associated with it, either by showing that there is a luminous connection between A and B or by showing that B is most unlikely to be close to A by chance, then we know the distance of object B; it is at the same distance from us as A.

The noncosmological red-shift hypothesis could thus be established if it could be demonstrated that objects with very different red shifts are physically associated. Then at least the difference in the red shifts must be intrinsic to the objects. On the other hand, the cosmological red-shift hypothesis would be confirmed if it could be established that objects which have the same red shift (to within small differences) are physically associated.

The latter hypothesis has been tested as follows: The fields around a number of quasars with small red shifts have been examined, and, where clusters of galaxies can be seen, attempts have been made to measure the red shifts of some of these galaxies. In three cases it has now been found that one or two of the galaxies in clusters or groups close to the quasars have red shifts (measured in velocity units) that are the same as those of the galaxies to within a few thousand kilometers a second. Provided that these associations are not due to chance, this is evidence that the red shifts of those quasars are of cosmological origin. However, the method used does not test the hypothesis that objects with *different* red shifts are associated together. This is because the quasars chosen for study were those with red shifts small enough so that if they were at cosmological distances, galaxies associated with them would just be detectable. The red-shift limit has been set at $z \simeq 0.3$ because for cosmological red shifts much greater than this galaxies would be too faint to be detected at all. Thus, for example, there has been no attempt to see if quasars with red shifts much greater than 0.3 are associated with galaxies with red shifts equal to or smaller than 0.3.

On the other hand, evidence that objects with very different red shifts are associated together has also been found. A study of the brightest radio quasars, of which there are not more than about 50 in the northern sky, shows that 5 of them lie very close, from a few minutes of arc to about 20 seconds of arc, to bright galaxies of which there are only a few thousand over the whole sky. The galaxies lie at distances typically less than about 300 million light-years (comparatively close by on the scale of the universe), while the red shifts of the quasars range from about 0.5 to 1.4. Such large cosmological red shifts would correspond to distances of billions of light-years. Extensive statistical analyses have led to the conclusion that there is less than one chance in 100,000 that this distribution is due to chance; that is, the statistical evidence favors the view that the quasars and the galaxies in this sample are physically associated. Further evidence in favor of this hypothesis is that the angular separations of the quasars from the galaxies are inversely proportional to the distances (the red shifts) of the galaxies, thus suggesting that in each case the quasar is at about the same distance from the galaxy.

This is not the only evidence for noncosmological red shifts. Over a number of years Halton Arp has shown evidence that suggests to him that many peculiar galaxies have ejected quasars and even other galaxies with very different red shifts. While the statistical arguments have often been weak, some of the cases shown are very convincing. For example, in Figure 5.5 we show the bright galaxy NGC 7603 that is joined by two arms to a small galaxy with nearly twice its red shift. Another case is the galaxy NGC 4319 with an object that is almost identical to a quasar with a very different red shift apparently lying inside one of its spiral arms. Probability arguments suggest that these two systems are physically associated. Arp has claimed to have found a physical connection between them in the form of a faint luminous bridge, but others have not been able to confirm its existence.

Finally, a number of small groups of galaxies have been found in which one member has a highly discrepant red shift. One such system is shown in Figure 5.6. Again probability arguments suggest that the objects are all physically associated, so that the galaxy with the discrepant red shift may have a comparatively large intrinsic red-shift component.

Thus, while there is evidence suggesting that some quasars have cosmolog-

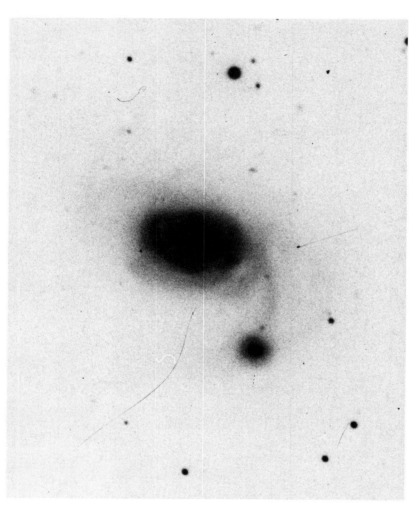

Figure 5.5. Photograph of the galaxy NGC 7603 and companion to which it is joined by two arms.

The companion has nearly twice the red shift of NGC 7603, although they must be at the same distance. The plate was taken by Dr. H. C. Arp with the Hale 200-inch telescope. We are grateful to Dr. Arp and to the Hale Observatories for giving us permission to reproduce this picture.

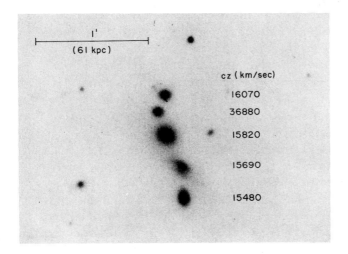

Figure 5.6. The chain of galaxies V-V (Vorontsov-Velyaminov) 172.
The red shifts of the galaxies (in velocity units) are marked, and it can be seen that one of them has a highly discrepant red shift though it is likely that they are all physically connected. We measured red shifts of two of the objects, and Dr. W. L. W. Sargent of the Hale Observatories measured the red shifts of the other objects, including the galaxy with a discrepant red shift. This plate was taken by Dr. H. C. Arp with the Hale 200-inch telescope. We are grateful to Dr. Arp and to the Hale Observatories for giving us permission to reproduce this picture.

ical red shifts, there is also evidence which suggests that noncosmological red shifts do exist.

If true, this is one of the most remarkable discoveries of the twentieth century. What does it mean? Some might want to argue that it casts doubt on the idea that is by now deeply imbedded in our thinking, namely, that the universe is expanding. We see no reason at all to take this point of view, since the evidence from normal galaxies is now stronger than it was when the discovery was made.

Conclusion

Where do we stand if the evidence just described is accepted at its face value? It appears likely that the active nuclei of galaxies and quasars can be understood only if something beyond our present theory is invoked. The alternative view is that the explosive events can be explained by conventional evolutionary schemes, and the anomalous red shifts as (local) Doppler shifts—involving the ejection of objects from galaxies at very high speeds, or as gravitational in origin. However, there are very severe difficulties associated with these latter ideas. If the shifts are due to ejection at very high speeds, there should be very many objects with blue shifts due to the fact that they have a large velocity component in our direction, and none have been found. The interpretation of the large red shifts as gravitational shifts also appears to be (exceedingly) implausible.

An entirely new suggestion made by Hoyle and Narlikar is that the masses of the fundamental particles are not everywhere the same—equivalent to supposing that the gravitational constant is not constant but varies with time and also with position. No well-defined theory lies behind this proposal, and to be accepted it will demand a radical rethinking of some of our current concepts in fundamental physics.

And what of the big bang? Certainly much of the cosmological evidence points to the view that there was a beginning. However, it may be that even this concept is far too simple. In any case, it appears it may be difficult to fit all of the observed facts into a scheme in which the conditions in a beginning were ultimately responsible for all of the complexities of the universe at present.

On these basic questions, the cosmologists are clearly divided at present. The majority feel that the evidence for an initial state is almost overwhelming, and somehow feel that it will be possible to fit all of the startling new discoveries into a picture in which it is supposed that the initial conditions were ultimately responsible for everything—galaxy formation and evolution, dense systems in nuclei, explosions, quasars, and the like. But there is a minority who feels that the new observations may cast doubt on the big bang, but even if there was a singular beginning, some of the observations require entirely new ideas that will have an impact on physics as great as some of the most remarkable discoveries of the first half of the twentieth century. It is this that may lead to yet another revolution of thinking in astronomy in this century.

Notes

1. Blackbody radiation is the technical name for radiation emitted by matter—solid, liquid, or gas—in which the radiation is completely specified by a single definite value of the temperature and in which the energy emitted at various wavelengths can be described by a mathematical formula in which temperature is the only determining quantity.

2. The red shift z is defined as $\Delta\lambda/\lambda_{lab}$ where $\Delta\lambda = (\lambda_{observed} - \lambda_{lab})$; λ_{lab} is the wavelength of a spectrum line measured in the laboratory, and $\lambda_{observed}$ is the wavelength measured in the spectrum of the galaxy or other extragalactic object.

3. The Schwarzschild radius is numerically equal to $2GM/c^2$, where G is the gravitational constant, M is the mass of the object, and c is the velocity of light.

4. Pulsars, discovered in 1967, are radio sources that vary or "pulse" in times of about one second, or even small fractions of a second. The only explanation found for such rapid yet regular variation is that it is produced by a very small massive object rotating very fast; a beam of radiation sweeps past the observer during the rotation. For an object to be so small, the material must be very compressed, stripped right down to its component elementary particles, and in fact must be mainly neutrons—a mass equivalent to that of the sun, or 2 million million million million million tons packed into a sphere of radius about 6 miles. Such stars would be at the end of their evolutionary lifetimes. Although regular variations have not been found in the nuclei of galaxies, they are known to be rotating fast, and it has been suggested that in some ways they might resemble giant pulsars—hence the name "spinars."

Part II

Biology

Summary

The number of quasi-Copernican revolutions in biology and also the depth of these revolutions appear to exceed those in many other domains of science. The first essay in this part gives a rapid overview of the fundamental changes in the thinking on the essence of life itself. The period covered extends from Leonardo da Vinci (1452–1519), through the time of the gradual demise of the so-called "vitalism," to the present day's musings about "genetic engineering." Subsequent essays concern selected particular episodes in this chain of events.

Two essays deal with related developments of the theory of evolution and of genetics, the study of heredity. The landmark in the theory of evolution is the opus of Charles Darwin, *On the Origin of Species* . . . , published in 1857. Its ideas of struggle for existence and of survival of the fittest reflect the general intellectual upheaval that swept Europe following the French Revolution. Genetics has its origin in the experimental work of Gregor Mendel. Published in 1865, Mendel's work was largely overlooked and then rediscovered at the turn of the century. It served as a complement to Darwinism.

Darwin's own studies were based on the anatomy of contemporary organisms and on fossil remnants, and had important limitations. The mid-twentieth century saw a new wave in the study of evolution, following the development of molecular biology and the discovery of the DNA molecule as the carrier of inheritable genes. The problems studied include the question of whether all life on earth started from a single distant primitive ancestor.

The evolution of species is paralleled by that of single cells, the building blocks of individual organisms. The cells of primitive organisms are relatively simple; those of higher organisms are much more complicated. A separate essay explains the idea that the complex cells of higher organisms resulted from such phenomena as the "invasions" of large primitive cells by many small primitive cells, each with its own inheritable properties. Then the invaded cell and the invaders establish a most interesting "cooperation" with a greatly expanded range of possible functions. A possibility seems to exist to produce similar "invasions" artificially, with intentionally directed changes in heredity of higher organisms.

The vision of a primitive ancestor of all life on our planet raised the question of the possible source of the big "biological" molecules, the building blocks of all living matter. Such molecules are normally produced through processes within living organisms. Also, as a result of considerable effort, some of these molecules can now be synthesized in laboratories. The interesting question is whether such molecules, or at least their important "submolecules," could have been produced spontaneously on earth. Twenty years ago this question was answered in the affirmative by a then graduate student in Chicago. His essay describes how it all happened.

The last essay in this part of the volume concerns neurobiology, the study of the brain and of the nervous system. Quite possibly, the neurons, the building blocks of the brain, are the most complicated cells of all. Even though a tremendous volume of facts has been accumulated regarding the structure of neurons and their interactions, both chemical and electrical, the understanding of such phenomena as memory and thought is still far away and might be brought about by a special revolution. The question is whether it will be a Copernican or a non-Copernican revolution.

Robert L. Sinsheimer

6 The Molecular Basis of Life

To early man the distinction between animate and inanimate matter was surely obvious. Animals moved, ate, sensed, slept, reproduced, and died; plants grew, flowered and reproduced, and died. Inanimate objects did none of these but merely persisted unless moved or rent by external force.

Living creatures could be observed in a striking variety of shapes and textures, with diverse spans of existence; so also were inanimate objects. It surely seemed plausible to surmise that the living creature was but an inanimate mass infused with a quick spirit of motion, a *vis vitae* which conferred to its host the properties of life and which simply departed upon death.

The motive power is the cause of all life.—Leonardo[1]

Hence the myth of Galatea: inanimate matter might be directly made animate if one knew the magic. And hence the doctrine of vitalism—that the substance of living matter differed throughout from the substance of ordinary matter by the addition of a mysterious power to each component particle—born in the most obvious observation and entrenched by centuries of tradition.

The life of the whole (animal or vegetable) would seem to be only the result of all the actions, all the separate little lives, if I may be permitted so to express myself, of each of those active molecules whose life is primitive and apparently indestructible.—Buffon[2]

Every individual particle of the animal matter then is possessed of life, and the least imaginable part which we can separate is as much alive as the whole.—Hunter[3]

The overthrow of this doctrine has been in truth a Copernican revolution and has provided a very different perception of the nature of the world around us. And indeed when coupled with an understanding of biological evolution, which has linked man directly to all the world of life, this perception has distinctly altered man's insight into his own inner nature.

Historical Development

The death of vitalism and the comprehension of life in terms of physical science have been the consequences of an extended series of major discoveries both expanding and complicating the concept of life on one hand, and both expanding and deepening the concepts of chemistry on the other.

The invention of the microscope and its early application by Anton van

Division of Biology, California Institute of Technology, Pasadena, California 91109.

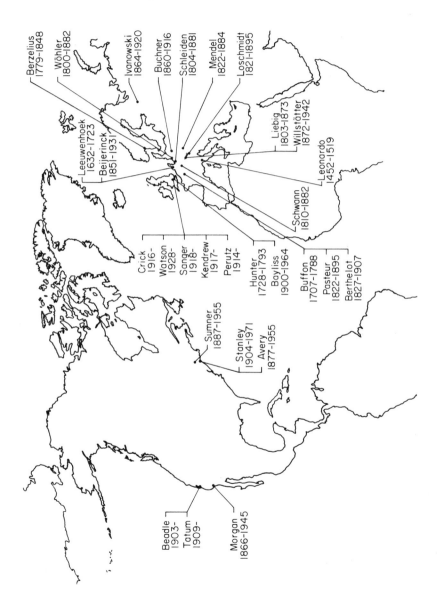

Figure 6.1. Geographical distribution of the scientists cited in this chapter.

Plate 6.1. Anton van Leeuwenhoek.
(Courtesy of The Bettmann Archive, New York.)

Leeuwenhoek immediately made evident a new world of previously invisible forms of life. These forms, too (in general), moved, grew, reproduced, and could die. Because of their ubiquity and their small size, their continual origin from inanimate matter seemed not to require great magic, and thus seemed not unlikely ("spontaneous generation") until finally disproved by Pasteur.

The discovery of the uniquely modular nature of all the larger living creatures, the cells, of Schleiden and particularly Schwann, brought a partial unity to the microscopic and the larger forms of life by revealing their common structural basis. At the same time, however, this discovery emphasized a new structural divergence between ordinary and living matter. The further discovery of common elements of structure, such as nuclei, plasmids, and mitochondria and membranes within most cells further strengthened the concept of a unity of life and its distinction from nonliving matter.

With the discovery of the viruses (tobacco mosaic virus by Iwanowski in 1892 and, independently, by Beijerinck in 1898) the concept of life was extended again, to yet smaller entities. (Beijerinck referred to the infectious agent as a "contagium vivum fluidum." But in the process the concept was confused by the loss of some of its salient characteristics. The viruses—submicroscopic, filterable—could evidently reproduce but only within other living (or recently living) forms. They appeared to lack an independent metabolism. Some researchers, therefore, thought the viruses to be the minimal units of life; others thought them to be the ultimate of obligate parasites.

While the concepts of life were thus extended at the cost of increased complexity and diversity, the domain and understanding and sophistication of the chemist were steadily growing. The fundamental list of natural elements grew and was given coherence in the table of Mendeleev. The importance of the light elements—carbon, nitrogen, oxygen, and hydrogen—in the composition of living matter was recognized.

However, while many chemists believed living or organic matter to be comprised of the same elements as compose inorganic matter, they also believed that the organic substances of living matter appeared only under the influence of "vital forces," able to unite their constituent elements in a special, complex way.

The cause of most of the phenomena within the animal body lies so deeply hidden from our view that it certainly never will be found. We call this hidden cause, *vital force*.—Berzelius[4]

In living Nature the elements seem to obey entirely different laws than they do in the dead . . . The essence of the living body consequently is not founded in its inorganic elements, but in some other thing, which disposes the inorganic elements . . . to produce a certain result, specific and characteristic of each species. . . .

This "something" which we call vital force is situated fully outside the inorganic elements and is not one of their original properties as are gravity, impenetrability, electrical polarity, and so on. . . .—Berzelius[5]

The historic synthesis of urea in 1828 by Wöhler was a major step toward the liberation of the chemistry of organic compounds from the mysticism of vitalism, although it did not silence vitalistic explanations of the mystery of the organization of living matter.

> The chemical force, under the dominion of heat, determines the form and properties of all the more simple groups of atoms, whilst the vital force determines the forms and properties of the higher order of atoms, that is, of organic atoms.—Liebig[6]

Of equally great importance for the molecular comprehension of life was the gradual chemical appreciation of the true dimensions of molecules. The first approximate determination[7] of "Avagadro's number" in 1865 by Loschmidt had profound implications for the biologist. A simple calculation could reveal that any ordinary cell, say 10 microns in diameter, could contain sufficient molecules to provide an immense diversity and chemical complexity.

That this potential molecular ferment could, in fact, generate and sustain the remarkable processes unique to life remained, however, to be demonstrated. With increasing skill the chemist proceeded to dissect the cell, to obtain and identify the strange and diverse molecules—the proteins, the carbohydrates, the nucleic acids, the lipids—unique to life. But their size, their complexity, their lability, their multiplicity were far beyond the means available for molecular separation and characterization.

The experiments of Edward Buchner were clearly not intended as a direct assault upon vitalism, yet in one sense they provided the first secure bridge between the domains of biology and chemistry, between the world of function and the world of molecules, and led to the eventual demonstration of the molecular nature of the enzymes.

Buchner's achievement was to demonstrate that the chemical reaction of fermentation, previously accomplished only in living yeast cells, could be accomplished as well by extracts of fragmented cells. The mysterious agents of these chemical conversions were called ferments (now called enzymes).

That catalysis was involved in the maintenance of life had been suspected soon after the discovery of inorganic catalysis.

> . . . thousands of catalytic processes take place between the tissues and the liquids and result in the formation of the great number of different chemical compounds . . . The cause will perhaps be discovered in the future in the catalytic power of the organic tissues of which the organs of the living body consist.—Berzelius[8]

> Among the phenomena that are related to the transformations of matter in living things, whether during their life or their death, there are few that do not involve fermentation to a greater or lesser degree.—Berthelot[9]

Plate 6.2. Edward Buchner.

(Reproduced, with permission, from *Noble Prize Winners in Chemistry 1901–1961*. Edited by Edouard Farber, Abelard-Schuman Limited, New York, 1961.)

Plate 6.3. Louis Pasteur.

(Photograph of Louis Pasteur from *Biographical Encyclopedia of Science and Technology,* by Isaac Azimov. Copyright © 1964 by Isaac Azimov. Reproduced by permission of Doubleday & Company, Inc., Garden City, N.Y., 1972.)

However, although "soluble ferments" (diastase, pepsin, invertase) were, in fact, discovered, the demonstration by Pasteur that the fermentation of sugar to alcohol (plus carbon dioxide) was accomplished by a living organism restored the vitalistic concept as a basic tenet of the chemistry of life.

My present and most fixed opinion regarding the nature of alcoholic fermentation is this: The chemical act of fermentation is essentially a phenomenon correlative with a vital act, beginning and ending with the latter.—Pasteur[10]

After Buchner, however, the arts of chemistry could be applied to the purification of the agents of fermentation and to those enzymes soon discovered in other cell extracts to effect other specific chemical reactions.

The problems which faced the contemporaries of Berzelius, Liebig and Pasteur have been solved. The differences between the vitalist view and the enzyme theory have been reconciled. Neither the physiologists nor the chemists can be considered the victors; nobody is ultimately the loser; for the views expressed in both directions of research have fully justified elements. The difference between enzymes and microorganisms is clearly revealed when the latter are represented as the producers of the former, which we must conceive as complicated but inanimate chemical substances.

We are seeing the cells of plants and animals more and more clearly as chemical factories where the various products are manufactured in separate workshops. The enzymes act as the overseers.—Buchner[11]

The chemical search for the molecular ferments was difficult. In retrospect we understand they were labile, present in small amount, admixed with many different but chemically similar molecules. Thus, although the protein nature of enzymes was frequently implied or claimed, other chemists would maintain that the observed reactions were not the consequence of a specific molecular structure in the extract but were rather the result of an "active group" carried upon or formed by an association of colloids.

The chemical nature of enzymes is probably very various. . . . Some appear to be complex systems of colloids with inorganic components, or other simple compounds.—Bayliss[12]

The protein therefore is no part of the enzyme . . . An enzyme seems to consist of a specific active group and a colloidal carrier.—Willstätter[13]

This question was resolved definitively with the crystallization by Sumner in 1926 of the enzyme urease and his association of the enzyme function

Plate 6.4. Richard Willstätter.

(Reproduced, with permission, from *Nobel Prize Winners in Chemistry 1901–1961*. Edited by Edouard Farber, Abelard-Schuman Limited, New York, 1961.)

Plate 6.5. James B. Sumner.

(Reproduced, with permission, from *Nobel Prize Winners in Chemistry 1901–1961*. Edited by Edouard Farber, Abelard-Schuman Limited, New York, 1961.)

Plate 6.6. Wendell M. Stanley.

(Courtesy of the Virus Laboratory of the University of California, Berkeley.)

with the physical and chemical characteristics of the protein molecules comprising the crystal.

. . . Sumner's results have now been accepted as verified and thus also accepted as
the pioneer work which first convinced research workers that the enzymes are substances
which can be purified and isolated in tangible quantities.—Tiselius[14]

The crystallization of other enzymes followed rapidly and then astonishingly, using essentially similar techniques, Wendell Stanley succeeded in the crystallization of the tobacco mosaic virus. He was able to demonstrate that these crystals contained a chemically reproducible species, inert in the crystal yet self-propagating in the plant, and that the characteristics of the chemical molecules comprising the crystal could be definitively associated with the characteristics of the infective agent.

As the domain of the chemist thus expanded, as the range of phenomena he could explain enlarged, the domain of the vitalist inexorably contracted. Increasingly the vitalist came to rely upon, and argue from, the wonderfully integrated character of life processes and thus upon the mysterious nature of the genes—for it had become clear that these obscure entities gave direction and character to the whole range of life forms and processes.

Who could dream that science would be able to penetrate the problems of heredity
. . . and find the mechanism that lies behind the crossing results of plants and animals;
that it would be possible to localize in these chromosomes which are so small that they
must be measured by the millesimal millimeter hundreds of hereditary factors which
we must imagine as corresponding to infinitesimal corpuscular elements. And this localiza-
tion Morgan had found in a statistic way! A German scientist has appropriately compared
this to the astronomical calculation of celestial bodies still unseen. . . .—Henschen[15]

The gene, the hereditary unit discovered by Mendel in 1865 and rediscovered in 1900, had remained a formal abstraction and, for some, a profound molecular riddle.

Curiously, the geneticists themselves had limited interest in the chemical nature of the gene.

What are genes? Now that we locate them in the chromosomes are we justified in
regarding them as material units; as chemical bodies of a higher order than molecules?
Frankly these are questions with which the working geneticist has not much concern
himself, except now and then to speculate as to the nature of the postulated elements.
There is no consensus of opinion among geneticists as to what the genes are—whether

Plate 6.7. Gregor Mendel.
(Courtesy of the Moravian Museum.)

Plate 6.8. T. H. Morgan.
(Courtesy of California Institute of
Technology.)

Plate 6.9. G. W. Beadle.

(Courtesy of California Institute of Technology.)

Plate 6.10. Edward L. Tatum.

(Courtesy of Stanford University Press.)

Plate 6.11. O. T. Avery.

(Reproduced, with permission, from *Biographical Memoirs of Fellows of the Royal Society*, Vol. 2, 1956, p. 35.)

they are real or purely fictitious—because at the level at which the genetic experiments lie it does not make the slightest difference whether the gene is a hypothetical unit or whether the gene is a material particle.—Morgan[16]

As a chemical structure, the gene had to be capable of precise self-reproduction through countless generations and also of manifold variation—mutation—with subsequent precise reproduction of the mutated form. Because of the great diversity and apparent complexity of the gene, only protein molecules seemed adequate to provide the requisite molecular diversity. Yet no plausible mechanism could be conceived which could endow a protein molecule with the known properties of the gene.

In addition, the experiments of Beadle and Tatum upon defective mutants of the mold Neurospora had established the concept that one major function of certain genes was the specification of the structure of specific proteins which performed structural or catalytic roles in the organism.

The identification, in 1944, by Avery, McCarty, and MacLeod of pneumococcal transforming factor—a transferable gene—as DNA inaugurated a new era in molecular biology.

If we are right . . . then it means that nucleic acids are not merely structurally important but functionally active substances in determining the biochemical activities and specific characteristics of cells—and that by means of a known chemical substance it has been possible to induce predictable and hereditary changes in cells which has long been the dream of geneticists.—Avery[17]

The subsequent resolution of the complementary structure of DNA by Watson and Crick in 1953 then led immediately to a simple explanation of the molecular basis of self-reproduction and of mutation, and to the identification of the gene as a tract of a DNA chain.

Detailed molecular understanding of the structures of protein molecules was forthcoming at nearly the same time as the result of the elaboration of methods of the chemical determination of amino acid sequence (initiated by Sanger) and for the determination by x-ray analysis of their three-dimensional structure (initiated by Kendrew and Perutz). With established structures known for both the DNA genes and their ultimate protein products, a precise formulation could be given of the "coding problem"—the unmasking of the code linking nucleotide sequence in the DNA gene to the amino acid sequence in the protein. The solution of this potentially thorny problem was greatly facilitated by the discovery that very simple nucleic acid

Plate 6.12. Left, Frances Crick; Right, James Watson.

(From James D. Watson, *The Double Helix*, Atheneum Publishers, New York, Copyright © 1968 by James D. Watson.)

Plate 6.13. Frederick Sanger.

(Reproduced, with permission, from *Nobel Prize Winners in Chemistry 1901–1961*. Edited by Edouard Farber, Abelard-Schuman Limited, New York, 1961.)

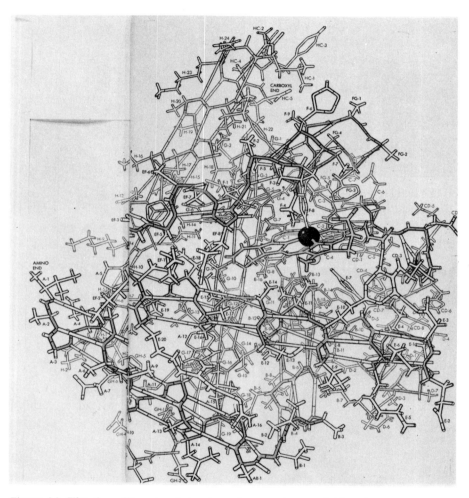

Figure 6.2. The three-dimensional structure of a protein, myoglobin.

This protein contains a heme prosthetic group with one iron atom (shown as a large, dark sphere) imbedded in the protein structure, a folded chain of 153 amino acids.

In Figure 2a, only the effective centers of each amino acid are shown as A1, B4, and so on. In Figure 2b the complete carbon nitrogen and oxygen framework of each amino acid is shown to illustrate more fully the three-dimensional complexity (note that, for clarity, the hydrogen atoms present in each amino acid are not shown).

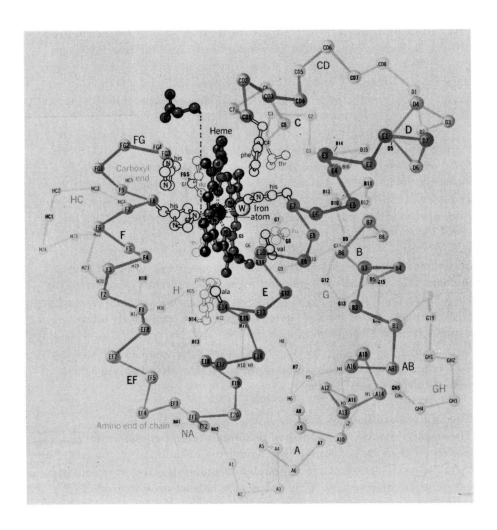

structures could, in *in vitro* protein-synthesizing systems extracted from cells, code for the synthesis of correspondingly simple "proteins"—far simpler than any now found in nature. Analysis of the correlation between the simple nucleic acid structure and the simple "protein" produced then generated the genetic code now known to be employed in all living forms and continuing testimony to the unity of all life.

The death of vitalism has a profound meaning for man. Life in all its manifestations is henceforth to be considered a part of the physical world to be understood in terms of natural law. Through the invention of a molecular memory, the gene, and taking advantage of the power of trial and selection, life has exploited the potentials inherent in organized matter. And, in achieving ever increased complexity, life has countered the universal thrust toward the increase of entropy and disorder.

Further, the death of vitalism signifies that the structural complexity required for the manifestation of life is not beyond the grasp of the human intellect, that man can comprehend the essential processes underlying life phenomena and can now begin in his own way to duplicate and even extend them. Proteins, genes, viruses—even, in time, cells—previously unknown in nature will soon be made to human design.

And since the Darwinian revolution joined man to all of life's forms, man's comprehension of the molecular basis of life includes then a comprehension of his own essence.

The Molecular Basis of Life

The essential ingredient of life—in the description provided by molecular biology—lies in the appropriately dynamic organization of the potential for chemical specificity of reaction, provided by molecular conformations of sufficient size and variety. This potential when adroitly expressed in a molecular assemblage of sufficient complexity confers the properties of self-contained self-reproduction.

Most of the chemical reactions observed in living cells will not take place at appreciable rates at the temperatures and in the solutions characteristic of these cells. These reactions are accelerated in highly specific directions by the creation of highly specific catalytic environments on the surface of (or in the crevices of) the large and specifically conformed macromolecules ubiquitous to life. By this means a great variety of chemical reactions—some

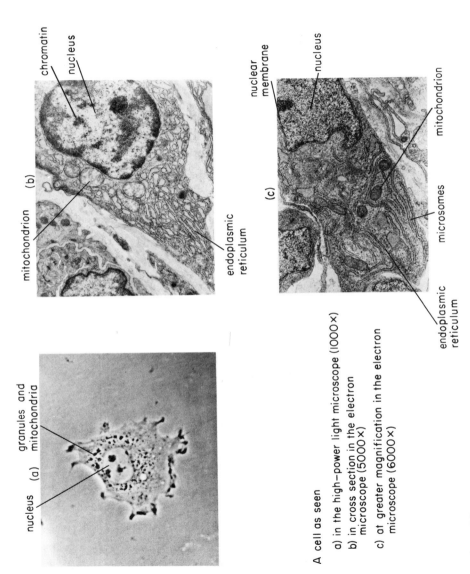

Figure 6.3.

A cell as seen

a) in the high-power light microscope (1000×)

b) in cross section in the electron microscope (5000×)

c) at greater magnification in the electron microscope (6000×)

synthetic, some degradative—are enabled to proceed within a highly confined environment with a minimum of interaction or conflict.

The complex molecular apparatus of a living cell is now seen to include the following elements as a minimum:

A molecular memory (the DNA genes) in which can be encoded the instructions for the design of the complex catalytic macromolecules previously described. This molecular memory must be one that can be precisely reproduced so as to permit the transmission of its vital information to progeny cells. At the same time it must be mutable, in such a way that the mutations can subsequently be reproduced, so as to permit the evolutionary development and trial of changes in the molecular assemblage.

A molecular mechanism to read out the information encoded in the molecular memory and to realize its expression in the form of the catalytic macromolecules.

(Obviously the memory must also contain within itself, among others, the instructions coding for the molecular readout mechanism so as to permit its increase.)

A mechanism for the precise reproduction of the molecular memory. (The memory thus must also contain within itself the instructions for this reproductive machinery.)

The maintenance of the living cell requires energy for the synthesis of the complex molecules and to counter the inevitable tendency to disorder implicit in the laws of entropy. Thus, the cell must have a metabolic machinery to provide this energy in a useful form. Ultimately this energy may be derived from light (in photosynthetic organisms), or by the uptake of energy-yielding nutrients.

Cells utilize a rich but limited variety of subunits as the building blocks of their macromolecules. The synthesis of these essential macromolecules then requires an adequate supply of these components. Therefore, among the catalysts of the cell there must be present some with the capacity to generate the subunits out of the compounds available externally. (Again, the molecular memory must contain the coded instructions for the design of these catalysts.)

The delicate specificity of the diverse reactions of the cell requires the maintenance of a rather specific internal chemical environment. To achieve this the cell must be bounded by a surface membrane that controls the ingress and egress of all materials in the cell. The presence of this membrane also can permit the achievement of molecular concentrations inside the cell much in excess of those in the environment. (Of course, the molecular memory must also contain the coded instructions for the structures of the major membrane components.)

In order to reproduce itself, the cell must have the capacity to divide, and in so doing to achieve an accurate partition of the reproduced molecular memory in order that both daughter cells have a complete set.

Possessed of this basic, integrated set of molecular systems, cells will, in an appropriate environment, increase in size and in number—they will be *alive*.

In addition to these minimal essential features, many cells have developed additional components found to be advantageous to their survival. These may include repair systems to cope with damage to the cellular membrane or even, within limits, to the molecular memory itself. Many cells have found motility advantageous and have developed means of self-displacement. Sensory detectors have been developed for a variety of important environmental parameters: light, gravity, various chemicals, and so forth.

Of exceptional importance has been the development of internal control mechanisms to keep the activities of the cell efficiently coordinated and integrated in an economical manner. Clearly, a cell will need different molecular proportions of the various subunits: macromolecules, particles, membrane components. And the proportions needed may vary dependent upon the external situation. The need of the cell for energy will vary. The timing and the succession of events in the division cycle must be accurately coordinated if the process is to be effective.

Thus the synthetic and the degradative reactions of the cell are almost invariably subject to regulation and control. This regulation, which may be achieved at various stages of the metabolic processes, is in general determined by feedback cycles; these again involve specific molecular interactions of various normal cell components that impact to varying degree on the control system, as their concentrations may fluctuate. Such self-contained processes serve to regulate the rates of the various reactions, to minimize the use of energy and other raw materials, and to maintain a coordination among the varied cellular activities.

In a multicellular organism many of the activities of a given cell must be subject to influence by the other components of the organism. Such cells therefore develop sensors of the internal state of the organism (which is the external environment of the cell) which are coupled directly to the cellular control mechanisms. In this way cells of a multicellular organism respond specifically and appropriately to such influences as surface contact with neighboring cells, circulating hormones, or nutrients, or the impact of a nerve impulse.

The Further Consequences of the Overthrow of Vitalism
The Copernican revolution accompanying the replacement of vitalism in human thought can already be seen to be spawning a series of additional revolutions in traditional human concepts—"aftershocks" of potentially even greater magnitude.

In Medicine
If the attributes of life are explicable in physical and chemical terms, then many of the ancient ailments of man should be explicable in physical and chemical terms. And in consequence, such ailments should be subject to alleviation by physical or chemical intervention. While this concept was implicit in medicine from early times, as in ancient herbal remedies, it now rests on a comprehensive understanding and extends to a far wider range of disorders. The successive discoveries and clinical applications of the vitamins, the hormones, the prostaglandins, the clotting factors, and so on, are but the first wave of this revolution.

Conceived as a mechanism, the human body should be subject to conscious maintenance and repair beyond its normal capacities, to an extent limited only by our understanding. This concept has increasingly spurred the development of large research programs intended to gain the knowledge to permit man to cope with the persistent disorders of the body, for the toll of these ills becomes increasingly intolerable as the ultimate certainty of a potential "cure" is envisaged.

The eventual scope of this revolution is difficult to foretell. Even the genetic and the psychological disorders (see later) are increasingly caught in its sweep. The potential changes in the human outlook that would accompany the virtual elimination of disease and a significant extension of the natural human life span are difficult even to imagine.

In Psychology

The successes of the nonvitalistic explanations of the properties of the cells and the body have led very naturally to the extension of such concepts to the interpretation and elucidation of the mind, that is, to the development of the proposition that the phenomena of the mind should be similarly explicable in physical and chemical terms. Very considerable progress has already been achieved in the description of the electrophysiological and biochemical processes accompanying varied psychological events. And this progress has, in turn, already generated the realization that emotional and psychological states are certainly at least influenced by the accompanying cellular processes. Reciprocally, then, such states can be affected by chemical or electrical intervention. This realization has already spawned the tranquilizers, the antidepressants, the shock therapies—and the fears of large-scale, socially directed, chemical thought-conditioning.

The ability of the neurosciences to provide mechanistic explanations of psychological phenomena is as yet only partial; nor is it certain that the detailed processes of the human mind are even in principle fully comprehensible to the human mind. But the potential for a great expansion of knowledge and understanding beyond that presently available is evident.

The impact of such insight into the nature of mental processes upon traditional concepts of free will, of motivation, of moral responsibility, of education, of insanity and lunacy and deviation and hysteria, and upon both the mystical and the rational theologies is bound to be revolutionary.

In "Genetic Engineering"

The mechanistic description of man encompasses an appreciation of the genetic bases of his most salient characteristics. Man is seen to be the product of his genetic inheritance as it develops and finds expression in his created culture. The relation between the inheritance of specific genetic characters and the development of specific traits is at present best realized in the tens of hundreds of genetically based clinical disorders now described. However, the relation between inheritance and many other distinctly human attributes is increasingly recognized and studied.

The realization of the great genetic heterogeneity of human beings is coming to have profound political and philosophical implications. And the biochemical understanding of the basis of such heterogeneity contains within it the potential for conscious intervention.

Serious discussion is already in progress as to possible means for the repair of those genetic "defects" that lead to clinical disorders. But with adequate understanding genetic intervention clearly need not be restricted to repair. In the evolutionary sense man is but the latest development in the whole continuous chain of life. His genetic endowment has derived by chance and selection from that of his immediate ancestors—and theirs from their ancestors. He is, ultimately, as surely defined and limited by his endowment as they by theirs.

But, remarkably, surely, unpredictably, man has for the first time among all the creatures on this planet achieved an understanding of his origin and his genetic endowment, and he has thus for the first time in evolution the capacity to modify this endowment as he may see fit, to transcend, in principle, his own genetic limitations, to initiate in truth a new stage of evolution.

To appreciate this concept is surely to alter man's entire view of his place in nature. A Copernican revolution indeed.

References

1. Leonardo da Vinci in *The Literary Works of Leonardo da Vinci*. Edited by J. P. Richter, 3rd ed. Vol. II, Phaidon Publishers, New York, 1970, p. 238.

2. As quoted in T. S. Hall, *Ideas of Life and Matter*. Vol. II, University of Chicago Press, Chicago, 1969, p. 8.

3. Ibid., Vol. II, p. 112.

4. As quoted in J. S. Fruton, *Molecules and Life*. John Wiley & Sons, New York, 1972, p. 70.

5. As quoted in B. S. Jørgensen, ed., *J. Chem.*, Vol. *42*, 1965, pp. 394–396. See also T. O. Lipman, "Wöhler's Preparation of Urea and the Fate of Vitalism." *Journal of Chemical Education*, Vol. *41*, 1964, pp. 452–458. See p. 455.

6. T. O. Lipman, ibid., p. 456.

7. R. M. Hawthorne, Jr., "Avogadro's Number: Early Values by Loschmidt and Others." *Journal of Chemical Education*, Vol. *47*, 1970, pp. 751–755.

8. Reference 4, p. 67.

9. Ibid., p. 52.

10. Ibid., p. 55.

11. Nobel Prize Lecture of E. Buchner, 1907. Reprinted in *Nobel Lectures, Chemistry, 1901–1921*. Elsevier Publishing Co., New York, 1966.

12. Reference 4, p. 156.

13. Ibid., p. 157.

14. A. Tiselius, Presentation Speech for Award of the Nobel Prize in Chemistry to J. Sumner in 1946. Reprinted in *Nobel Lectures, Chemistry, 1942–1962*. Elsevier Publishing Co., New York, 1964.

15. F. Henschen, Presentation Speech for Award of the Nobel Prize in Physiology or Medicine to T. H. Morgan in 1933. Reprinted in *Nobel Lectures, Physiology or Medicine, 1922–1941*. Elsevier Publishing Co., New York, 1965.

16. Nobel Prize Lecture of T. H. Morgan, 1933. Reprinted in *Nobel Lectures, Physiology or Medicine, 1922–1941*. Elsevier Publishing Co., New York, 1965.

17. Quoted by L. C. Dunn in *Genetic Organization*. Edited by E. W. Caspari and A. W. Ravin. Vol. I, Academic Press, New York, 1909, p. 63.

R. C. Lewontin

7 Darwin and Mendel—The Materialist Revolution

The first half of the nineteenth century was a period of great tension in
many domains of thought, a tension produced by the steady development
and intensification of a fundamental contradiction in world views. The political and economic organization of the Western world had been overturned
by the successful bourgeois revolution. Beginning with the revolutions of
1776 and 1789 and culminating in the Reform Bill of 1832 and the continen-
tal revolutions of 1848, the middle classes had succeeded in overturning the
static world order in which each man from peasant to king was fixed in his
position by the grace of God. In its place was established a dynamic social
order in which individuals and their children could rise, by aggressive and
competitive behavior, from peasant to peer. The high Tory Prime Minister,
Sir Robert Peel, was the grandson of a Midland peasant turned hawker, and
Josiah Wedgwood, Darwin's maternal grandfather, began in life as a potter's
apprentice and ended as convenor of the powerful General Chamber of
Manufacturers. This change from the static, fixed, aristocratic society to a
dynamic, shifting, bourgeois world, had inevitable consequences for the general development of philosophical and scientific conceptions. Men reason
by analogy from the conditions of their own lives to the conditions of the
universe. Before the eighteenth century, ideas of the physical and biological
world were consonant with the rigid aristocratic social structure. The universe had been created in its known form by the grace of God, the earth was
fixed in its place, and the species of living organisms were immutable and
unchanging. A static society saw a static universe.

The bourgeois revolution of the late eighteenth and early nineteenth century produced in its turn a bourgeois universe, a universe in which change
and mobility are the dominant properties. If there is any world view that
characterizes the social, philosophical, and scientific outlook of Europe between the first and second French revolutions, it is *evolutionism*, the view
that constant change is the essential characteristic of all natural systems and
human institutions. "Tout change, tout passe, il n'y a que le tout qui
reste." [1] Evolutionary cosmology was founded by Kant in his *Metaphysical
Foundations of Natural Science* in 1786 and by Laplace in his nebular hypothesis of 1796. In geology, catastrophic theories like deluvianism, which
postulated a series of extraordinary floods to explain sediments and fossils,
were replaced by Hutton's uniformitarian principle that the same forces of

Museum of Comparative Zoology, Harvard University, Cambridge, Massachusetts 02138.

continuous and gradual change were always in operation. Charles Lyell had already carried the day for uniformitarianism in geology in his *Principles of Geology* of 1830, and from that time geology was thoroughly evolutionist. Lyell at first resisted carrying over his evolutionism to organic life but later became a supporter and sponsor of Darwin. Evolutionary thermodynamics was founded in 1824 by L. N. Sadi Carnot and fully developed by William Thomson in 1851 when he formulated the second law of thermodynamics. According to this law the total energy in the universe is in a constant process of redistribution, with the differences between parts growing less and less, so that eventually the distribution will become completely uniform and no mechanical or chemical process will any longer take place.

Evolutionism as a general intellectual current was widespread outside of the natural sciences. Herbert Spencer in 1857 wrote in *Progress: Its Law and Cause,* that

> It is now universally admitted by philologists, that languages, instead of being artificially or supernaturally formed, have been developed. And the histories of religion, of philosophy, of science, of the fine arts, of the industrial arts, show that these have passed through stages.[2]

The ubiquity of the consciousness of change and instability is shown by the recurrence of the themes of evolution and chance in literature. Around 1840 Tennyson wrote in *In Memoriam* that nature was not careful to preserve "the type," but rather that

From scarpéd cliff and quarried stone
She cries, "a thousand types are gone:
I care for nothing, all shall go." [3]

and in the *Idylls of the King* he legislates that "The old order changeth, yielding place to new." [4] The novels of Dickens, like *Dombey and Son* (1846), *Bleak House* (1852), and *Hard Times* (1854), chronicle the sweeping away of old social relations by railroads and ironmasters.

By the time of the publication of the *On the Origin of Species by Means of Natural Selection* in 1859, the evolutionary view had already made progress in biology, but resistance was great among leading natural historians, at least in part because the evolution of organic life, including man, was perceived to be directly threatening to Christian religious belief, just as Copernicus's heliocentric and infinite universe was a threat to the established religion of his day. The eighteenth-century naturalist Buffon had, in his *Natural History,* proposed the transformation of species including the descent of man and apes from a common ancestor. His evolutionary outlook brought Buffon in direct conflict with Linnaeus, whose scheme of universal classification of plants and animals was the culminating scientific development of the static world view. Charles Darwin's grandfather, Erasmus Darwin, in *The Temple of Nature* and *Zoonomia,* published around the turn of the century, presented a thoroughgoing evolutionary view, if a romantic one, of the development of organic life. The most influential evolutionary theories before Darwin were those of Lamarck and Geoffrey St. Hilaire, who between 1794 and 1830 developed theories of the transformation of species which were to be taken seriously by biologists even into the twentieth century. Popularizers and demiphilosophers like Herbert Spencer and Robert Chambers asserted the obvious truth of organic evolution, and Chambers's *Vestiges of the Natural History of Creation* (1844) was the most influential work on natural history of the time until the publication of the *Origin of Species.* Even universities, those strongholds of entrenched intellectual reaction, recognized evolutionism, the University of Munich offering a prize for an essay on "The Causes of the Mutability of Species." [5]

If evolutionism so thoroughly permeated the intellectual life of the early nineteenth century, including the idea of the evolution of life on earth, was the "Darwinian revolution" not already an accomplished fact before Darwin? No, for Darwin's revolution was a second revolution, brought into being by the first, and necessitated by a deep contradiction between the new evolutionary world view and an older strain in philosophy that still permeated nineteenth-century thought but that was incompatible with it. This old strain was Platonic and Aristotelian idealism or "essentialism." More than merely a self-conscious "philosophy," idealism is a total way of looking at the world. It regards the real objects of the world as imperfect reflections of underlying ideals or essences. It supposes that "we see through a glass, darkly," and that the real variations between real objects only confuse us in our attempts to see the essential nature of the universe. According to this view, it is the task of philosophy and science to sweep away the irrelevant distortions that characterize actual realizations of the underlying ideals, so that the "true" relationships between the ideal categories can be perceived. A concomitant of this view is that the differences between objects that belong to the same category are of a different nature, and have a different origin, than the differences between the categories themselves. The differences between objects within a category are "disturbances" and are ontologically unlike the differences between the ideals, which are a revelation of the essential structure of the universe.

In the particular case of living organisms, each species is regarded as a type, so that the individual members of each species are seen as conforming more or less closely to the type that expresses the true nature of their species. Indeed the "type" is a concept still used in taxonomy. It is a single specimen upon which the description of the entire species may be based (although it may in fact turn out to be very "atypical") and which is deposited in a museum for further comparison. The variations between individuals are an inconvenience for such typological taxonomy and, under an idealist view of nature, such intraspecific variation has different causes and a totally different significance for our understanding of nature than the differences between species.

So long as evolutionists held to a typological view of nature they were confronted with a problem that they did not solve successfully. If species change one into another, this means that they must change from one *type* into another. But what mechanism exists for a change in type? The problem is made more complicated by the assumption that there is a continuous chain of life with all living organisms coming from previously existing ones. Either an individual itself must change its type or else an organism of one type must give birth to an offspring belonging to a different type. Two solutions were offered before Darwin. The first, by Lamarck, was that individuals themselves changed type through the use or disuse of particular organs. If a giraffe stretched its neck slightly to reach high leaves, the slight increase in neck length would be passed on to its offspring, which would in turn stretch their necks, and so on. Their type changed slowly and imperceptibly in response to the adaptive alterations of individuals. An alternative solution was Geoffroy's, that the change in type was discontinuous, large in magnitude, and occurred at the production of offspring. An individual of one type gave rise to one of a different type. Neither of the solutions was convincing to a wide group, although both reappear persistently in the nineteenth century as rivals to the Darwinian theory of evolution, and neither was effectively abandoned until the rediscovery of Mendel's work in 1900. The failure of evolutionism to become the dominant ideology in biology before Darwin must, in large part, be laid to the very weak position that early evolutionists found themselves in because they were unable to provide a satisfactory explanation of change of type.

Darwin's revolution lay in turning his attention away from the type of the species and concentrating on the actual individuals that made it up. Rather than regarding the variation between individuals as obscuring the essential difference between species, he took the actual variation between individuals to be the proper object of study. Variation became the thing-in-itself. Instead of regarding individual variation and differences between species as ontogenetically distinct and opposed to each other, he took them to be directly related causally. *Darwin's revolutionary theory was that the differences between organisms within a species are converted to the differences between species in space and time.* Thus, the differences between species are already latent within them, and all that is required is a motive force for the conversion of the variation. That force is *natural selection*.

While Darwin dated his concept of natural selection from his reading of Malthus's essay on population in 1838, the concept flows ineluctably from Darwin's study of individual variation, as does the metaphor "natural selection" itself. The *Origin of Species* begins with a chapter on Variation under Domestication, a subject to which he devoted an entire two-volume work in 1868. He observes the very great difference between breeds, varieties, and individuals within domestic species is much greater than is observed between organisms in wild forms. He explains the origin of these divergent breeds as chiefly the result of man's selection of naturally occurring variation.

The key is man's power of accumulative selection: nature gives successive variations; man adds them up in certain directions useful to himself.[6]

Thus, the idea of variation and of selection comes together out of the consideration of domesticated plants and animals. In order to transfer them to nature, a mechanism is required whereby "Nature" rather than man exercises "selection" among the variants. Darwin found this mechanism in the overproduction of young, beyond the numbers that can be supported by the food supply, so that only the best adpated can survive on the average. A similar principle was extended by him to sexual selection, on the analogy with the competition between individuals for marriage partners in man.

The demonstration by Darwin of a simple mechanism by which "Nature" could "select" among variations, without the intervention of divine will or a mysterious vital force stands, together with his concentration on variation as the essential reality, as a distinctive feature of his system. To see this we must differentiate between the *mechanism* proposed by Darwin and the *explanations* proposed by Lamarck and Geoffroy. The Darwinian theory is based on three easily verified observations about the organic world:

1. Individuals within a species vary one from another in morphology, physiology, and behavior.
2. Variation is in some part heritable so that variant forms have offspring that resemble them.
3. Different variants leave different numbers of offspring.

From these three observations it follows *mechanically* that evolution will take place since the differential reproduction of different heritable variants must lead to a change in population composition and, thus, evolution. The schemes of Lamarck and Geoffroy were, on the other hand, a priori explanations that offered no demonstrated mechanism. Lamarck's theory of the inheritance of acquired characters, while it would have led to evolution had it been true, was a pure postulate of convenience rather than an observation of processes in nature. Geoffroy's idea of saltation had the virtue that occasionally grossly different (usually monstrous) forms were in fact produced in the offspring of normal individuals. But the appearance of these variant *individuals* played no role in Geoffroy's theory, which postulated unobserved changes in type from the evidence of homology.

What is extraordinary in the Darwinian system is the total lack of inferred but unobserved entities or forces whose existence is necessary to the explanation. There are here no metaphysical constructs like Newton's ideal bodies moving in ideal rectilinear paths from which actual bodies departed more or less. The Newtonian revolution of the seventeenth century in which such metaphysical constructs played an essential role was totally opposed in spirit and method to the Darwinian revolution of the nineteenth century that required the sweeping away of such constructs and the concentration on the actual and individual. The intellectual force required for this revolution was quite as great as that needed by Copernicus when he made the earth and sun change places.

The Problem of Heredity and Variation

Darwin gave simple natural historical causes for the third element in his system, differential reproduction. He found these causes in the competition of individuals for resources in short supply, including mates, and in the differential ability of different forms to survive climatic rigors. He was, however, unable to cope successfully with the underlying mechanism of his first two elements, variation and heredity. What was the origin of the variation between individuals? What was the mechanism that led to the similarity of parents and offspring? How can we reconcile the creation of individual variation and the heritability of characters, both of which must occur, and both of which underlie evolution? This is a problem with which Darwin struggled unsuccessfully, as did every other naturalist of the nineteenth century. The irony of the history of this problem is that Darwin's failure lay in his inability to apply the same materialist method to the study of heredity that he applied to the theory of evolution. In fact, he accepted tentatively in his first edition (1859), and more explicitly in his fifth edition (1867) Lamarck's theory of the origin of variation from environmental induction.[7] Even more curious is his "Provisional Theory of Pangenesis" in the *Variation of Animals and Plants under Domestication*. This theory postulated unseen and unobservable entities, *gemmules,* which were of different sorts in each organ of the body and gave that organ its specificity, and which were later collected together in the reproductive organs in order to be passed on to the next generation. The theory of pangenesis is a characteristically pre-Darwinian construct and one that partook in no way in the revolution that Darwin had made.

The problem of heredity in the nineteenth century was closely linked to the problem of the transmutation of species. For the pre-Darwinian (and anti-Darwinian) evolutionists, the leading problem was how the "type" was inherited and how in the course of inheritance organisms might be altered to a new type. Because of this preoccupation, much of the early work on heredity was carried out by crossing species or widely divergent varieties, chiefly in plants where wide crosses are possible. The outcome of such crosses, where the hybrids were not sterile, was a vast and unanalyzable diversity in later generations with a tendency in any case to revert over the course of generations to one or the other of the parental types. The hope of experimenters like Gärtner, whose *Researches on Plant Hybridization* appeared in 1849, and of Wichura (1865) was to show that new species were formed from hybrids between old ones. Even those like Naudin (1856) who carried out crosses on closely related varieties, worked with species crosses as well, and Nägeli who was Mendel's long-time correspondent, writes that he resumed the study of heredity in 1864 "when Darwin's writings had made the problem of species a burning one." It was Nägeli's aim to provide an alternative mechanism to natural selection for the origin of species. Because the aim of so much experimental breeding was to elucidate the relation between species, the eventual acceptance of Darwin's theory of natural selection removed the motivation for this work. Bateson[8] observed that "The *Origin* was published in 1859. During the following decade, while the new views were on trial, the experimental breeders continued their work, but before 1870 the field was practically abandoned." The irony of this development was that, far from solving the problem of heredity, Darwin's theory had exacerbated the underlying paradox of heredity and variation.

The problem of inheritance is to explain the fact that offspring resemble their parents, yet they differ from their parents, and these differences themselves are to some extent heritable. There are, then, two antithetical properties that characterize the process of reproduction. The solution of the nineteenth-century breeders and naturalists to this contradiction was to *separate* the two conflicting elements and to concentrate on the similarities, while regarding the variations as ontogenetically different. This separation had three consequences. First, it created a dichotomy between species differences and the variation within species as being of different natures. The differences between species, that is, between types, were the objects of interest because the types themselves were the objects of interest. The variations among individuals were those disturbances in spite of which the truth might be seen. Second, it focused attention on group description rather than on individual description in the outcome of experimental crosses. The results of several generations of breeding were more often than not described by the degree to which the progeny as a whole resembled one or another parent, and variation among individuals was a quality of the population, rather than being given a quantitative expression. Third, the investigation of cellular mechanics was with the object of discovering the physical basis of exact heredity. While this search for the physical carriers of hereditary similarity was in the end a necessary part of a fully mechanistic explanation of heredity, it could not in itself reconcile the contradictory aspects of heredity and variation. The great cytologists of the 1880s and 1890s, Boveri, Hertwig, Roux, Strassburger, and Weismann inter alia, had succeeded in showing that there were rodlike bodies, the *chromosomes,* in the nucleus of every cell, that these chromosomes were in characteristic numbers and shapes in each species, that they maintained their individuality during the process of cell division, that they were duplicated when cells divided, each cell receiving one copy of each chromosome, and that at the fertilization of an egg the chromosomes from the paternal gamete and the maternal gamete came together in the nucleus. From these observations it was inferred by Weismann, but not observed, that during the formation of gametes, the number of chromosomes must be cut in half. All of these observations and inferences, which are such a striking parallel to the Mendelian system, did not, however, allow the cytologists to produce a theory of genetics because the observed behavior of the chromosomes could be related only to the phenomenon of heredity and not to the phenomena of variation.

To understand the failure of the pre-Mendelians, we must first see how Mendel's own method and viewpoint succeeded. Mendel carried out his crosses using true breeding varieties of the edible pea, *Pisum sativum,* which differed from one another in one or a few distinct characters. Thus, one variety might have full round seeds, while the other would have wrinkled seeds, or one variety would be tall while the other was dwarfed. He was extremely careful to nurture each seed from a cross so that he could get complete information on the number of each type of offspring that resulted from a cross. Finally, he kept separate records on the outcome of breeding of separate individuals. Whether or not two plants looked the same, *their offspring were separately collected and separately analyzed.* This last aspect of his method was, in fact, the crucial one.

When two varieties were crossed, the hybrids (or F_1 generation) were all alike with respect to the trait being investigated, resembling either one parent or the other which was then said to show the *"dominant"* character. The character that disappeared in the F_1 was the *"recessive"* character. So, tall is dominant to dwarf, since all F_1 plants of a cross between true-breeding tall and true-breeding dwarf varieties were tall. If these F_1 hybrid plants were then crossed with each other, their offspring (the F_2 generation) were of two types, tall and dwarf, so the recessive character reappeared. A further generation of self-fertilization showed that the F_2 plants with the recessive trait now bred true, just as the original parental dwarf plants did. Thus, the determinant of the dwarf trait, although apparently disappearing in the F_1, *reappeared uncontaminated and unchanged* in the F_2. Moreover, the dominant F_2 plants were of two sorts, as revealed by their own further breeding. Some, when self-fertilized, behaved like the F_1 hybrids, producing both dominant and recessive forms in their offspring. But other individuals, when self-fertilized, bred true for the dominant trait. Thus, the mixing of dominant and recessive "factors" (as Mendel called them) in the F_1 hybrid did not in any way contaminate or alter the factors themselves, because both pure-breeding dominant and pure-breeding recessive plants appeared in the F_2. Moreover, those F_2 plants that were not pure breeding but behaved like the F_1 hybrids were not the results of "contamination" since in subsequent generations (Mendel went on to the F_7) they continued to segregate out pure breeding dominants and recessives.

If the two varieties originally crossed differed in two traits, say plant height and pod shape, each of the characters separately showed the same phenomena of segregation, so that in the F_2 of a tall, full pod plant crossed with a dwarf constricted pod plant, both original varieties appeared, but also the new combinations: *tall, constricted pod* and *dwarf, full pod*. Not only were the alternative factors for the *same* character segregating in the formation of eggs and pollen of F_1 hybrids, but factors for different characters could be studied separately and had no necessary relation to each other.

From the observations so far described, the essentials of Mendelian inheritance can be deduced. Traits are inherited as if they were determined by discrete, particulate alternative factors which come together in a hybrid organism but which separate again from each other when the hybrid forms reproductive cells, without any contamination from their contact in the hybrids. As Mendel wrote to Nägeli, ". . . in each successive generation, the two primal characters issue distinct and unaltered out of the hybridized form, there being nothing whatever to show that either of them has inherited or taken over anything from the other." Each individual organism is formed from the union of a male and female reproductive cell or gamete. Each gamete carries one of the alternative "factors" for a trait so a fertilized zygote has two factors for each trait, one inherited from its male parent and one from its female parent. These two alternative factors that are reproduced in all the cells of the growing organism then separate from each other again at the time of the formation of gametes, so completing the cycle of segregation, union, and segregation as each generation follows.

Other regularities were deduced by Mendel, but these turn out to be less general than the universal mechanism of segregation of particulate factors. From the ratios of contrasting characters in various crosses, he deduced that there are equal numbers of the two types of gametes produced by a hybrid, and that gametes fused at random in producing new zygotes. Designating by A the dominant factor and by a the recessive factor, the self-fertilization of a hybrid Aa resulted in an F_2 generation one-quarter of which was aa, true-breeding for the recessive trait, one-quarter AA, true-breeding for the dominant trait, and one-half Aa, that are again segregating hybrids. So, too, a hybrid Aa crossed with a true-breeding recessive aa produced half true-breeding recessive plants aa and one-half segregating hybrids Aa. In fact it is now known that some hybrids do not produce equal numbers of each kind of gamete and that fertilization is not always at random. The complete dominance of one factor over its complement in hybrids is not the general rule, and often the hybrid is intermediate in form. Had Mendel worked with such traits, his results would have been even more transparent in their interpretation. Finally, Mendel's observation that factors for two different traits, such as plant height and pod shape, segregate completely independently of each other is the exception rather than the rule. The probability of choosing, as Mendel did, seven different characters at random that would segregate independently is extremely low, and the source of Mendel's remarkable "luck" in this respect remains an open problem.

To what did Mendel owe his successful solution of the problem, and why did his work, published in 1865 remain totally unnoticed and unassimilated until 1900? The answers do not lie, as is often suggested, simply in Mendel's experimental material nor in his observations of segregation. Seton and Goss in 1822 and Knight in 1823 made crosses of green and pale varieties of garden pea, Mendel's very material, and observed segregation of green and pale seeds in the second and backcross generations. Moreover, they observed that some plants produced pods in which all the seeds were green, while other pods were mixed. Gärtner in 1849, Verlot in 1865, and Haacke in 1893 observed segregation in F_2 plants from varietal crosses. Naudin (1861–1865) postulated the "disjunction of specific essences" in advanced generations of hybrids, coming close to Mendel's principle of the purity of gametes. Nor was individual breeding a sufficient element of method. Vilmorin (1856) kept separate records of individual crosses, which he regarded as the cornerstone of his method and counted the forms in the F_2 of crosses and recorded obvious 3:1 ratios but made nothing of them (1879).

What was common to all the observers before 1900, with exception of Mendel, was that they concentrated on the group of progeny *as a whole*, rather than focusing on the variation among the progeny as of fundamental interest. Experimental breeders of the nineteenth century, like the evolutionists, still viewed the world as a collection of types and individual organisms as imperfect manifestations of those ideals. Such a world view effectively prevented them from searching in the variation among organisms for the laws of regularity of inheritance. Their Platonism led them to take the similarity between members of a group and a description of the group as a whole as the primary objects of attention, regarding variation between individuals as ontogenetically different. Naudin's "specific essences" were Platonic essences characteristic of the organism as a whole. Vilmorin's crosses were individual, but his characterizations of the results were group descriptions. Even Haacke, who came closer than any breeder before 1900 to the Mendelian laws, gave only summary descriptions of the successive generations.

The most important consequence of the idealist viewpoint was that inheritance and variation belonged to two different causal chains. It is ironic that Darwin, who based his explanation of evolution on the variation between individual organisms, was unable to apply this same outlook to his consideration of inheritance. *The Variation of Animals and Plants under Domestication* contains extensive inquiries into the nature of inheritance and reports the results of some of Darwin's own experiments, which were quite Mendelian in method but not in outlook. For example, he found a 90 to 37 ratio of normal and distorted flowers in the F_2 of a cross between a normal and a distorted variety of snapdragon. But Darwin's general discussion of inheritance makes clear that he held inheritance and variation to be fundamentally different phenomena. "Hence," he writes in his Summary on inheritance, "we are led to look at inheritance as the rule, and noninheritance as the anomaly." [9] Mendel's unique contribution was his realization that both inheritance and noninheritance were manifestations of the *same* underlying phenomena and that the study of the pattern of variation among the individuals of a cross would lead to an understanding of the patterns of regularity. He took the apparently antithetical and contradictory observation that offspring resembled their parents, yet differed from their parents, and synthesized them into a single explanatory scheme that encompassed both inheritance and noninheritance. The possibility of this synthesis was created by concentration on individual differences. Thus, Bateson, who did more than any other of the early twentieth-century geneticists to defend, spread, and augment the Mendelian system, wrote in 1909, "For the facts of heredity and variation unite to prove that genetic variation is a phenomenon of individuals. Each new character is formed in some germ cell of some particular individual at some point of time. More we cannot assert." [10]

Nowhere is the difference on point of view about similarity and variation more clearly illustrated than in the contrast between Mendel's system and the work of Francis Galton, toward the end of the nineteenth century. Galton was among the founders of modern statistics, and he placed strong emphasis on the importance of studying not only the mean value of a measurement but also its variation. He was scornful of those who looked only at averages, likening them to the man who thought Switzerland would be fine if all the mountains were filled into all the lakes, making a more equable landscape. Galton's approach to the study of inheritance, however, was to compare averages of parents with averages of offspring.[11] So, for example, he compared the height of Cambridge undergraduates with the average height of their parents by taking a large number of individuals who were, say, 62 inches tall and recorded the average height of all their parents, the same for 63-inch students, and so on through all the height classes. When the average height of offspring was plotted against the average height of their parents, height class by height class, the parents fell almost exactly on a straight line, but with a slope less than 45 degrees. That is, very tall parents had offspring who were on the average not so tall as they were, and very short parents had offspring who were less extreme than they were. This phenomenon that extreme parents produced offspring closer to the mean of the entire population, he called "regression" and established the "Law of Filial Regression" as a universal rule of inheritance. Despite Galton's insistence that variation among members of the total population was worthy of study, his Law of Regression made no use of the variation between brothers, or between parents, or between parental pairs whose offspring belonged to the same height class, or between offspring whose parental pairs belonged to the same height class.

A second aspect of Galton's theory of inheritance was his attempt to partition the hereditary influences on an individual into contributions from various ancestors. Beginning with the undoubted fact that each parent contributes only half of its own hereditary information to an offspring, Galton concluded by a heuristic and alogical argument that value of an individual's character was determined half by his parents, one-quarter by his grandparents, one-eighth by his great-grandparents, and so on. One of the chief appeals of this theory to Galton was that the sum of the geometric series $\frac{1}{2} + \frac{1}{4} + \frac{1}{8} + \frac{1}{16} + \ldots$ is unity, so that, in some sense, all of an individual's inheritance was accounted for. A much more sophisticated and rather different form of this theory was elaborated by the great mathematical statistician, Karl Pearson, who named his own theory "Galton's Law of Ancestral Inheritance." [12] In either case, the viewpoint of Galton could not predict differences between offspring of the same cross since an individual's heredity was a sum over all his ancestors, so two individuals with the same ancestry ought to be the same.

A consequence of Galton's blending theory of inheritance, and of his averaging over ancestry, was that he rejected Darwin's theory that evolution was based upon the selection of small individual variations. Galton quite logically reasoned that such small continuous variation could not lead to evolution if the nature of inheritance were indeed continuous. Rather, he asserted, against Darwin, that natural selection operated on rare "sports," large discontinuous changes in form, whose laws of inheritance were quite different, but unknown. Thus, the failure to concentrate on individual variation led Galton to an incorrect view of both heredity and evolution, the one leading to the other. Again we see how critical was a concentration on variation itself as the only proper object of study.

In contrast to Galton, Mendel derived his system entirely from a consideration of the pattern of variation between offspring of a given cross. It was precisely the level of variation that Galton averaged out which for Mendel was critical. Nowhere in Mendel's paper do we find a separate characterization of the average appearance of offspring and of the variation in their appearance. Means play no role at all. For Galton, means and variances were ontogenetically separate. For Mendel, the variation was the only reality.

Mendel's work lay unnoticed for thirty-five years except for a passing reference in 1881. Then, in 1900, in the March, April, and June issues of the *Reports of the German Botanical Society*, there appeared three papers reporting results and interpretations identical to those of Mendel, and all citing Mendel as the first discoverer. The experimental work on which these papers were based was carried out independently during the last decade of the nineteenth century by H. De Vries, C. Correns, and E. Tschermak, and each apparently rediscovered Mendel's paper independently while writing his own report. While their reading of Mendel's paper must have had some influence on their own interpretations, the nature of their experiments and the observations they made, and their perception of the close relevance of Mendel's own results to their own, made De Vries, Correns, and Tschermak as much the discoverers as the rediscoverers of the fundamental rules of hereditary transmission.

That we talk of "Mendelism" is in part a recognition of the greater intellectual problem faced by a monk of the middle of the nineteenth century, than by the established professionals of the beginning of the twentieth, but also reflects a preoccupation with priority and personification that is a characteristic of the social structure of our intellectual life. The simultaneity of discovery and rediscovery of genetics (although not of the publications which were induced one by the other) by De Vries, Correns, and Tschermak leaves us in little doubt that intellectual trends had come to a conjunction that made modern genetics inevitable, although no satisfactory explanation of this problem of historical determination has yet been made. It is not surprising, then, that Mendelism was very rapidly accepted after 1900. In 1902 Bateson published *Mendel's Principles of Heredity: A Defence*.[13] By 1909 he could write a textbook, *Mendel's Principles of Heredity*. There was nothing left to defend. Mendelism had swept the field.

The revolution begun by Copernicus was not completed until the work of Kepler sixty years later. The revolution in thought begun by Darwin and Mendel was not completed in 1900 or 1909 nor is it really completed in

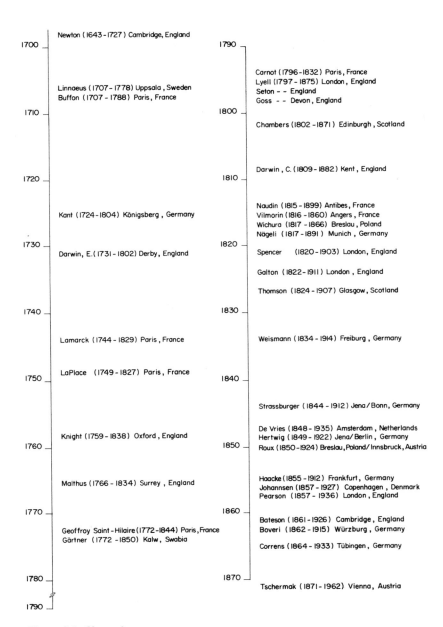

Figure 7.1. Chronology.

1973. Indeed, Darwin and Mendel themselves could not carry their revolutions in thought beyond a certain stage. In order to establish laws of inheritance, Mendel needed to distinguish between internal "factors" that were passed from generation to generation and the external appearance of the organism. This distinction was given its modern form by Johannsen's (1903) concepts of *genotype,* the internal, inherited genetic constitution of an organism, and *phenotype,* the outward manifestation of the genotype. This necessary and correct distinction leads in turn to a somewhat shakier distinction between variation that is inherited and variation that is nonheritable, a distinction already stressed by Darwin. It is a short step from seeing *population variation* as arising from two different underlying causes to seeing *individual variations* as due either to heredity or environment. Thus arises the nature-nurture controversy that poses questions like "Is heredity or environment more important in determining behavior?" Beginning with the vital distinction between genotype and phenotype, we end with the false and scientifically destructive distinction between hereditary and environmental determination of characters. This strain can be seen even within scientific genetics where textbooks speak of the phenotype as the "sum of the genotype and the environment." But modern developmental genetics has shown that the "sum" is a false metaphor. *The phenotype is the result of a unique interaction between gene and environment,* an interaction that destroys the distinction between gene and environment as causes of the state of an organism. Until that interaction of gene and environment in determining organism is fully integrated into scientific and social thought, the Darwinian and Mendelian revolutions will remain incomplete.

References

1. Denis Diderot, *Le Rêve de d'Alembert*, 1769.

2. Herbert Spencer, *Progress, Its Law and Cause. Westminster Review*, 1857.

3. A. Tennyson, *In Memoriam*, Part 56, Stanza 1.

4. A. Tennyson, *The Idylls of the King*, "The Passing of Arthur," line 408.

5. G. Himmelfarb, *Darwin and the Darwinian Revolution*. Peter Smith, Gloucester, Mass., 1959.

6. C. Darwin, *On the Origin of Species by Means of Natural Selection*. 1st ed., 1859, p. 30.

7. C. Darwin, *On the Origin of Species*. A facsimile of the first edition with an introduction by Ernst Mayr, Harvard University Press, Cambridge, Mass., 1964.

8. W. Bateson, *Mendel's Principles of Heredity*. C. P. Putnam and Sons, New York, 1913.

9. C. Darwin, *The Variation of Animals and Plants under Domestication*. 2nd ed., revised, Vol. 2, p. 59. D. Appleton and Co., New York, 1883.

10. W. Bateson, *Mendel's Principles of Heredity*. Cambridge University Press, 1909, p. 289.

11. F. Galton, *Natural Inheritance*. Macmillan, London, 1889.

12. K. Pearson, "Mathematical Contributions to the Theory of Evolution. On the Law of Ancestral Heredity." *Proceedings of the Royal Society* Vol. A *62*, 1898, pp. 386–412.

13. W. Bateson, *Mendel's Principles of Heredity: A Defence*. Cambridge University Press, 1902.

E. Margoliash

8 Informational Macromolecules and Biological Evolution

The subject of this presentation is the revolutionary impact that the newly developed understanding of genetic phenomena at a molecular level is having on our ability to decipher the fascinating history of biological evolution. This natural process started possibly some 3 billion years ago, and from most primitive beginnings has led to the profusion of life forms that grace our planet today. These developments are at least in part the end products of evolutionary changes in the structures of those large molecules that constitute the totality of the biological information specifying every organism—the so-called "informational macromolecules," *proteins* and *nucleic acids*. Our recently acquired knowledge of this relationship and our even newer ability to examine the chemical structures of these large and complicated molecules, have led to the realization that more precise quantitative information on evolution than previously available can be obtained from comparisons of such molecules in many living species. Such studies are in the process of causing a revolution of Copernican dimensions in our capacity to unravel the course of biological evolution.

Our curiosity about biological evolution reflects one of the most fundamental questions man asks in viewing himself against the background of the surrounding universe, living and nonliving, namely, "Where do I come from and what is my relation, if any, to all the other forms of life which surround me?" Two of the most fruitful upheavals in biological thought of modern times, that wrought by Charles Darwin's concept of the evolution of species by natural selection, and that initiated by Gregor Mendel's *Experiments on Plant Hybrids,* reported in 1865, gave for the first time a resolute and satisfactory answer. These will be merely succintly defined here, insofar as they bear directly on the genesis of our immediate subject. Their backgrounds and implications are considered in detail in the essay by Richard Lewontin in the previous chapter of this volume.

The Classical Approach to Evolution and Genetics

Charles Darwin defined evolution as "descent with modification," namely, the derivation of new species from previously existing ones. All species reproduce in much larger numbers than the number of individuals that can actually survive. Populations nevertheless remain relatively constant in size because only a proportion of offspring survives. Offspring that differ from

Department of Biological Sciences, Northwestern University, Evanston, Illinois 60201.

their parents by having characteristics that fit them better for survival in their particular environmental niche will survive longer and in larger numbers than others less fit, and in turn have correspondingly a larger number of surviving offspring. Thus, the population will gradually tend to contain a larger and larger number of individuals carrying the favorable characteristics that will have been *selected for*. Similarly, unfavorable characters will be *selected against* by the same mechanism, termed *"natural selection."* This entire simplified version of the Darwinian concept of evolution was contained in the title of his famous 1859 book: *On the Origin of Species by Means of Natural Selection or the Preservation of Favoured Races in the Struggle for Life*. From this relatively simple start to the present-day so-called "classical," "synthetic," or "neo-Darwinian" theory of evolution which extends into many branches of biology, the primordial importance of natural selection has remained a constant feature. Without it the evolutionary fixation of new characters in an evolving population is likely to be a rare event that can take place only under very special conditions. As is the case with every other living being, man is conceived as being descended from some nonhuman ancestral species, and by extension, to be more or less closely related to every other living creature.

Darwin's concepts provided no basis for understanding how variations occur and are inherited among individuals, variations without which natural selection has no way of operating. The lead came from Gregor Mendel's work on the inheritance of characters in the offspring of hybrids of pea plants. He recognized that pairs of traits, such as short and tall, or green versus yellow seeds, correspond to pairs of "elemente" (later termed *genes*), present in the cells of the plant, and which directed the appearance of the related characters. By the time the Austrian monk's work was rediscovered independently by three scientists in 1900 and his 1865 paper exhumed from its resting place in the *Transactions of the Brunn Natural History Society*, genetics was on its way to its historical confluence and intertwining with cellular and later molecular biology. A veritable mass of work and workers contributed to that development. They are far too many to cite here, and, without some detail of how the work of each one of these scientists fitted in with that of the others, the reader could not get a useful impression of the way in which the magnificent scientific edifice of classical genetics was put together. Suffice it, therefore, to say that by the mid 1930s genes, which had been defined as the bearers of the influences that lead to the inherited biological traits of the individual such as sex, eye color, and so on, had been shown to be located in definite linear orders on the *chromosomes,* bodies located in a specialized structure called the *nucleus* of the cell. Each chromosome carries its own complement of genes. The relative linear positions of genes along chromosomes had been mapped by purely genetic experiments. It was eventually even observed under the microscope in the particularly favorable case of the giant chromosomes of the salivary glands of the fruit fly, *Drosophila,* a favored object of genetic experimentation. These chromosomes, as seen in Figure 8.1 show a clear pattern of bands, and the relative positions of these bands were found to correlate directly with the relative positions of genes, as determined genetically. Finally, with the study of *mutations* (events, spontaneous or induced, such as by radiations, mutagenic chemicals, and so forth, which lead to a permanent change of the genetic constitution of a cell by affecting one or more genes), classical genetics provided the background for an understanding of the variation of individual characteristics required by Darwin's concept of evolution. More recently, the science of the genetics of populations, by focusing attention on the totality of the genes available to a population and on the mechanisms by which variations occur in them, has made it possible to study evolution at the level of the real biological unit in which the phenomenon takes place, namely, the population as a whole, not the individual.

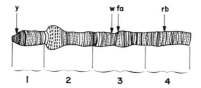

Figure 8.1. Drawing showing the banded appearance of tip section of the giant X chomosome of the salivary gland of the fruit fly, *Drosophila melanogaster.*

The fly has four chromosomes that are arbitrarily divided into 102 divisions. This drawing shows the first four divisions of the X chromosome (marked 1 to 4), the whole chromosome consisting of divisions 1 to 20. Each division is divided into six subdivisions and each subdivision contains a varying number of bands. As many as 159 different genes have been located in the sections pictured in the figure, and these represent only a fraction of the genes they contain. Examples are the gene for yellow body color located at the position indicated by the arrow marked "y," the gene for white eye color located at "w," the gene that affects the arrangement of facets in the compound eye located at "fa," and the gene for ruby eye color at "rb."

Thus, by the 1940s it was obvious that, though the conceptual bases of biological evolution were well established, any description of what had actually occurred in a historical sense during the descent of species could be achieved only from the study of fossils, those preserved remnants of past living forms. A series of fossils can indeed show us directly stepwise descent from an ancestral species to a more recent form, and at the same time let us know, through the application of suitable dating methods, when these evolutionary transitions took place. But the fossil record has two crucial limitations. First, it is so incomplete that only a small fraction of total evolutionary history can be followed in it. In fact, fossils are more and more poorly represented as we delve further back into the times at which the major cellular evolutionary events took place and even further back near the time when life originated. Among such early events one can list the first formation of the nucleus of cells, the structure that contains the genetic apparatus; the appearance in evolution of the structure that carries out photosynthesis, the process by which green plants harness the energy of sunlight to build the compounds they require from carbon dioxide and water; and also the appearance of the structure responsible for cellular respiration, without which the majority of the chemical energy needed to drive cell reactions would not be available to nonphotosynthetic organisms such as animals. It must be emphasized that the development of these cellular structures represents crucial steps in the evolution of life as we know it, without which its past and current multitudinous diversity could never have been attained. Second, even when fossils are available they tell us little about the physiological activities of the organisms when they were alive and give hardly any information about the molecular machinery they employed.

The Rise of Molecular Biology

For all of its brilliant development, genetics had not succeeded to give much flesh to its fundamental concept, the gene, which even by the 1940s was still rather near the abstract realm of Mendel's postulated "elemente." One knew that genes were linearly arranged along the chromosome and that they had two characteristic properties. First, during cell reproduction they replicated with the utmost precision so that the new copies were identical to the original, and when a mutation had changed the gene, the changed or mutated gene was also faithfully reproduced. Second, genes directed the physiological activities of cells in some very essential manner since they caused the appearance of the huge variety of traits by which they were distinguished. It remained for molecular biology to discover the material bases for these properties.

The first essential question was about the substance of which genes are made. The chain of events which eventually led to the correct answer started as far back as 1869 when Friedrich Miescher in Tübingen, analyzing cells that had a large nucleus, such as pus cells or salmon sperm, discovered a new substance, later named nucleic acid. By the 1920s two varieties of nucleic acid had been found. They are now labeled DNA (technically, deoxyribonucleic acid) and RNA (ribonucleic acid). These molecules are long chains of units known as nucleotides. There are five nucleotides involved; three of them, A, G, and C, appear in both DNA and RNA. Of the other two, the nucleotide T is found only in DNA. Instead of T, the nucleic acid RNA has the nucleotide U.

Important experiments of O. T. Avery and his colleagues at the Rockefeller Institute had demonstrated by 1944 that DNA extracted from one strain of bacteria could transform the genetic constitution of another strain. This implied that the genes transferred in this process were DNA. For a variety of reasons, this conclusion was not generally accepted until in 1952 Alfred Hershey and Martha Chase showed that a type of virus, known as bacteriophage or phage and which multiplies in bacterial cells, reproduced itself by injecting nothing else but its DNA into the bacterium. The original phage, with an overall volume of about 1/1000 of the bacterial cell, consists of a head that is a mass of DNA within a protein envelope and a tail. The introduction of the DNA of one phage into a bacterium results within 30 minutes in the production of one hundred complete phages, heads, tails and all, demonstrating that the injected DNA carried the entire genetic information of the phage and specified not only its own hundredfold replication but also the production of all the proteins needed to assemble the phage.

The Hershey-Chase experiment had finally convinced the scientific world that genes were indeed DNA, and within a year James Watson and Francis Crick at Cambridge University had solved the structure of that master molecule by building a model that fits the x-ray diffraction data obtained by M. H. F. Wilkins and his colleagues at King's College, London. The Watson and Crick paper appeared in the April 25, 1953, issue of *Nature* and suggested that DNA was made up of two intertwining chains of nucleotides arranged as a double helix like a two-stranded rope. The strands were strictly complementary, every A on one corresponded to a T in the other, and every G on one being opposite a C on the other, the pairs being held together by hydrogen bonds that could form only as long as the complementarity was maintained. Thus, if the sequence of nucleotides of one strand was specified, it was automatically also fixed for the other, explaining the then three-year-old observation of Erwin Chargaff of Columbia University that in all samples of DNA for every molecule of A there was one molecule of T and for every molecule of G there was one molecule of C, even though the ratio of G to A varied among the DNA's of different species.

This structure of DNA strongly suggested a mechanism for one of the two fundamental properties of the gene, that of precise self-replication. If the two strands are separated by unwinding the double helix, each strand can act as a template and proceed to build on itself a precisely complementary chain, as specified by the allowed pattern of bonding of A to T and of G to C. Similarly, if a mutation results in the substitution of one nucleotide for another in one strand, on replication the same mechanism dictates the replication of the appropriate complement, thereby faithfully reproducing the mutation. This is the so-called *semiconservative mode of replication* in which a parental strand becomes distributed so that the two immediate daughter double helices each contain one parental strand and one newly synthesized strand. That this did in fact occur was demonstrated by Matthew Meselson and Frank Stahl in 1958 (see Figure 8.2).

There remained the far more complex problem of how genes direct the physiological activities of cells. A start on solving this question was made by George Beadle and E. L. Tatum in the 1940s. Their studies of mutants of the bread mold, *Neurospora,* led them to the hypothesis of "one gene— one enzyme," implying that one gene controls the formation of one enzyme, and in this manner exerts its influence on cellular functions. Indeed, enzymes are the substances which carry out the numerous chemical reactions that occur in living organisms. But enzymes are proteins, and proteins are made of chains of building blocks known as *amino acids* (see Table 8.2). The twenty different kinds of amino acids are arranged in a definite and unique sequence for each protein. That proteins do indeed have definite

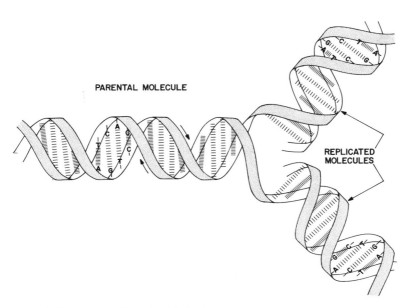

Figure 8.2. Diagram of the double helical structure of DNA and of its semiconservative mode of replication.

Two turns of the double helix of the parental DNA molecule are shown to the left. The flat bands represent the continuous backbone of each strand, while the banded bars represent the hydrogen-bonded nucleotide pairs that keep the two strands in the double helix conformation. In some of the bars letters indicate the nature of the nucleotides, showing that a G on one strand is always opposite and hydrogen bonded to a C on the other, while an A on one strand is always similarly related to a T on the other. To the right of the drawing the two parental strands are shown separated one from the other and a new strand is being synthesized on each parental strand used as a template. Here also hydrogen-bonded pairing decides the order of the nucleotides in the new strands. It is clear that when the synthesis is completed each of the two daughter DNA molecules will contain one parental strand and one newly synthesized strand.

amino acid sequences was first demonstrated by Frederick Sanger in the early 1950s when he determined the amino acid sequence of the protein insulin. Thus, if genes are molecules of DNA carrying their information in the sequence of their four nucleotide bases, while a single gene controls the formation of a single protein, then it is not difficult to conclude that what the gene specifies is the amino acid sequence of the protein chain. Such chains fold in space into very precise three-dimensional structures, termed the *native conformation* of the protein. It is only in this folded state that proteins can carry out the myriad physiological functions entrusted to them in the cell, and as the only information transferred from gene to protein is the amino acid sequence, then the essential folding into the active structure must be a spontaneous process. This was first shown to be the case in the early 1960s by Christian Anfinsen of the National Institutes of Health for the enzyme ribonuclease.

The question of the mechanism whereby the sequence of nucleotides in DNA can specify a sequence of amino acids in a protein chain was gradually solved in the late 1950s and early 1960s. In the first place, since there are only 4 nucleotides and 20 amino acids, the code, termed *genetic code,* used in the translation of one to the other must contain at least 3 nucleotides per code word, termed a *codon.* Two nucleotides per codon and 4 different nucleotides would yield only $4 \times 4 = 16$ codons, while a triplet genetic code has $4 \times 4 \times 4 = 64$ codons (Table 8.1). That such a triplet code was actually in operation was first conclusively demonstrated in genetic experiments with the bacteriophage T4 by Francis Crick and his collaborators in 1961. The working out of which nucleotide triplets correspond to which amino acid residues was carried out in a variety of ways, and the names of Marshall Nirenberg, Severo Ochoa, and Har Gobind Khorana are prominently associated with this work. Though it was first deciphered using components derived from the bacterium *Escherichia coli,* the genetic code turned out to be universal, applying equally to other bacteria, plants, and animals, including higher vertebrates. All 64 possible triplets are used. Three specify no amino acid at all, namely cause the termination of the synthesis of the protein chain (*chain terminating codons:* UAA, UAG, and UGA). There are also many synonyms. For example, the amino acids leucine, arginine, and serine are each coded for by 6 different codons and there are 5 amino acids that have 4 codons each. The genetic code is displayed in Table 8.1.

The genetic code was found to function by a complex though efficient series of reactions that can rapidly assemble the particular amino acid sequences of the chains of proteins, varying in length from under 100 to over 500 amino acids, and most remarkably without making errors. The first step is the *transcription* of the DNA chain of the gene into its complementary RNA sequence, the DNA serving as the template for the formation of what is termed a *messenger* or *m-RNA*, just as in gene replication, one strand of DNA serves as the template for the formation of a new strand. The m-RNA

Table 8.1. The Genetic Code.

SECOND NUCLEOTIDE OF CODON

		U	C	A	G	
FIRST NUCLEOTIDE OF CODON	**U**	Phenylalanine	Serine	Tyrosine	Cysteine	U
		Phenylalanine	Serine	Tyrosine	Cysteine	C
		Leucine	Serine	C. T.	C. T.	A
		Leucine	Serine	C. T.	Tryptophan	G
	C	Leucine	Proline	Histidine	Arginine	U
		Leucine	Proline	Histidine	Arginine	C
		Leucine	Proline	Glutamine	Arginine	A
		Leucine	Proline	Glutamine	Arginine	G
	A	Isoleucine	Threonine	Asparagine	Serine	U
		Isoleucine	Threonine	Asparagine	Serine	C
		Isoleucine	Threonine	Lysine	Arginine	A
		Methionine	Threonine	Lysine	Arginine	G
	G	Valine	Alanine	Aspartic acid	Glycine	U
		Valine	Alanine	Aspartic acid	Glycine	C
		Valine	Alanine	Glutamic acid	Glycine	A
		Valine	Alanine	Glutamic acid	Glycine	G

THIRD NUCLEOTIDE OF CODON

The first nucleotide of each codon is indicated in the left-hand column, as marked, the second across the top of the table, and the third on the right-hand column. Thus, for example, the amino acid phenylalanine at the topmost left corner is coded for by either of two codons, UUU and UUC, while the amino acid glycine at the bottom right corner is coded for by any one of the four codons, GGU, GGC, GGA, and GGG. The three codons marked C. T. (UAA, UAG, and UGA) are chain terminating, namely, they result in the termination of the synthesis of the protein chain at the point at which they occur.

is picked up by a *ribosome,* a small intracellular body which is the site of protein synthesis. A cell of the bacterium *Escherichia coli,* for example, contains some 15,500 of them, and a single ribosome can form a protein chain at the rate of approximately 15 amino acids per second at 37° centrigrade. The molecular details of this process and the complex structure of the ribosome have in good part been established, but any description of them would take us far beyond the confines of the present outline.

In summary, the rise of molecular biology has clearly led to a most profound revolution of our views of the living world. One can truly qualify it as Copernican in scope. Gone are the great unknowns of genetics. Even further, a good start has been made on the molecular mechanisms by which proteins function. Though the full mystery of life has certainly not been solved, such a giant step forward has been made in that direction that biology today is a thoroughly different science than it was just twenty years ago. Furthermore, the concepts and methods of molecular biology are in the process of engendering a full-blown revolution in our approach to the problems of biological evolution. This fascinating interaction is described now.

Macromolecules and Evolution
The macromolecules, or large molecules, referred to here are those substances, nucleic acids and proteins, which constitute the biological information carried by each organism. Indeed, as soon as it was deduced that the amino acid sequences of proteins were merely translations of the sequences of nucleotides of the corresponding genes according to the simple and universal rules of the genetic code, it also became obvious that the amino acid sequences of the proteins of present-day species of living beings were the current molecular end products of the totality of the variations undergone by their genes in the course of evolution. Genes, as we noted earlier, are the hereditary material transmitted from one generation to the next and which contain, as it were, the complete biological blueprint of the organism. Since changes occurring in the course of evolution are the result of the replacement in an evolving population of changed, mutated, more favorable genes for the older less favorable genes, and that genes dictate the amino sequences of proteins, such sequences will bear the traces of every evolutionary event that has taken place since their emergence in biological evo-

lution. This fundamental relation was eloquently stated by Francis Crick in 1958, as he wrote:

Biologists should realize that before long we shall have a subject which might be called "protein taxonomy"—the study of the amino acid sequences of the proteins of an organism and the comparison of them between species. It can be argued that these sequences are the most delicate expression possible of the phenotype of an organism and that vast amounts of evolutionary information may be hidden away within them. (Symposium *Soc. Expermtl. Biol.*, *12*, 138, 1958.)

(Taxonomy is the science of classification of plants and animals. The phenotype is the sum total of the bodily characteristics of an organism, as contrasted with the genotype which is the genetic constitution of the organism, namely the assortment of genes received from the parents.)

The problem was to find the key to unlock these vast stores of evolutionary information. This enterprise was made possible to a large extent by the fact that the great majority of evolutionary changes occurring in protein structure are due to substitutions of nucleotides in the triplet codons of the corresponding gene, and not to other possible types of mutational events. The resulting changes in the amino acid sequence of the protein will be according to the genetic code. Conversely, given that two proteins differ by one amino acid, one can calculate from the genetic code what is the minimal number of single nucleotide replacements in the codons for these amino acids required to change one amino acid into the other. For example, from the genetic code (Table 8.1) it can be seen that the codon GCU codes for the amino acid alanine, while the codon GUU codes for the amino acid valine. If two proteins differ by the replacement of a valine by an alanine, the evolutionary fixation of a single nucleotide change, from GCU to GUU, in the corresponding codon will be enough to explain the evolutionary transformation of one protein into the other. If two proteins differ by more than one amino acid, one can calculate for each pair of amino acids compared how many single nucleotide replacements are required, and sum them for the entire protein to obtain the *minimal replacement distance* of one protein to the other. The more similar the proteins, the smaller will be their replacement distance, and the more closely related will they be from the evolutionary standpoint; the more different the proteins, the larger their replacement distance and more distant their evolutionary relation. It is this sort of calculation that has provided the basis for examinations of evolutionary relationships on a molecular basis, and it is evident why such an approach was not even conceivable before the successes of molecular biology described in the previous section.

Another problem is that the determination of the amino acid sequence of a protein is still an experimentally tedious and demanding procedure, in spite of considerable advances in recent years in regard to the automation of various phases of this work. The efforts of numerous investigators have resulted to date in knowledge of the amino acid sequences of several dozen different sorts of proteins, each type represented in most cases by the protein from one or at most a few species. However, the use of such information to establish evolutionary relationships and history requires knowledge of the amino acid sequences of proteins of the same type from numerous different species. Sufficiently large sets of such sequences from different species are presently available in only a very few cases. The most abundant of these is for cytochrome c, a protein which is involved in the complex process of cellular respiration and for which the amino acid sequences are known for more than fifty different species, covering the entire biological world (see Table 8.3). Most of the examples given will be based on cytochrome c as the author is particularly familiar with it. However, it should not be overlooked that other proteins, such as the red oxygen-carrying protein of blood, hemoglobin, and fibrinopeptides which represent a small segment of the protein that causes the clotting of blood, fibrinogen, as well as other proteins, have been examined for a fairly large number of species and have also yielded interesting evolutionary information. For a survey of such data the reader is referred to the *Atlas of Protein Sequence and Structure* by M. O. Dayhoff, 1972.

Finally, a protein cannot function when it is in the form of a mere long string of amino acids but only when the chain has folded into the precise native three-dimensional structure that is its prime characteristic. Since evolutionary selection for or against it will depend on how well the protein functions in its own environment, to understand evolutionary changes that have occurred as a result of natural selection, one must know with great precision how changes in the amino acid sequence affect the folding of the protein in space and its functional activities. This sort of knowledge is gradually accumulating for many different proteins, but there is at this point not enough understanding of structure-function relations for any one protein to comprehend fully the action of natural selection on it. Whatever evolutionary changes do occur must at all times and in all intermediate stages maintain a functional protein, otherwise the species, except in very special circumstances, is unlikely to survive at all to continue the evolutionary process. This requirement for functional continuity severely limits the changes of amino acid sequence which are fixed in evolution since most of the structural changes that occur as a result of random mutations are likely to be deleterious and lead to the eventual extinction of the line of descent bearing the mutated gene. Though, as we will see later, there is some question as to whether natural selection is at present importantly involved in improving the functional performance of proteins, it is certain that natural selection is most effective in eliminating proteins that have become defective as a result of damaging mutations.

These difficulties and limitations notwithstanding, fundamental strides have been taken in the use of protein amino acid sequences to describe the process of biological evolution in a completeness of detail and a precision far beyond those which could conceivably have been attained by the classical approaches. The aspects of the subject of molecular evolution described later will document the revolutionary reversal it is having on our concepts of how far it is possible actually to describe the major events that took place in the dim past history of biological evolution. The vast exciting area of investigation opened up by these concepts and this new methodology is only now beginning to be explored, and a truly quasi-Copernican revolution is in the making in this biological domain.

Statistical Phylogenetic Trees
If one examines the amino acid sequences of the same protein from two different species, it very quickly becomes apparent that there are extensive similarities between them. For example, Table 8.2 presents the sequences of the cytochromes c of man and baker's yeast. Notwithstanding the enormous evolutionary distance between them spanning the entire biological world, 65 amino acids out of the total of 104 in the human protein and 108 in the yeast protein are the same. This is far beyond what could be expected by chance. A possible reason for such a degree of similarity is that the corre-

Table 8.2. A Comparison of Baker's Yeast and Human Cytochromes *c*.

Thr-Glu-Phe-Lys-Ala-   -Ser-Ala-Lys-   -Ala-Thr-Leu-   -Lys-The-Arg-   -Leu-Gln-
             Gly   Lys-Gly   Phe   Cys   Cys-His-
-5   Acetyl- -Asp-Val-Glu-   -Lys-Lys-Ile-   -Ile-Met-Lys- -Ser-Gln-
            I   10   HEME

            -Pro-   -Val-   -Ile-   -His-Ser-
Thr-Val-Gly-Lys- Gly-Gly   His-Lys   Gly-Pro-Asn-Leu-His-Gly Phe-Gly-Arg
                  -Lys-   -Thr-   30   -Leu-   -Lys-Thr-
     20   40

       -Gln-   -Asp-   -Ile-Lys-   -Asn-Val-Leu- -Asp-   -Asn-
Gly-Gln-Ala   Gly-Tyr-Ser-Tyr- Thr   Ala-Asn   Lys   Trp   Glu
      -Pro-   -Ala-   -Lys-Asn- -Gly-Ile-Ile- -Gly-   -Asp-
                            50   60

Asn-Met-Ser-   -Thr-   -Ala-   -Gly-
          Glu-Thr-Leu   Asn-Pro-Lys-Lys-Tyr-Ile- Pro-Gly-Thr-Lys-Met   Phe
Thr-Leu-Met-   -Glu- 70   80 -Ile-   -Val-

  -Leu-   -Glu- Lys-Asp-   -Asn-   -Thr-   -Cys— D —
Gly   Lys-Lys   Arg   Asp-Leu-Ile   Tyr-Leu-Lys-Lys-Ala   GluCOOH
  -Ile-   -Lys- Glu-Glu- -Ala-   -Ala-   - Thr-Asn-
                      90   100   104

This table merely represents the sequence of the amino acids in both proteins, indicated by their three-letter abbreviations (see below) in a continuous line as in writing a sentence. The way in which these chains are folded in space to give the active functional forms of the proteins is *not* indicated. The top line lists the sequence of the 108 amino acids that make up the baker's yeast cytochrome *c* chain, the bottom line the 104 amino acids of human cytochrome *c*. When the same amino acid occurs in both proteins in the same location, it is marked once between the two lines in large type. The numbers under the line indicate the amino acid positions starting at glycine 1. The two amino acid sequences are aligned by placing together the two cysteines (abbreviated Cys) in each sequence (amino acids 14 and 17) which in both proteins are fixed to the heme (marked "HEME" in the table). This is the iron-containing group that is essential for the activity of the protein but is not composed of amino acids. The first amino acid in human cytochrome *c* (amino acid 1) is glycine, but as in the case of all the cytochromes *c* of vertebrates it is blocked by reaction with acetic acid. This is indicated in the table by the "Acetyl" marked before it. In yeast cytochrome *c* this "acetyl" is replaced by a sequence of five amino acids (amino acids −1 to −5). Extending from the glycine to the other end of the sequence the yeast protein is 103 amino acids long, while the human protein is 104. The "D" marked at position 103 of the yeast cytochrome *c* indicates that a space or "deletion" has been introduced in order to maximize the similarity between the two proteins. The three-letter abbreviations for the names of the amino acids are Ala, alanine; Arg, arginine; Asn, asparagine; Asp, aspartic acid; Cys, cysteine; Gln, glutamine; Glu, glutamic acid; Gly, glycine; His, histidine; Ile, isoleucine; Leu, leucine; Lys, lysine; Met, methionine; Phe, phenylalanine; Pro, proline; Ser, serine; Thr, threonine; Trp, tryptophan; Tyr, tyrosine; Val, valine.

sponding genes are *homologous,* namely that they had a common gene ancestor at some time in distant past evolutionary history, and that the differences that now exist between them are the result of mutations that have accumulated since they first diverged from their common ancestor. It must not be overlooked, however, that even if the degree of similarity observed cannot be ascribed to chance this does not necessarily mean that the proteins are homologous. One could, for example, conceive that the genes for baker's yeast and human cytochrome *c* arose from separate evolutionary ancestors, but that natural selection caused them to become more and more similar, namely *converge,* in order to provide proteins having similar functions. In other words, these two cytochromes *c*, even though they originated from different ancestors, could have acquired in the course of evolution the degree of similarity made necessary by their similarity of function. In such a case one would call these genes *analogous* rather than homologous.

On the basis of only two such amino acid sequences it is not possible to tell whether the observed similarity is the result of a convergent evolutionary process or merely what remains of the original ancestral form following a divergent evolutionary process. However, similarities of the sequences of the cytochromes *c* exhibited in Table 8.3, corresponding to some 50 different species from the lowest to the highest, indicate strongly that all cytochromes *c* did indeed arise from a common ancestral form! Cytochrome *c* being essentially universally distributed among all species, this result is particularly interesting. To the present author, it constitutes the strongest evidence yet obtained that all living forms on earth as we know them today, including man, are the result of a single occurrence and descend from a common ancestor.

The amino acid sequences for a set of homologous proteins also provide an extremely powerful tool for determining the ancestral relationships (commonly termed the *phylogeny*) of the species making these proteins. The basic measure used is the minimal replacement distance between any two sequences, as was defined in the preceding section. A statistical procedure is used to develop a tree depicting the evolutionary relations between the species examined. Such a statistical phylogenetic tree for the cytochromes *c* of 29 species is shown in Figure 8.3. The numbers along the branches of the tree are estimates of the numbers of single nucleotide replacements (mutations) that occurred in the gene for cytochrome *c* in each line of descent

Table 8.3. Composite Amino Acid Sequence of the Cytochromes c of Known Structure.

```
                        Tyr                              Thr
                   Ser  Phe  Lys Lys           Ile Ala  Ser                  Val
          *    Lys Glu Pro Thr Pro Val Glu Gln  *   Ser Val Glu Lys      Lys Thr Leu
     Ala - Ala - Ser - Phe - Ser - Glu - Ala - Pro - Pro - Gly - Asn - Pro - Asp - Ala - Gly - Ala - Lys - Ile - Phe -
      -9            Thr Ser Ala Gly        Ser Ala  1  Asp Ala Lys Asn      Glu Asn Thr 10
                            Asn             Ala            Ser Thr
            Ala                                  Ala   Gln      Glu              Val
     Glu   Glu                   Gly              Glu  Gln      Lys Val Thr Asn
     Thr   Gln  Arg  Ala   Ser  Leu        Gly  Cys  Gly  Gly  Asn  Leu  Pro  Gln  Ser      Thr
     Lys - Thr - Lys - Cys ∘ Ala - Gln - Cys - His - Thr - Val - Asp - Ala - Gly - Ala - Gly - His - Lys - Gln - Gly - Pro -
     Ile   Met        Glu  Glu             Ser  Ile  Glu  Asn  Ala  Gly  Lys            Ile    30
     Val   Ser     └────HEME────┘           20       Lys            Ser            Val
            Gln  Val                  Thr              Glu  Val                              Thr
            Asn  Ser                  His                    Glu                              Glu
     Ala    Ser  Ile  Ile  Ser       Lys  Thr       Ser  Val  Gln       Phe  Thr       Thr  Asp
     Asn ∘ Leu - His - Gly ∘ Leu - Phe - Gly ∘ Arg - Gln - Ser - Gly - Thr - Thr - Ala ∘ Gly - Tyr - Ser ∘ Tyr - Ser - Ala -
     Ser    Tyr  Phe  Tyr                 40         Gln  Ala  Asp            Ala            Asn
            Trp                     Val        Lys              Pro       His              Pro
            Lys  Met               Ile        Asn  Gln  Glu    Lys   Arg  Val              50
     Gly  Asp  Ile  Arg  Ala  Asn  Ile  Leu  Gln  Asn  Pro  Asn  Met  Ser  Ile  Phe      Thr
     Ala - Asn - Lys - Asn - Lys - Ala - Val - Glu - Trp - Glu ∘ Glu - Asn - Thr - Leu - Tyr - Asp - Tyr - Leu - Leu - Asn-
            Gln  Arg  Gly       Thr        Ala  Asp  Asp  Ser       Met  Glu            Glu  70
            Ser                 Lys        Asp       Lys  Asp       Phe  Lys
            Ala                 Gln        Gly       Gln
                                Asn        Thr            Val                          Lys  Ala
                                           60                                          Thr  Asp
            Val        Lys                      Ile       Thr       Ile  Ser  Ala  Glu  Gly  Glu
     Pro - Lys ∘ Lys - Tyr - Ile - Pro - Gly - Thr - Lys - Met - Val - Phe - Pro ∘ Gly - Leu - Lys - Lys - Pro - Gln - Asp -
                                           80   Ala       Ala       Phe       Ser  Asp  Thr  90
                                                          Gly                      Ala  Ser
            Lys                                                D                         Glu
            Gln                          Val  Asp  Ser       Lys Lys                    Asn
            Thr  Asn  Ile  Val  Thr  Phe  Met  Leu  Glu  Thr  Ala  Asn  Ala
     Arg - Ala - Asp - Leu - Ile - Ala - Tyr - Leu - Lys - Lys - Ala - Thr - Ser - Ser - Glu
            Glu  Val  Ser              Thr  Ser  Lys  Cys  Ala  Glu
            Gly                        His  Thr  Leu  Ser  Glu  Gln
            Asn                        Gln  Val       Asp
            Val                        Ile
                                       100
```

The three-letter abbreviations for the names of the amino acids are listed in the legend of Table 8.2. The asterisk at position −8 indicates that this amino acid is acetylated in the higher plant cytochromes c, while the asterisk at position 1 similarly denotes the acetyl present in all vertebrate cytochromes c. The cytochromes c tabulated are those of 15 mammals, 5 birds, 2 reptiles, 1 amphibian, 4 fish, 1 cyclostome, 4 insects, 6 fungi, 11 plants, and 2 protists. These last are a large group of organisms, mostly composed of one cell only, and which are intermediate in characteristics between plants and animals. References to the species and to the original work on the structures of these proteins are given in Margoliash, *Harvey Lectures*, Vol. *66*, 1972, p. 177.

during the particular period of evolution represented by that line. Summing the numbers along the branches connecting any two species yields an estimate of the replacement distance between their cytochromes c. Out of the very many possible trees, that tree was considered best for which all the replacement distances reconstructed from the lengths of the branches most closely match the minimal replacement distances calculated directly from the comparison of the amino acid sequences of the cytochromes c, as described in the preceding section.

It is indeed remarkable that notwithstanding this very restricted basis of the molecular evolutionary events occurring in a single small gene, the resultant phylogeny is so excellently in accord with classical scientific opinion. Strictly the only information given the computer that derived this tree was the amino acid sequences of the cytochromes c of the species listed, the genetic code, and a very simple set of statistical instructions. Nevertheless, the groups of species are properly separated, and with only two or possibly three exceptions the relationships between species are exactly those expected from previous zoological knowledge. Furthermore, the results obtained with one gene can be checked independently by similar studies with any number of other genes. Choosing the proper proteins for such studies is of major importance. Cytochrome c varies rather slowly in evolution, a single mutation being fixed on the average every 20 million years in the cytochromes c of two diverging lines of evolutionary descent. It is useful, therefore, for examining phylogenetic relations of major groups of species over long periods of evolutionary history but not useful for determining the relations of rather closely related species that are likely to show few or no differences among their cytochromes c. Thus, men and chimpanzees have identical cytochromes c, as have goats, sheep and pigs, and also chickens and turkeys. For more closely related species, one would best examine a gene product that has varied more rapidly in the course of evolution, such as hemoglobin, which accepts mutations about six times more rapidly than cytochrome c, or fibrinopeptides, which fix mutations some fourteen times faster than cytochrome c.

Ancestral Amino Acid Sequences
Once a phylogeny has been decided upon, one can use it to infer the approximate sequences of nucleotides of the gene that coded for the protein

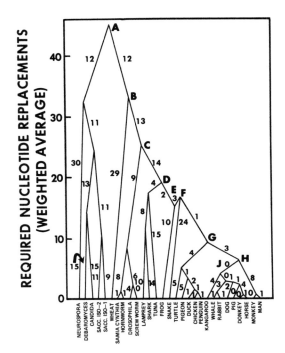

Figure 8.3. Statistical phylogenetic tree based on the minimal replacement distances between the cytochromes *c* of the species listed, according to Fitch and Margoliash, *Science,* Vol. *155,* 1967, p. 279.

Each number on the figure is the replacement distance along the line of descent as determined from the best fit of the data so far found. Each peak is placed at an ordinate value (height in the diagram) that is the average of the sums of all nucleotide replacements (mutations) in the lines of descent from that peak, weighting equally the two lines descending from any one peak.

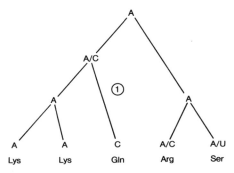

Figure 8.4. Determination of the ancestral nucleotide for the first nucleotide of the codons for the amino acids marked in the bottom row, given the indicated topology of protein relationships.

The procedure is described in the text. The circled 1 denotes that one mutation in the marked branch is the minimum required to derive the ancestral nucleotide.

in any one of the ancestral forms. These ancestral sequences are represented by the various branching points of the phylogenetic tree such as those marked A, B, C, and so on in Figure 8.3. They can be reconstructed from the sequences of the present-day proteins. To do so one derives that ancestral nucleotide sequence which can give rise to the genes of the present-day proteins in the smallest possible number of nucleotide substitutions, following the topology of relationships depicted by the phylogenetic tree. Figure 8.4 gives a simple hypothetical example.

The phylogenetic tree drawn in Figure 8.4 consists of five amino acid sequences. In a certain position these proteins carry the amino acids lysine, lysine, glutamine, arginine, and serine as indicated in the bottom row of the figure. Let us consider the first nucleotide of the triplet code words for these amino acids. These are listed above the amino acids, A, A, C, A or C, and A or U, respectively. The reconstruction rule is simply that a peak will have the nucleotide that is common to both descendant forms, yielding A in the rightmost and leftmost peaks. However, in the case of the intermediate peak marked A/C, there is no common nucleotide in the immediate descendants, hence both nucleotides are put down. This now permits us to decide that the topmost peak is A, thus deriving the nature of the common ancestral form for this particular coding position for this particular phylogenetic tree. A further examination now shows that only a single nucleotide replacement or mutation is required to explain the descent from the common ancestral form A to all the five present descendant forms if we assume that the peak marked A/C is in fact A. That mutation will be placed in the A/C to C branch, as indicated by the circled 1. If peak A/C were considered to be C, two mutations would be required to explain the descent. That possibility is therefore rejected.

Carrying out a similar process for all the nucleotides in all the codons of cytochrome c according to the statistical phylogenetic tree yields the codon sequences and hence the amino acid sequences for all intermediate ancestral forms, including the one at the topmost peak of the tree. This sequence is listed in Table 8.4 and represents an approximation of the amino acid sequence of the cytochrome c of one of the creatures that first utilized the type of molecular apparatus for cellular respiration which is in general use today.

The estimation of ancestral sequences has a number of important consequences. First, the time is fast approaching when the laboratory synthesis of protein chains will be a fairly routine procedure. Knowledge of ancestral amino acid sequences will then make it possible to obtain such proteins synthetically and test their properties and functional activities, just as the properties and functional activities of present-day proteins are examined. This will have the far-reaching consequence of turning evolution into an experimental science, employing the proteins of species long extinct to explore experimentally the mechanisms and consequences of transformations that took place in eons past.

Second, these procedures give a fascinating approach to the very earliest times near the period when life began its evolutionary plunge into the

Table 8.4. Amino Acid Sequence of the Ancestral Form of Cytochrome *c* at the Topmost Peak of the Phylogenetic Tree in Figure 8.3.

	Ala	Glu			Thr				Pro												
Ala -9	Lys [Glu] [Thr]	Ser [Ala]	Ser Phe	Ala Ser	Gly	Val Phe	Ser [Thr] [Ala]	Ala Pro	Gly 1	Asn	Ala	Lys Glu	Lys	Gly	Ala	Lys Asn	Leu Ile	Phe 10	Lys		

Thr -	Lys Arg	Cys	Ala	Gln	Cys	His	Thr	Val	Glu	Glu Gly Ala	Gly	Gly	Thr	Lys [Arg]	His	Lys	Val	Gly	Pro	Asn 30
		└── HEME ──┘						20												

							Ser													
Leu -	His -	Gly -	Leu -	Phe -	Gly -	Arg -	Lys Gln 40	Thr -	Gly -	Gln [Pro] -	Ala -	Glu Ala -	Gly -	Tyr -	Ser -	Tyr -	Thr	Asp -	Ala 50

Asn -	Lys -	Lys Asn	Lys -	Gly -	Val Ile	Lys Thr	Trp -	Glu Asp 60	Glu -	Asn -	Thr -	Leu -	Phe -	Glu -	Tyr -	Leu -	Glu -	Asn -	Pro 70

								Ala		Gly				Glu		
Lys -	Lys -	Tyr -	Ile -	Pro -	Gly -	Thr -	Lys -	Met - Val 80	Phe -	Ala	Gly -	Leu -	Lys -	Lys -	Pro Ala [Gln]	Lys -	Asp -	Arg 90

Glu										Ala Glu
Ala										
Thr -	Asp -	Leu -	Ile -	Ala Thr	Tyr -	Leu Met	Lys -	Lys -	Ala -	Thr - Ser - Ala
Lys						100				[Thr] Ser
Asn										104

Any of the alternative amino acids shown would permit the evolution of the 29 descendant cytochromes *c* in the minimum number of 366 nucleotide replacements, assuming the topology shown in Figure 8.3. Amino acids in brackets have not yet been observed in any present-day cytochrome *c* in these positions.

future. Indeed, the examination of numerous sets of proteins will yield re-
constructions of the structure of the ancestral form for each set. Having in
this way eliminated the mutations that have accumulated between the time
these early species existed and the present, it is quite possible that the ancestral sequences will exhibit similarities that are not visible in their present-
day descendants. One could then work out a tree based on these ancestral
sequences and thus estimate the structure of the ancestral sequence of these
ancestral sequences. A similar approach could be used also with nucleic
acids. The second-order ancestral sequences of both varieties of biological
molecule should correspond one to the other and take us back to near the
original replicating molecule at a time, near the borders of chemical and
biological evolution, at which the genetic code and the genetic machinery
first came into being.

Finally, the reconstruction of ancestral sequences yields a straightforward
way of deciding whether the members of a set of proteins have descended
from a common evolutionary origin, or whether they originated from several such sources. For example, this has been done by comparing the
cytochromes c of the fungi, such as yeast, and the other cytochromes c of
known sequence. It was shown that the most recent common ancestral form
of the cytochromes c of fungi was far more similar to the most recent common ancestral form of all the other cytochromes c than are the present-day
proteins. Thus, those two sets of proteins are clearly not converging from
different evolutionary sources and becoming more similar as they evolve.
On the contrary, they are undergoing a divergent evolutionary process
having come from a single source in the past and becoming more different
as they evolve. This is the evidence upon which was based the statement of
the common origin of all living beings given in the preceding section.

Concluding Remarks

The foregoing account briefly describes how the understanding of the fundamental genetic mechanisms that came from the advance of molecular biology in the last two decades has yielded an entirely new approach to the
study of the history and mechanisms of biological evolution. This approach
is in the process of extending our knowledge of the actual occurrences of
evolutionary history to areas and events that could not possibly have been
observed through classical methods and classical concepts and is yielding

new and unexpected insights into basic molecular evolutionary processes. It is already clear that one can obtain a thorough description of the evolutionary relationships of all biological macromolecules that carry genetic information, such as proteins and nucleic acids. This can take the form of a master statistical phylogenetic tree for a large number of different genes, namely a gene phylogeny, and for each gene a phylogenetic tree of the gene variations in many species, namely, species phylogenies. In addition to providing precise phylogenies of the ordinary zoological type, these will make it possible to decide important aspects of evolutionary history that would otherwise have remained forever buried in the debris of the distant past. Among them are the identification of the invertebrate group from which vertebrates arose; the precise evolutionary relations of species that have no cell nuclei, such as bacteria, and those that carry nuclei and represent the majority of species; the evolutionary origin and history of intracellular structures such as those involved in respiration, photosynthesis, and so on. Tracing biological evolution even further back, the molecular approach may eventually lead to an approximation of the structure of the original duplicating biological macromolecule existing at the time when the genetic code was being established, and thus provide an understanding of how this actually happened. Finally, and even more importantly, the ability to derive ancestral amino acid sequences may well make it possible to turn molecular evolution into an experimental science through the laboratory synthesis of ancestral proteins. This would result in the application of modern techniques of structural and functional study to molecules and events resurrected from the dim distant past. Such a revolution in the making is undoubtedly Copernican in scope and impact.

Acknowledgments
The author is grateful to Professor Grier Davis for reading the manuscript and Professor Robert King for supplying the material for Figure 8.1.

General Supplementary Readings

F. C. H. Crick, "The Genetic Code." *Scientific American*, October 1966, p. 55.

M. O. Dayhoff, *Atlas of Protein Sequence and Structure*. National Biomedical Research Foundation, Silver Spring, Md., 1972.

R. E. Dickerson, "The Structure and History of an Ancient Protein." *Scientific American*, April 1972, p. 58.

W. M. Fitch and E. Margoliash. "The Construction of Phylogenetic Trees." *Science*, Vol. *155*, 1967, p. 279.

E. Margoliash and W. M. Fitch, "The Evolutionary Information Content of Protein Amino Acid Sequences," in *Homologies in Enzymes and Metabolic Pathways*. Edited by W. J. Whelan, North-Holland Publishing Co., Amsterdam, 1970, p. 33.

E. Margoliash, "The Molecular Variations of Cytochrome *c* as a Function of the Evolution of Species." *Harvey Lectures*, Vol. *66*, 1972, p. 177.

C. Yanofsky, "Gene Structure and Protein Structure." *Scientific American*, May 1967, p. 80.

Emile Zuckerkandl, "The Evolution of Hemoglobin." *Scientific American*, May 1965, p. 110.

Seymour S. Cohen

9 On the Origins of Cells:
The Development of a Copernican Revolution

For many years I have been interested in the chemical differences among organisms. In a general way we can ask if the differences between man and the microbes are to be ascribed primarily to the differences in the organizational complexity of their common chemical units or if the chemistry of higher organisms, such as man, has also evolved in significant respects from that of the bacteria. If the latter were true, we would have chemical approaches to the study of the evolution of living things, to the nature of our species. Inevitably this question raises the problem of the origin and nature of the first living things; inevitably we must ask about the nature of creation.

In concluding that the earth was not at the center of the universe and indeed spun around the sun, Copernicus denied the accuracy of the Bible in matters of cosmology. Inferentially he raised doubts concerning the utility of this book and the knowledge of its guardians as guides in other scientific matters, including biology. In concluding that the earth was moving to a greater degree than the sun, Copernicus was also challenging the apparent evidence of his senses, the testimony of common belief, and the authority of his teachers and of various officials of the Church. Concomitantly he had adopted the attitude of looking closely at natural phenomena and of developing closely reasoned conclusions from his observations. His conclusions and this approach to the problems of the real world have had the broadest consequences, reaching even into the biochemical question I wish to explore.

The Pace and Difficulties of a Copernican Revolution
However, this Copernican transformation of intellectual attitudes has been very slow, taking hundreds of years. The problem of the origins of living things has only recently become a focus for active scientific investigation, although it has lurked behind the Copernican discovery for centuries. Furthermore, it is obvious that the rule of reason is far from complete, that even now a Copernican approach to many problems has numerous competitors.

For example, I have just returned from a symposium held in South America; our topic was "Fundamental Approaches to the Improvement of Agriculture." The meeting took place in a city founded by a conquistador in the year 1533, ten years before the publication of Copernicus's book on "The Revolutions of the Heavenly Spheres." This city of a million possesses a giant airport, factories, television, a medical center, university and agriculture

University of Colorado, School of Medicine, Denver, Colorado 80220.

stations, but it operates in the midst of hunger, disease, poverty, and sullen and recent rebellion. Not even Bolivar and his successors have been able to resolve the continuing struggle between the heirs of the conquistadors and of Copernicus.

In the last day of our meeting, papers were given on the control and obliteration of insect pests, the natural processes of repair of human cells subjected to damaging radiation, and the development of new species of plants. In unspoken defiance of Joyce Kilmer and his Heaven-sent "Trees," three groups of scientists from Hawaii, France, and Long Island had isolated single cells from sugarcane or tobacco and during growth in glass vessels had eliminated certain stabilizing or complicating chromosomes, unwanted packages of genes from the cell nuclei. Following this manipulation, whole complex plants, sugarcane or flowering tobacco of normal appearance, were grown from these single more manipulable cells, as a first step in producing richer crops, more resistant to parasites or to adverse weather conditions. As I followed the presentations, I thought that surely this bold manipulation of the biological world was one of the consequences of the giant step of Copernicus, who had timidly challenged dogma and the inertia of the traditional scholarship of his time. Nevertheless, I had only to step from the building to see that his influence had not been greater than that of his contemporaries, the conquistadors.

It is clear then that the giant strides of recognition of scientific and social truth in defiance of rigid tradition, which we celebrate as Copernican, are not instantaneous achievements. Copernicus was handed a printed copy of his book when on his deathbed in 1543. It is known that for about thirty years he had been in possession of the observations and calculations which led to his revolutionary theses. Despite his eminently respected position as physician, financial adviser, and doctor of canon law, he was somewhat hesitant to publish his work and conclusions in astronomy. He suspected, quite correctly, that there would be some undesirable repercussions to the thesis that the planetary residence of man was somewhat off center. Indeed, both Calvin and Luther indicated their preference of Scripture to Copernican theory.

After the appearance of his book, a less cautious philosopher, Giordano Bruno, lectured on a number of off-beat notions, as well as on this work. Indeed, he made a significant contribution to the spread of the Copernican world view. For these indiscretions he was burned at the stake in 1600 for heresy. Galileo obtained confirmatory and even more extensive data, published these and his conclusions supporting the Copernican thesis, and we know the trouble he had with his employers. Having advanced his affirmation of Copernican astronomy in 1616, he was forced to promise not to discuss these matters again. The reaffirmation of his position in a classic volume in 1632 led to the banning of his book, and he was "detained" under threat of more serious treatment. Galileo never openly discussed the Copernican theory again until his death in 1643. The deliberate censorship of this verifiable scientific conclusion was imposed until 1758, and only in 1828 was the prohibition explicitly withdrawn by the Church. Most traces of serious opposition, at least in non-Catholic countries, were swept away in 1687, the date of appearance of Newton's great work, *Mathematical Principles of Natural Philosophy,* which integrated the Copernican system within Newton's theory of the physical world. Nevertheless, as we know full well, the Old Testament and Mosaic cosmogony are still upheld as guides to the teaching of geology and biology in many American states.

Is Man Alone?

Why should the fact that man is not at the center of the universe have seemed so threatening to anyone? Perhaps a questioning of Scripture (as Luther indicated in detail), or even worse, a correction of dogma, of authority, would lead to other much worse aberrations, such as questions about the divine right of kings and clergy. Was the creation of man, of living things, a peripheral or central event in the creation of the universe? Was it possible that, as Bruno (the heretic), Tycho Brahe, and Johannes Kepler believed, man and other living creatures could exist on heavenly bodies far from earth? Had the drama of the birth of Adam then been reenacted more than once? Was there actually even one creation of Adam requiring a heavenly midwife? Where did life and man on all these places come from anyway?

I suspect that Kepler might have been slightly disappointed (as indeed were many biologists) and various seventeenth-century clergy might have been somewhat relieved if they had known of the results of space flights of this past decade. Man is not quite at the center of the universe, but no sign of a comparable higher organism (or lesser one either) can be detected outside of earth. Life and its language-speaking, tool-wielding, weapon-brandishing culmination seem to be all alone within a rather extensive (by some criteria) volume of space. Most of the scientists involved in the space efforts seem agreed that it is unlikely that life as we know it on earth can survive (or develop in) the extraterrestrial environments we can explore in our time.[1]

How Did Life Appear on Earth?

As noted earlier, these biological problems did not become urgent for centuries. Everyone knew that the creation of life was a simple thing and that it occurred frequently, almost under our very eyes. Vermin of all kinds were particularly easy to generate, in the kitchen, in dark dusty corners, stables, and so on. This was not a real problem, although the variety of organisms was not as generally representative as thoughtful philosophers might have liked. Of course, babies sometimes appeared in doorways, but this was well understood and was not quite the same problem.

With the discovery of microorganisms in the seventeenth century and their observed growth in even boiled liquids, it was clear that these tiny living things could form in the simplest of media and under the most ordinary contemporary environmental conditions. The experiments of Spallanzani in 1777 had demonstrated that various infusions heated to kill germs would remain sterile if air was excluded, but no one paid much attention to this, even though we now credit him with the inadvertent discovery of the preservation of food.[2] Nevertheless, it took the furious battles of Pasteur in the sixties and seventies of the nineteenth century to convince scientists that the spontaneous generation of microorganisms, of living things on earth at this time, simply did not occur, that life came from life in modern times. Thus, it had taken three centuries before a corollary to Copernican theory, that we ought also to be concerned with the origins of the first living things, had emerged as a commonly accepted problem. Where did life on earth come from anyway; how did it originate?

The English physiologist Huxley thought that at the bottom of the sea he had discovered the primordial ooze that had originated from inorganic matter and from which life had evolved. Many workers tried to instill life into complex substances, and it was claimed that radium would do so if added to a solution of gelatin. Lord Kelvin treated these reports with magisterial disdain.

A very ancient speculation, still clung to by many naturalists (so much so that I have a choice of modern terms to quote in expressing it), supposes that, under meteorological conditions very different from the present, dead matter may have run together or crystallized or fermented into "germs of life" or "organic cells" or "protoplasm." But science brings a vast mass of inductive evidence against this hypothesis of spontaneous generation. Dead matter cannot become living without coming under the influence of matter previously alive.

On the Late Late Show
Another approach to the problem of the origin of life was that of "panspermia," which proposed that life-giving seeds or "cosmozoa" were drifting about in space. Such a view had been formulated by Anaxagoras among the early Greeks, but only a century ago the famous botanist, Ferdinand Cohn, and the authoritative Lord Kelvin were also active proponents of this theory, the latter stating in his presidential address to the British Association:

I am fully conscious of the many objections which may be urged against this hypothesis. I will not tax your patience further by discussing any of them on the present occasion. All I maintain is that I believe all to be answerable.

Svante Arrhenius, one of the founders of modern physical chemistry and Nobel laureate in 1903, published *Worlds in the Making* in 1908, developing the theory of panspermia in great detail. He believed that life is ubiquitous and that spores, emitted from myriads of habitable worlds, are continually traversing space; a few such surviving organisms will light on a potentially habitable world, and in germinating provide the seed for the development of life on the fallow planet. It is obvious that this hypothesis is exciting science fiction but is unlikely to be easily testable. However, even more serious is the displacement of the problem to some other part of the universe. How did a first "spore" arise in any part of the universe?

Scientific Beginnings
The Darwinian revolution made it clear that present forms of living things on earth had evolved from simpler forms. For some scientists uncramped by religious dogma, it was natural to ask if the initial living things were not the product of chemical evolution expressed over millions or even billions of

years of geological time. Nevertheless, the development of biochemistry was not yet adequate to orient the chemists on the nature of the compounds that must have evolved. As the splendid recent book of historical essays by J. Fruton (*Molecules and Life,* John Wiley & Sons, New York, 1972) makes clear, a good working knowledge of the detailed nature of the crucial cellular components, the proteins and nucleic acids, has developed only in the period after the Second World War. Some philosophic individuals may have considered the problem of the origins of living things, but the time was not yet ripe for the necessary experimental chemical work. As it turned out, a good start came from astronomy and cosmology, from the rapidly developing knowledge of the development of suns and planets, the very discipline to which Copernicus had contributed.

We enter the modern era of thought on the origin of life in the 1920s with the independent theoretical work of the Russian, A. I. Oparin, and the "eccentric" Englishman, J. B. S. Haldane. Both of these innovators posed the problems of the origin of life on earth in a new way, as a function of the chemical and geological conditions prevailing during the development of the third planet. It is probably no accident that both of these men were Marxists, who in a revolutionary era attempted to solve our problems here and now. Even if the existing science and technology were not yet ripe for such attempts, these men believed that the problem of the origin of life must be understood within the context of the development of our own planet, and not someone else's. "Panspermia" as formulated by Arrhenius and his predecessors was obviously an evasion of the responsibility of scientists, who had shunted the problem to some other corner of the universe. It is of great interest that the Oparin-Haldane theories that seemed so remote and experimentally untestable at the time of their formulation nevertheless focused on the real capabilities of several sciences that would possess such capabilities within a mere three decades.

By 1930 the key papers were published and had disappeared from view. Oparin's new and far more comprehensive volume *The Origin of Life,* published in Russian in 1936 and in English in 1938, still did not mention Haldane's little-known paper. Haldane, writing in 1927, did not know of Oparin's first essay published in Russian in 1924.[3]

Oparin had promulgated the view that substances rich in hydrogen (compounds such as methane, ammonia, water, and hydrogen itself) must have

been the primary precursors for the synthesis of the complex organic molecules found in cells. Haldane, then interested in metabolism and the catalytic
(enzymatic) control of metabolic events had been impressed by the lack of
requirements for oxygen by many organisms in the early steps of processing
glucose. He had suggested then, that the generality of this anaerobic metabolism in the handling of a substance so obviously useful as the common
sugar, glucose, which is a primary synthetic product of light-utilizing plants,
indicates a conservation of a primitive mechanism. Indeed, he thought that
this mechanism may have reflected the primitive earth conditions in which
metabolism originated. And so he wrote of life developing anaerobically in
a primitive ocean with "the consistency of a hot dilute soup" of organic compounds, substances whose anaerobic formation had indeed been presaged by
Oparin only a few years earlier.

It was only after these prophetic papers, in 1929, that H. N. Russell observed hydrogen as the predominant constituent of the solar atmosphere.
In his turn, R. Wildt in 1931 recorded the presence of methane and ammonia in the atmosphere of Jupiter. It is remarkable how generally accepted
these insights of Oparin and Haldane are today, although the problem is
still debated of the exact amount of oxygen which might have been contributed to the primitive atmosphere of earth as a result of the decomposi-
tion of water by radiation.

Synthesizing the Building Blocks
Oparin's seminal volume in 1936 (*The Origin of Life*) described many early
efforts to approximate the conditions prevailing on our primitive planet.
In 1953 a graduate student, S. L. Miller, working under the direction of the
physical chemist, H. Urey, who had become interested in the origin of
planets and moons, made a signal breakthrough in synthesis under anaerobic
conditions. Using a glass apparatus provided with an electrical arc as an
energy source to simulate the possible effects of lightning, and containing a
mixture of water, methane, ammonia, and hydrogen, he was able to produce
some of the constituents of proteins, some amino acids. As one adds new
constituents to this system, for example the gas hydrogen sulfide, it is possible to detect the formation of new amino acids, such as the sulfur-contain-
ing amino acids. Other precursors treated under similarly simple conditions
in slightly modified gadgets permit the synthesis of nucleic acid constituents

in astonishingly high yield. It has been possible to make many if not most of the simple organic building blocks of living cells in this way, and slightly more energetic conditions, for example, higher temperatures, produce associations of these molecules into relatively large polymers, that is, regular arrangements of linked subunits. Although the mechanisms of these syntheses are far from being understood, it has been apparent for two decades, since the early fifties, that given the expected components of the primitive atmosphere and an energy source, the synthesis of the components of living things can occur at significant rates abiogenically, that is, in the absence of life, and that such potentially prebiological syntheses are fully amenable to laboratory study and experimentation.

What Did Viruses Tell Us?

The experiments of Miller and Urey, the concepts of Oparin, Haldane, and others began to titillate scientific neurons at a particularly appropriate period in the history of biochemistry. In the mid 1950s and the early 1960s, we began to know how viruses were organized and operated, and we began to understand the roles of their key components, the nucleic acids and proteins, as well as how they were synthesized. This understanding posed a new and as yet unresolved problem in the development of our knowledge of biopoiesis, the origin of life. Given two different key substances, that is, the nucleic acids and proteins, the scientist wishes to know which one came first, or whether they were not both required simultaneously.

The 1930s were a period in which proteins and their catalytic representatives, the enzymes, were being purified and characterized quite actively. In this country the Rockefeller Institute was a major center for such studies, as well as for the exploration of infectious disease, including virus disease. At the Department of Animal and Plant Pathology of the Institute in Princeton, New Jersey, John Northrop and Moses Kunitz were crystallizing various enzymes with surprising ease and were exploring the chemical structure and catalytic activity of these proteins. In 1934, a young chemist, Wendell Stanley, was working in the same institute (Figure 9.1) on the problem of the nature of plant viruses. He asked whether the causative agent of a representative virus disease, that causing the tobacco mosaic disease, would not be similarly amenable to the techniques of protein isolation. It is of interest that viruses had first been discovered in the study of this disease late in the last

Figure 9.1. Some modern Copernicans at the Rockefeller Institute, Department of Animal and Plant Pathology, Princeton, 1942: (a) J. Northrop, (b) M. Kunitz, (c) W. M. Stanley.

century; these infectious entities had been shown to be smaller than the smallest cells, that is, bacteria, by the Russian, D. Iwanowski and the Dutchman, M. Beijerinck. Nevertheless, viruses were able to multiply (duplicate) inside cells and to evoke characteristic physiological responses, for example, induce the synthesis of specific new proteins; these qualities of the viruses were those of the chromosomal units of inheritance, of genes. Using the techniques of protein isolation, Stanley proceeded to isolate the tobacco mosaic virus in aggregates posessing the two-dimensional regularity (Figure 9.2) describable as paracrystalline.[4] These aggregates were assemblies of rod-shaped particles, which surely contained protein. In one of the triumphs of protein chemistry the dimensions of a single virus particle, as measured in Stanley's laboratory, were later shown to be almost identical with the dimensions observed directly in the electron microscope. These particles were not only readily "crystallized," but also they were stable and uniform in size (Figure 9.3a), and the compositions of various preparations of virus were quite reproducible. Thus, the virus particles possessed qualities used to define "molecules." Stanley had isolated an infectious virus, with many qualities ascribed to living things, and it proved to be a relatively simple regular chemical substance.

The Nucleic Acids Seem Important

In 1936 English plant virologists, F. C. Bawden and N. W. Pirie, found a nucleic acid containing the sugar, ribose, that is, *ribose nucleic acid,* or the familiarly known RNA, in the tobacco mosaic virus (Figure 9.3b). In the same year a German refugee working in London, M. Schlesinger, isolated a bacteriophage, a virus that infects bacteria, and found it to contain a nucleic acid of the *deoxyribose* type, DNA (Figures 9.3c and 9.3d). It took until the 1950s to discover that the nucleic acid of a virus determines the genelike properties of these agents. The nucleic acids were in fact the units of inheritance; one or the other type directed an infected cell to make the proteins and enzymes, to replicate the virus nucleic acid in the form of the original and to make more proteins, including the virus proteins that coat and protect the nucleic acid as a fully packaged infectious agent.

In viruses the directing nucleic acid can be of either the RNA or DNA type; in cells the genetic units of the nuclear chromosomes are exclusively of the DNA type. The arrangement of subunits in these tapelike strands of

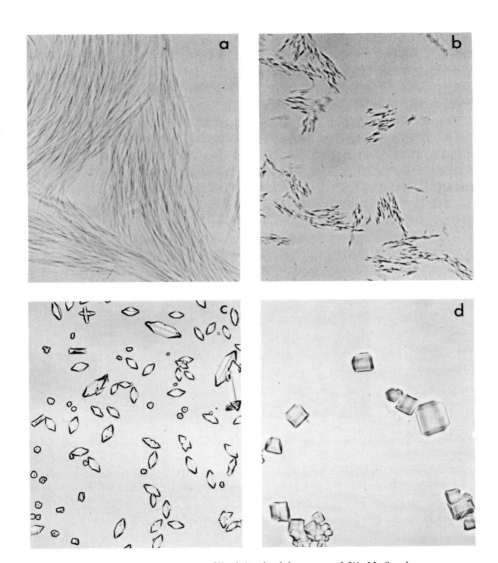

Figure 9.2. Some plant viruses crystallized in the laboratory of W. M. Stanley.

(a) Tobacco mosaic virus; (b) green aucuba virus; (c) tomato bushy stunt virus; (d) tobacco necrosis virus. The rod-shaped viruses presented in (a) and (b) are in paracrystalline arrays, unlike the spherical viruses that aggregate in three-dimensional regularity, as in the crystals in (c) and (d). Heparin method.

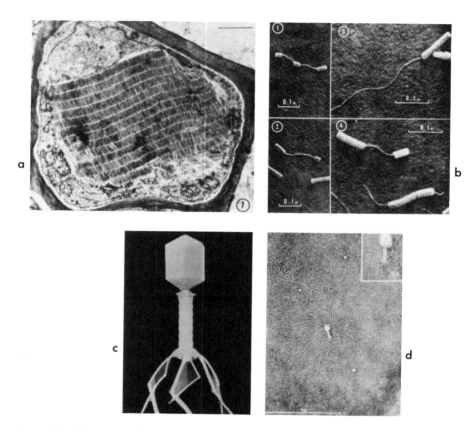

Figure 9.3. The morphology of two intact and disrupted viruses examined in the electron microscope.

(a) Monolayers of stacked rods of tobacco mosaic virus (TMV) within an infected plant cell, courtesy of H. E. Warmke and J. R. Edwardson; (b) partially unraveled TMV lacking some protein subunits and revealing uncoated filaments of nucleic acid (RNA), courtesy of M. K. Corbett; (c) the very complex T4 bacteriophage, a virus that infects certain intestinal bacteria. The picture is of a model exhibited at the Seattle World's Fair, courtesy of L. M. Kozloff; (d) T4 bacteriophage with a disrupted head, liberating a single thread of nucleic acid (DNA), courtesy of A. K. Kleinschmidt.

nucleic acid determines, via an elaborate translation mechanism, the quality of the protein synthesized, that is, the arrangement of many different amino acid subunits in linked linear sequences that are characteristic of particular proteins. The nucleic acids can be said to determine the synthesis of proteins (and enzymes); the enzymes are in their turn instructed by the nucleic acids to duplicate the latter. How did all this begin; how were cells first made? As noted earlier, we must ask: Did the nucleic acids or proteins come first? Oparin had developed his theories before the knowledge of the crucial informational role of the nucleic acids had been developed, when it was thought by biochemists generally that these substances were too simple to have such a role. His major work relied on the potentialities of the proteins, whose abilities to aggregate and to interact with other substances in salt solutions demonstrably led to the formation of tiny sacs containing differentiated protein complexes. Such spherules can even be shown to undergo budding and fission. Oparin devised elaborate schemes to show that these tiny sacs would survive if they assembled like units and grew within their boundaries, and following such synthesis, multiplied by division. Survival and stability required such dynamic synthetic behavior, and only those spherules or cells would survive which actually selected enzymatic proteins capable of synthesis and replication.

Trying to Put Cells Together
How did such cells develop their information-rich hereditary systems? Oparin and Fox have suggested that metabolism in the spherules led to the evolution of the nucleic acids, the information repositories, within the tiny cells-to-be. The analogy has been given that giant enterprises usually begin with small factories; the development of a leading administration and board of directors is most often a late evolutionary stage, at least in the United States. However, the younger generation of working biologists, brought up with the primary informational role of the nucleic acids in their bones, doubts that protein systems and duplication could have begun independently of the nucleic acids. These, they say, must have been first; after all, the start and building of giant enterprises in Oparin's homeland do not begin without the instructions of a Central Committee. I have been amused to note that the emblem of the Biochemistry Congress in Moscow in 1961 was a tiny cell, stuffed with enzymes and smaller substances processed in metabolism but lacking

a nucleus and its key components, the nucleic acids, whereas the emblem of the 1964 Congress held in New York exclusively celebrated the informational molecule, double-stranded DNA. I do not know if these emblems, reproduced in Figure 9.4, were consciously intended to reflect the geographic centers of the two major theories mentioned earlier. More importantly, we have not had much progress on this apparently key question in the past decade. In 1963 Haldane postulated the polymerization or zipping together of nucleic acid subunits as primary events and suggested a coupling of such subunits with amino acids by a simple polymerization mechanism. "I want to suggest that the initial organism may have consisted of one so-called 'gene' of RNA specifying just one enzyme, a very generalized phosphokinase, which could catalyze all the above reactions."

The Oparin hypothesis makes use of some long-known, remarkable properties of the interactions in water of protein, nucleic acids, and some other large, charged molecules, such as the sugar-containing gums. Not only do these aggregates separate into tiny bounded sacs, but the sacs have numerous smaller sacs of different compositions within them. These artificial products, known to early chemists as "coacervates," indeed begin to look like plant and animal cells, which also contain tiny membrane-bounded sacs or "organelles." I wonder if this superficial resemblance between the artificial "co-

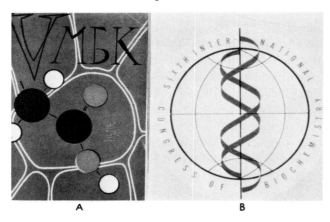

A   **B**

Figure 9.4. The emblems of two successive Biochemistry Congresses.

(a) Moscow, 1961—a cell without nucleus and nucleic acid; (b) New York, 1964—doublestranded helical DNA without a cell.

acervates" and the natural cells has not served to obscure the problem of the origin of cells on the one hand, and on the other, to mask the real problem of the natural origin of the sacs within true cells, the origins of the organelles. The true nature of the latter problem has only begun to emerge within the past decade, and I wish now to consider this matter, which may give us some new insights concerning the nature of man.

The Discovery of Cells and a Cell Organelle, the Nucleus

The concept that plants and animals are made of tiny cells, a theory usually associated with the names of T. Schwann and M. J. Schleiden who worked in the mid-nineteenth century, stems from a long history of research. This began in 1664 with the work of Robert Hooke, who described the outlines of residual cells seen in petrified wood. Some ten years later, just about three hundred years ago, Leeuwenhoek began his description of the use of the microscope. In the 1830s several decisive discoveries emerged from work with the microscope, among which were the discovery of cells within animal tissues and the recognition of the existence of nuclei within plant and animal cells. The latter, of course, provided the basis of the discussion (and much work) on the role of this organelle, on the problems of cellular organization and of the division of labor within cells. As we know, this discussion culminated in the recognition of the nucleus as the site of the genes, whose molecular nature as DNA provides the store of information for the synthesis of proteins.

Some Smaller Organelles in Animal and Plant Cells

Plant and animal cells also contain other saclike inclusions, or organelles, outside the single, generally central, relatively large nucleus. One of these types of inclusion is called a mitochondrion, and several thousands of these tiny particles, which resemble bacteria in size at least, may be present in cells. This is exemplified in the Figure 6.3 of Sinsheimer's chapter. These microscopic organelles have two layers of membranes, the inner of which supports the proteins which appear to fulfill the main function of the mitochondria. This function is that of the respiration of the cells and the useful handling of the energy liberated in the stepwise oxidation of foodstuff. The utilization of oxygen by this organelle is an obviously crucial activity in the life of plant and animal cells; without it they would be largely anaerobic and relatively inefficient in generating energy for synthesis and growth.

Plant cells uniquely contain an additional organelle, the chloroplast, which uses its green pigment to take up light energy and to convert carbon dioxide and water into glucose, liberating oxygen in the process. Indeed, it is believed that most of the oxygen in our atmosphere stems from this source. The primitive organisms were probably anaerobic bacteria and eventually evolved a mechanism for photosynthesis and oxygen production. Bacteria possessing the ability to produce oxygen in this way are called blue-green algae. As we shall see later, the evolution of these forms and the production of oxygen established the conditions for the metabolic transformation of bacteria into the so-called higher plant and animal cells.

It has been found in the past few decades that bacteria do indeed differ chemically from higher cells in important ways. For example, their synthetic mechanisms function almost entirely without atmospheric oxygen, presumably as the residue of their primitive origins, as noted by Haldane. Not only are they simpler in structure but they also lack many substances whose synthesis requires oxygen. After the development of blue-green algae and atmospheric oxygen, the latter substance was used to make many new compounds found only in higher cells, substances that are building blocks in the many new structures of higher cells. To invoke Lucretius,

For nothing is born in the body in order that we may be able to use it.
But, rather, having been born, it begets a use.

Chemical evolution resulted in the formation of primitive cells, of which present-day bacteria are possibly the closest biochemical survivors. Biochemical evolution among the bacteria led to the development of photosynthesis, one form of which, among the blue-green algae, culminated in the production of oxygen. This latter metabolic event eventually produced important geological changes, such as a modern, planetary atmosphere with oxygen. The availability of this reactive gas has in its turn permitted new directions in biochemical evolution, permitted the formation of new compounds capable of organizing in new ways, capable of assembling into the first prototypes of higher cells. But how did this final step occur; how did the higher cells appear with all the organelles that are missing among bacteria and blue-green algae?

Do Bacteria and Algae Live within Higher Cells?

For decades biologists considered that this required the relatively direct
conversion of a bacterium that lacks a nucleus, mitochondria, and chloroplasts into a nucleated higher cell, containing the extra nuclear organelles.
However, new knowledge concerning the biochemical differences between
bacteria and nucleated cells, as well as data on the composition of the organelles, has confronted us with a major evolutionary alternative. It was discovered early in the 1960s that the chloroplasts of plant cells and the mitochondria of both plant and animal cells contain their own DNA, their own
chromosome. Several thousand scientific papers later we know that this
organelle DNA determines the synthesis of numerous organelle components,
including the machinery for synthesizing proteins within the organelle. The
fact is these well-organized sacs begin to look like bacteria (in the case of
mitochondria) and blue-green algae (in the case of chloroplasts). Is it possible that they actually represent the modern remnants of bacterial and algal
invasion in the formation of higher cells, that in this evolutionary development several kinds of cells have learned to live together in a single cellular
format to the mutual benefit of the various individual cells, that is, sym-
biotically? Is it possible that in this sense all higher cells are similar to the
lichens, the dual organisms that are composed of an alga and a fungus? If
this were true, how shall we think of man, who has no longer evolved to his
pinnacle merely by direct descent, but is also comprised of the combined
bodies and genetic experiences of several of the most lowly organisms, which
are still detectable within him?

When mitochondria were detected at the end of the last century by R. Altmann, he described them as respiratory particles resembling bacteria. In fact
he suggested that they contain genetic information and that they divide.
The debate over his proposal that mitochondria are derived from the early
ingestion of primitive symbiotic bacteria was sufficiently acrid and damaging so that some thought it contributed to the early death of this talented
scientist. We can note that the dissection of organisms as an approach to
the understanding of the details of the biological and biochemical division
of labor has been attacked at every stage in this history of biology, as a "re-
ductionist" destruction of the meaning of the whole organism. In the first
half of this century, the tools were not ready to explore Altmann's vision
of the nature of these organelles. At this time, biologists and biochemists

have been in hot pursuit for a decade, picking up obvious trails and developing new leads. The chase has proved to be unexpectedly complex and difficult, as we shall see.

The Complexity of Animal and Plant Cells

It has been possible to relate the size of DNA and RNA to the size and numbers of proteins that the nucleic acids determine. It appears that the proteins found in mitochondria and chloroplasts are far too numerous for the amount of DNA present in these organelles. As a matter of fact, it has been shown that at least 80 percent of the proteins present in the organelles are determined by nuclear genes and are constructed in large part outside of the organelle. On the other hand, modern biochemical techniques have demonstrated that major portions of the internal structure, that is, various nucleic acids and some other proteins, are determined by the organelle DNA, as mentioned earlier.

Most strangely, some of these proteins, essential to the performance of respiration in the case of the mitochondria, or essential to the performance of photosynthesis by the chloroplasts, were found to be determined by the cell nuclei. Thus, the organelles and their most characteristic functions are determined by at least two spatially separated different sets of genetic units. Normal cell physiology must then be thought of as the interaction of several discrete packages of genes. To the virologist this is relatively old hat, because, whenever we study a virus-infected cell we are studying the interplay of viral and cellular chromosomes. Indeed some virus-induced proteins are comprised of subunits, some of which are determined by the host and some by the viral chromosome.

This is all very interesting, but how was this confusing accommodation developed? If invasion and symbiosis had occurred, it appears that some bacterial or algal genes essential to the microbial functions were somehow transferred to the nucleus. On the other hand, if we think that the evolution of a symbiotic association does not explain our higher cells, we can imagine that the original complement of nuclear genes may have split off bits of itself. This fragmentation and subsequent self-enclosure must have provided advantages to the cell, advantages that have been perpetuated.

And there we are. The cell is far more complicated than we had imagined, and the problems of the origins of modern cells, although they can now be posed in fairly clear terms, will require the systematic interaction of many approaches: genetics, biochemistry, and many other disciplines required for the study of cells before we can resolve the possibilities we have now delineated.

How Do Copernican Revolutions Develop?

Our editor has defined a Copernican revolution as the perception of a new truth in the face of tradition and resistance. We have, in the course of our compressed history, seen all too many instances in which large and influential groups of scientists themselves have asserted the position of tradition and demonstrated resistance to new ideas, new approaches, new forms of experimentation. We have seen this in the experiences of Pasteur and Altmann, in the reactions to descriptions and chemical explanations of virus structure and behavior, of hereditary units, and so on. It appears that few among us have examined our own heritage of data and assumptions, indeed of "revealed law," provided by our schools and teachers. Even fewer among us have developed new compelling and incontrovertible experiments to sweep unwarranted belief into the dust heap of history.

In developing his calculations, Copernicus included some data that he had collected himself. We can suggest then that Copernican revolutions themselves have had an evolution, in which the possibilities of discerning truth contrary to tradition have depended more and more on the acquisition of new data, on new kinds of experiments, on the detection of facts that cannot be denied or buried. Whereas pre-Copernican man believed there was little point to examining the earth and the stars in great detail, post-Copernican man, embodied in some few singular scientists, has evolved a belief in the importance, the necessity, of doing precisely the opposite, has evolved the attitude that a question is to be answered both by an examination of the real world and by an experiment. Thus, in our own era, we have many potential Copernican revolutions in biology, with many questions hanging fire, with many experiments needed to tell us who we are and how we came to be.

Notes

1. However, some others become quite exercised at the suggestion that we have given the search for extraterrestrial life a fair test, even in our initial ventures to the moon. Indeed, they resent the neglect of many important biological qustions by the space program. If our medieval fathers wished to be central, many of us do not wish to be alone, to be as unique as that implies. The idea that there are probably earthlike planets containing earthlike, even manlike, life associated with other suns, receives general lip service, but it may take a very long time before we can obtain some data relevant to this question.

2. In a treatise of domestic medicine and cookery, edited in 1824 by my favorite biochemist, Thomas Cooper, I find an essay by M. Appert first published in 1810 by the French Ministry of the Interior:
"We have just seen, from all that has been said, that alimentary substances, in order to be preserved, should be without exception, subjected to the application of heat in a water bath; after being rigorously excluded from all contact with air, in the manner, and with the precautions already indicated.
"This method is not a vain theory. It is the fruit of reflection, investigation, long attention and numerous experiments, the results of which for more than ten years, have been so surprising, that notwithstanding the proof acquired by repeated practice, that provisions may be preserved two, three, and six years, there are many persons who still refuse to credit the fact.
"Medicine will find in this method the means of relieving humanity, by the facility of meeting everywhere, and in all seasons, animal substances, and all kinds of vegetables, as well as their juices, preserved with all their natural qualities and virtues; by the same means it will obtain resources infinitely precious in the production of distant regions, preserved in their fresh state.
"From this method will arise a new branch of industry, relative to the productions of France, by their circulation through the interior, and the exportation abroad, of the produce with which nature has blessed the different countries.
"This method will facilitate the exportation of the wine from many vineyards: wine which can scarcely be kept a year, even when not removed from the spot, may hereafter be preserved many years though sent abroad.
"Finally this invention cannot fail to enlarge the domain of chemistry, and become the common benefit of all countries, which will derive the most precious fruits from it.
"So many advantages, and an infinity of others which the imagination of the reader will easily conceive, produced by one and the same cause, are a source of astonishment."

3. In the ensuing decades both theoretical and experimental work developing from the Oparin-Haldane theory led to many symposia, culminating in the first meeting of both of these scientists at a small conference in Wakulla Springs, Florida, in 1963. The meeting had been sponsored by NASA, whose Exobiology representative opened the discussion with an economic judgment disproved within a decade: "The beginning of life has an obvious and compelling fascination for all humanity: All wish to know and, I believe, are prepared to pay the price."
   At this meeting in which a number of eminent biochemists participated actively ("Prebiological chemistry has not always been regarded as fully respectable by all conventional biochemists."—S. W. Fox), Haldane contributed his last theoretical paper on this subject, attempting to blueprint the minimal components for the first organism. Haldane, who was a central figure in modern biology for fifty years and had made numerous important contributions to genetics, evolutionary theory, biochemistry, physiology and biometry, died in India approximately a year later.

4. A British x-ray crystallographer, J. D. Bernal, who first clarified the problem of the crystalline arrangement of preparations of this virus, later turned his attention to problems of the origin of life. He has presented the idea that certain ordered inorganic substances, such as clays, which themselves can organize spontaneouly in relatively specific crytalline arrays, might have, while sitting at the bottom of a warm, evaporating, reactive soup, that is, a pond, directed the syntheses of the specific chemical units found in living things.

Stanley L. Miller

10 The First Laboratory Synthesis of Organic Compounds under Primitive Earth Conditions

It is now generally accepted by scientists that life arose on the earth early in its history. The sequence of events started with the synthesis of simple organic compounds by various processes. These simple organic compounds reacted to form polymers,[1] which in turn reacted to form structures of greater and greater complexity, until one was formed that could be called living.

This is a relatively new idea, first expressed clearly by A. I. Oparin in 1938 [2] (a Russian version of his book appeared in 1924 but received little notice). The first synthesis of organic compounds under primitive earth conditions was not done until 1953. There had been a great deal of speculation prior to this on the origin of life, but much of it was incorrect, incomplete, or not convincing. An example of an incorrect hypothesis was the proposal that a living organism arose spontaneously by an extremely improbable event while the atmosphere of the earth was essentially the same as now; such an organism would have had to be like an alga and to be able to synthesize all its constituents from carbon dioxide, water, and light. The reason for making such a proposal was that there was no known mechanism for making organic compounds in the earth's atmosphere except by photosynthetic organisms. When a mechanism became available for synthesizing organic compounds in the atmosphere of the primitive earth, proposals of this type involving extremely improbable events were no longer seriously considered.

It is not strictly correct to say that the 1953 experiments were the first organic compound synthesis under primitive earth conditions. Many of the reactions previously studied by organic chemists have turned out to be important primitive earth synthetic reactions. This includes the first synthesis of an organic compound from inorganic materials,[3] the synthesis of urea from ammonium cyanate by F. Wöhler in 1828. But the motivation of such studies was to synthesize organic compounds and not to understand what happened on the primitive earth. In addition, a number of scientists had attempted to produce organic compounds under primitive earth conditions, but they had assumed the wrong primitive atmosphere and they obtained no organic compounds at all or in extremely small yield. So the first successful prebiotic synthesis experiment was done in 1953, and I will explain how these experiments came about.

Department of Chemistry, University of California, San Diego, La Jolla, California 92037.

At the beginning of my senior year (1950) at the Berkeley campus of the University of California, I made the rounds of the professors I knew to find out which graduate schools in chemistry were the good ones. Perhaps this was earlier than necessary, but the choice of a graduate school is an important one. I received some frank evaluations of various graduate schools and was advised to go to the best all-around chemistry department unless I was interested in a particular field of chemistry or professor. But there was also the essential factor of financial support. At that time, almost the only source of support for graduate students was a teaching assistantship. Of the universities with good chemistry departments, only the University of Chicago and the Massachusetts Institute of Technology had an adequate teaching assistantship. I applied to both these places, with the University of Chicago being my first choice, and I was very pleased when a telegram arrived in February 1951 saying I was accepted there.

I arrived in Chicago in September 1951, registered for the usual courses, and naturally attended the chemistry department seminars. About the middle of the fall semester, Urey gave a seminar on the origin of the solar system, in which he pointed out that the solar system is reducing (that is, there is an excess of molecular hydrogen) except for the earth and the minor planets (Mars, Venus, and Mercury), which are more oxidized, with the earth's atmosphere being highly oxidized. The earth's atmosphere contains carbon dioxide (CO_2), molecular nitrogen (N_2), water (H_2O), and molecular oxygen (O_2); it is highly oxidized because of the presence of molecular oxygen. It seemed very likely to Urey that the earth was also reducing when it was first formed. A reducing atmosphere would contain methane (CH_4), ammonia (NH_3), water (H_2O), and molecular hydrogen (H_2); this is just the atmosphere present on Jupiter and Saturn, except that the water has been frozen out, and on Uranus and Neptune, where both the water and ammonia have been frozen out.

A reducing atmosphere would be a favorable place to synthesize organic compounds, Urey thought, and the organic compounds synthesized would form the basis to make the first living organism on the earth. The present oxidizing atmosphere is not a favorable place to synthesize organic compounds, and a number of experiments in other laboratories which attempted to synthesize organic compounds from the constituents of the earth's atmosphere (usually from carbon dioxide and water) were total failures or gave very disappointing results. Urey suggested that someone should do an experiment to test the feasibility of organic compound synthesis in a reducing atmosphere.

Urey's point immediately seemed valid to me. For the nonchemist the justification for this might be explained as follows: it is easier to synthesize an organic compound of biological interest from the reducing atmosphere constituents because less chemical bonds need to be broken and put together than is the case with the constituents of an oxidizing atmosphere. Alternatively one can say that methane is already an organic compound; therefore, it should be easier to synthesize a larger organic compound from methane in most, but not all, cases than to synthesize it from carbon dioxide, which is inorganic, as a starting material. This point is very obvious today, but it was not so clear then.

After this seminar someone pointed out to Urey that in his book Oparin had discussed the origin of life and the possibility of synthesis of organic compounds in a reducing atmosphere. Urey's discussion of the reducing atmosphere was more thorough and convincing than Oparin's; but it is still surprising that no one had by then done an experiment based on Oparin's ideas.

It was about this time that I was thinking about what type of chemistry I would do for my dissertation. As an undergraduate I had done an experimental senior research project and found experiments to be time-consuming, messy, and not as important, I then thought, as theoretical work. The synthesis of organic compounds from the reducing atmosphere constituents promised to be a very messy and time-consuming experiment. So I decided to look for a theoretical thesis topic.

I went around to the professors doing theoretical work and listened to a great many interesting projects, but the one that seemed most attractive and fundamental was suggested by Edward Teller. This was to figure out how the elements could be synthesized in very hot stars. After about six months of looking at this problem and just as I was beginning to understand it and make a little progress, Teller announced to me one day that he was leaving the University of Chicago to set up a laboratory in Livermore, California. Although Teller said I could continue the project in Chicago while he was in California, this did not seem a good way to do things, and the chemistry department advised against this. Subsequently others worked on the problem of element synthesis in stars, and the generally accepted method for their synthesis is along the lines of the project I had just started.

It was now September 1952, and I was confronted with the problem of finding a new thesis topic. I now remembered Urey's seminar and the more I thought about it, the more reasonable it seemed, and the more fundamental. The problem was not just how to synthesize organic compounds on the primitive earth, it was the first step in understanding how life started on the earth. This was every bit as fundamental as the origin of the elements. So I went to Urey's office and said that I wanted to do an experiment on organic synthesis in a reducing atmosphere.

Urey seemed reluctant to go along with this and suggested that measuring the amount of the element thallium in meteorites, which was important for reasons I can no longer remember, would be a good thesis topic. I explained that thallium was very nice, but I was more interested in organic compound synthesis and hinted that it was either organic compound synthesis or I would go to another professor for a thesis project. Urey realized I was determined and then explained his reservations about the project I wanted to do. It was not that he did not believe in it, the problem was that it was an unusual project, and the chances for success were quite small. A professor has the responsibility to give a student a project that can be finished in a reasonable length of time (two to four years) so that he can finish his Ph.D. He then agreed that I could try the organic compound synthesis for six months or a year. If nothing came of the project by this time, I would have to switch to a more conventional project with a greater probability of success.

He suggested that I read the paper on which his lecture had been based,[4] Oparin's book, and a book on biochemistry of which I knew next to nothing. A few weeks later I went back to Urey to discuss what to do next, and we quickly decided that ultraviolet light and electric discharges must have been the most abundant energy sources on the primitive earth, and that an electric discharge would be best to try for the first experiments. So I repaired to the library to find out what was known about electric discharge reactions. I found considerable work had been done on electric discharge reactions of

methane alone, but little had been done on mixtures of methane with nitrogen or oxygen compounds. Some work had been done producing hydrogen
cyanide from the methane and ammonia and from methane and nitrogen,
but this did not make an impression on me then.[5] The action of electric discharges on methane seemed to give a random mixture of hydrocarbons, so I
was convinced that I would at least get hydrocarbons as products and probably a random mixture of organic compounds containing nitrogen and/or
oxygen; and this random mixture of organic compounds would contain,
hopefully, a trace of amino acids and other compounds of biological interest.

I went back to Urey and reported what I had found in the literature and
my feelings about a random buildup of amino acids. We decided that amino
acids were the best group of compounds to look for first, since they were the
building blocks of proteins and since the analytical methods were at that
time relatively well developed. We then designed a glass apparatus that contained a model ocean, an atmosphere, and a condenser to produce the rain.
We finished our conversation by reassuring each other that we ought to be
able to find at least traces of amino acids from the action of the electric discharge on the model primitive atmosphere.

I went back to the lab and drew out the apparatus carefully enough for
the glassblower to build it. The apparatus is shown in Figure 10.1. I then
looked at the spark I was going to use—a Tesla coil used for detecting leaks
in vacuum lines—and read the instructions that said it produced 60,000 volts
of high-frequency current. As I looked at the glass apparatus I began to have
second thoughts about its design. I felt that this high voltage in the presence
of the water vapor might be dangerous and decided to interchange the condenser and spark from that shown in Figure 10.1. It took about a week to
get everything ready. It was particularly important to make sure that there
were no leaks in the apparatus because air and hydrogen or methane form
explosive mixtures. I filled the apparatus with the postulated primitive atmosphere, water, methane, hydrogen, and ammonia, turned the spark on
and let it run overnight. The next morning there was a thin layer of hydrocarbons on the surface of the water, and after several days the hydrocarbon
layer was somewhat thicker. So I stopped the spark and looked for amino
acids by one-dimensional paper chromatography (Figure 10.2a).

Paper chromatography is a simple technique for separating compounds
by differences in the rate of migration of compounds, in this case amino

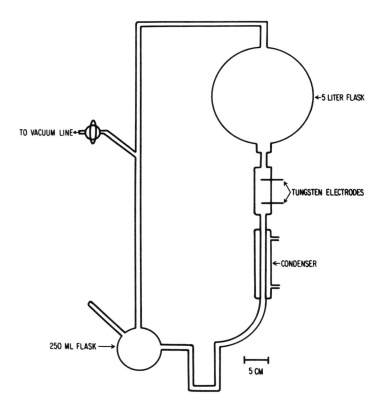

Figure 10.1. The apparatus used in the first electric discharge synthesis of amino acids. It is made entirely of glass except for the tungsten electrodes. (From S. L. Miller, "A Production of Amino Acids under Possible Primitive Earth Conditions." *Science*, Vol. *117*, 1953, p. 528.)

acids, on a piece of filter paper. The amino acids are colorless, but most of them give a purple color when they react with a compound called ninhydrin.

I was not particularly disappointed when no amino acids showed on the paper when sprayed with ninhydrin, since only traces at best were expected. I then thought about the interchanging of the spark and condenser and decided that this had been a mistake. The spark was not really dangerous. It has a high voltage but low current and can be held against one's finger with only a tingling sensation. So I took the apparatus back to the glassblower, and he put it back into its original design.

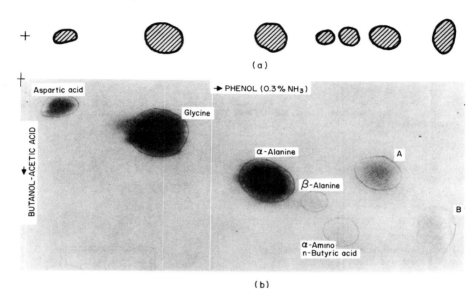

(a)

(b)

Figure 10.2. (a) Drawing of a one-dimensional paper chromatogram.

One-dimensional paper chromatography involves putting a drop of an amino acid solution at the origin (+). The solvent (phenol saturated with 0.3% aqueous ammonia in this case) is allowed to run down the strip of filter paper to the end. Different amino acids travel at different rates relative to the solvent, thereby separating them.

(b) Photograph of a two-dimensional paper chromatogram.

Two-dimensional paper chromatography involves putting a drop of amino acid solution at the origin (+) and allowing the solvent (butanol-acetic acid) to run down the sheet of filter paper as in one-dimensional paper chromatography. The paper is then dried, rotated clockwise through 90°, and the second solvent (phenol) is allowed to run down the paper, separating the amino acids over the sheet of paper. (From S. L. Miller, "A Production of Amino Acids under Possible Primitive Earth Conditions." *Science*, Vol. *117*, 1953, p. 528.)

Again after about a week's work getting everything ready, I filled up the apparatus with the same mixture of gases and turned the spark on, keeping the heating coil on the 500-ml flask at a low heat. The reason for the low heating rate was that a closed system was being heated, which can build up dangerous pressures. After two days I decided to see what had been produced. This time there were no visible hydrocarbons, but the solution was a pale yellow. I concentrated the solution and ran a paper chromatogram. This time I found a small purple spot on spraying with ninhydrin that moved at the same rate as glycine, the simplest amino acid.

The next day Urey returned from a lecture tour, and I went to him with the paper chromatogram. Before I could show him the chromatogram he told me about one of his lectures. He had described his ideas about the early solar system and the primitive reducing atmosphere and mentioned that I was doing an experiment. During the question period someone in the audience had asked him in a voice full of doubt, "and what do you expect to get?" Urey replied "Beilstein." *Beilstein's Handbuch der Organische Chemie,* now over 100 volumes, contains a list of properties and reactions of every organic compound that has ever been synthesized. Urey's reply meant that the electric discharge would be expected to produce a little of everything.

I then showed him the paper chromatogram with the faint glycine spot and commented that maybe I was doing better than "Beilstein" (a random mixture of organic compounds) although it was too early to tell. He was extremely pleased with this result. I said that I would repeat the experiment, this time letting the spark run for a week.

I set the apparatus up again and this time boiled the water more vigorously, making sure dangerous pressures did not build up, and started the spark. In the morning when I looked at the apparatus the solution looked distinctly pink. Then I noticed that the heater, which had an exposed coil, was glowing red and that the color probably came from the reflection of the coil. I removed the heater and the solution was still pink. My immediate thought was porphyrins (the red color in blood is due to an iron porphyrin), and I rushed over to Urey and brought him back to see the color, which he viewed with as much excitement as I did. As the sparking continued, the solution became at first a deeper red and then a yellow-brown which obscured somewhat the red color.

At the end of the week, I removed the solution and did a little processing on it and then ran a two-dimensional paper chromatogram (Figure 10.2b). Two-dimensional paper chromatography involves letting one solvent separate the amino acids as in one-dimensional paper chromatography, followed by letting a second solvent run at right angles to spread the amino acids over the entire square sheet of filter paper. This time seven purple spots showed up on spraying with ninhydrin. Three of these amino acids were strong enough and in the correct position to be identified as glycine, α-alanine, and β-alanine. Two spots were considerably weaker in color but corresponded to aspartic acid and α-amino-n-butyric acid. I was less confident of these identifications and decided to refer to them as tentatively identified. The two remaining spots did not correspond to any of the amino acids that occur in proteins or any known amino acids I then had on hand, so I simply referred to them as A and B.

Paper chromatography is not a very quantitative procedure, but I estimated that I had at least 10 mg of amino acids (subsequent quantitative experiments showed the presence of 110 mg of amino acids). To play it safe, I stated that the total yield of amino acids was estimated to be in the miligram range.

I showed all this to Urey and said that I thought I had enough to write a short paper on the results and he agreed. It was now December 1952 and I had been working on the project only three and a half months. The results were strikingly successful, and there was no question about my being able to continue on the project and to write my dissertation on it. The experiment needed repetition at least once and preferably twice before publication, but I could do this during the writing and during the time between the submission of the paper and its appearance. So I wrote up a short paper putting Urey's name as a coauthor and brought it to him for approval. He first said that I should take his name off since I had done this work largely by myself, and if his name was on it I would receive little or no credit. This was extremely generous of him since in chemistry the research director almost always appears on the papers resulting from the student's thesis.

We then decided that *Science* would be the best journal for the paper rather than the *Journal of the American Chemical Society*, the most prestigious chemical journal, because of the paper's interest to several fields of science. I made the necessary revisions of the paper and sent it off a few days later.

I had by this time repeated the experiment once and began the second repetition of the experiment. The results in both repetitions were the same as before, so I was extremely confident of the results. After four months the paper appeared in *Science* (May 15, 1953).[6]

I had expected that this paper would generate considerable interest, and that *Time* magazine would probably have a piece about it in their science section since they liked this sort of thing. But the reaction to the paper startled me. It began with an article by Earl Ubell in the *New York Herald Tribune* and an article and editorial by Walter Sullivan in the *New York Times,* and spread to most newspapers across the country, including considerable mention in the local Chicago newspapers. There was even a Gallup poll asking whether people thought it possible "to create life in a test tube." (The results were 9 percent yes, 78 percent no, and 13 percent no opinion. The results of a poll today would probably give a much higher percentage of yes.) After a while the furor died down and I was able to devote full time again to the experiments.

There was a lot to do, and I spent another year and a half at Chicago completing my dissertation and a year after that at Cal Tech finishing up the experiments. The story would be too long to go into complete detail on everything I did, but I will discuss some of the highlights.

The red color observed the first day of the "successful" experiment turned out to be elusive, and the color has never been identified. It is the only aspect of the experiment that has not proved reproducible. I was able to observe the color in about six experiments, but when I rebuilt the apparatus the color no longer appeared. The red color was adsorbed on the silica that was leached from the glass. The Pyrex glass of the first apparatus was particularly rich in trace elements, and this is apparently why the red color appeared only in the original apparatus. Professor Klüver at the University of Chicago looked for porphyrins in the discharge products, especially the red color, but he could not detect any.

One day Urey came to me and said that he was concerned that the amino acids may have been produced by bacteria. We had discussed this possibility before, but it was clear that someone had expressed his concern and doubts to Urey about the experiment. Urey said that if I had made a mistake it would be far better for me to find this out and retract the *Science* paper than if someone else found this out.

I was not really concerned about bacteria since most of the apparatus was too hot for bacteria to live in, the amino acids were not the kind and distribution obtained from bacteria, and it was evident organic compounds were being made by the spark (some of the yellow color could be seen to be made on the tungsten electrodes). But it was best to settle the bacteria problem once and for all. I filled up the apparatus with the reduced gases, completely sealed the apparatus and heated it in an autoclave for 18 hours. I then ran the spark and showed the yield was the same as the runs where the apparatus was not autoclaved. Since 15 minutes heating in an autoclave is usually sufficient to kill all bacteria, the 18 hours was overdoing it a bit, but the bacteria issue has not been raised since.

It was important to obtain quantitative values for the amino acids as well as to find out what else was present in the mixture of organic compounds produced. To do this an elaborate separation scheme involving various ion exchange resins was developed. I was able to separate the different amino acids and was also able to separate out the aliphatic and hydroxy acids. The scheme was so thorough that I would have found any single compound present in greater than 0.5 percent yield (greater than about 10 mg).

The problem of positive identification of the compounds arose. Urey sent me to one of the professors of organic chemistry (Weldon Brown) to make sure I was doing things correctly. He approved of what I had done so far but inquired if I had obtained any melting points of my amino acids. At that time the usual method of positively identifying organic compounds was to obtain a melting point of the compound or a derivative of it, show that an authentic sample of the compound has the same melting point, and most important, that the unknown and authentic compound when mixed has the same melting point. (Two different compounds both melting at, for example, 130° will melt at a considerably lower temperature when mixed.) I replied that I had not done this and that it would be difficult because of the small amounts of material available. Professor Brown said that in spite of any difficulties I should get the melting points.

This was several months work, but it was worth the effort, since the identifications were then firmly established. In subsequent work by others in this field such care in identifying compounds was not taken, and a number of embarrassing errors have been made.

I was able to prove unequivocally by melting point the presence of glycine, alanine, and β-alanine, which I had "identified" by paper chromatography. The α-amino-n-butyric acid, which I had tentatively identified in the *Science* paper, was proved to be present, but the spot I had tentatively identified as aspartic acid turned out to be iminodiacetic acid (aspartic acid was present, but in much smaller amounts). Compound A turned out to be sarcosine (N-methyl glycine), and compound B turned out to be, in part, N-methyl alanine.

During the isolation and attempt to crystallize a derivative of sarcosine, I ran into some trouble. A compound precipitated that was not the sarcosine derivative, even though I knew that the sample must have been relatively pure sarcosine. The sarcosine derivative was eventually induced to crystallize, but there was the problem of the first set of crystals. A look at the melting point tables and a good guess showed that these crystals might be α-aminoisobutyric acid, and I was able to prove this guess correct. So this amino acid was found by an accident during the crystallization process when I did not even suspect that it was present.

The search for amino acids was very thorough in terms of the techniques then available. Nine amino acids were identified, but about twenty-five others were shown to be present but not identified, mainly because the yields were too small to prepare derivatives for identification. Between 1957 when I reported the final details of these experiments[7] and 1972, the experiment was repeated many times by others using many variations. No one was able to find any amino acids that I had not found, with one exception (serine), although many invalid claims were made. One laboratory spent five years trying to find additional compounds without success. In 1972 I repeated the experiments and using more modern and sensitive techniques, it was possible to identify 33 of the amino acids produced by the spark discharge.[8] It was also surprising that the yields of amino acids from these first experiments are the highest so far reported in any prebiotic experiment of this type.

There was always the question of whether these experiments were an adequate model of the primitive earth. For example, one objection would be that the input of electrical energy was far higher than possible in the primitive atmosphere. This is a complicated topic, but a striking confirmation of the general validity of these experiments came from the analysis of a meteorite.

On September 29, 1969, a meteorite fell in Murchison, Australia. This meteorite was a carbonaceous chondrite, a meteorite that contains organic compounds. Amino acids had previously been reported in other carbonaceous chondrites, but these amino acids were generally believed to be contamination. In the case of Murchison, the meteorite was a freshly fallen one, and the techniques used were the most modern and showed convincingly that the amino acids were indigenous.

The amino acids reported in Murchison in the first paper[9] were glycine, alanine, sarcosine, glutamic acid, α-aminoisobutyric acid, all of which I had found in the electric discharge, the α-aminoisobutyric acid by accident. Valine and proline, which I had not found among the electric discharge products, were also reported. In the second paper[10] on amino acids in Murchison, among the amino acids reported were N-methyl alanine, β-alanine, aspartic acid and α-amino-n-butyric acid. In other words, most of the amino acids I had found in the electric discharge synthesis were in the meteorite. I was stunned at this result. It looked as if the Murchison amino acids were produced by an electric discharge reaction. However, a number of amino acids were found in Murchison that I had not found in the discharge. But it was easy to show on repeating the discharge experiments and using modern analytical techniques that all the amino acids found in the meteorite could also be produced in the electric discharge apparatus.

The synthesis of organic compounds under primitive earth conditions is not, of course, the synthesis of a living organism. We are just beginning to understand how the simple organic compounds were converted to polymers on the primitive earth, but how these polymers organized into the first living organisms is completely unknown. Nevertheless we are confident that the basic process is correct, so confident that it seems inevitable that a similar process has taken place on many other planets in the solar system. The question that is discussed is whether millions of planets in our galaxy, as well as other galaxies, have some form of life on them, or whether it is just a few planets. It is to be expected if there are millions of planets with life that some of them will have civilizations more advanced than our own and may be trying to communicate with us. Radio telescopes have been used to listen for such signals, so far without success, but this listening program is just beginning. We are sufficiently confident of our ideas about the origin of life that in 1976 a spacecraft will be sent to Mars to land on the surface with the primary purpose of the experiments being a search for living organisms.

Notes

1. An example of a polymer is Nylon, in which many molecules of an amino acid ε-amino caproic acid) are linked together forming long chains. Hemoglobin, the protein in blood that binds oxygen, is also an amino acid polymer, but there are twenty different amino acids in this polymer and they are linked together in an exactly specified sequence.

2. A. I. Oparin, *The Origin of Life*. Macmillan, New York, 1938; reprinted by Dover, 1953. A translation of the 1924 pamphlet, *Proiskhozhdenie Zhizni, Moskovskii Rabochii*, Moscow, 1924, is in J. D. Bernal, *The Origin of Life*. Weidenfeld and Nicolson, London, 1967.

3. An organic compound means a compound containing carbon and usually hydrogen and may contain other elements as well. Diamond, carbon monoxide, and carbon dioxide are usually considered to be inorganic compounds. The word organic compound originally meant compounds that occur in living organisms or derived from them. It was thought that organic compounds were fundamentally different from inorganic compounds. Wöhler's synthesis of urea showed that there was no fundamental difference since an inorganic compound could be converted to an organic compound.

4. H. C. Urey, "On the Early Chemical History of the Earth and the Origin of Life." *Proceedings of the National Academy of Science* (U.S.), Vol. *38*, 1952, pp. 351–363.

5. A year later when I made a more thorough search of the literature I found a paper by Löb in 1913 (W. Löb, "Über das Verhalten des Formamids unter der Wirkung der stillen Entladung. Ein Beilrag zur Frage der Stickstoff-Assimilation." *Berichte der Deutschen Chemischen Gesellschaft*, Vol. *46*, 1913, pp. 684–697) that reported the synthesis of an amino acid (glycine) by sparking a mixture of carbon monoxide and hydrogen. This paper had been missed by workers in the field of electric discharge reactions.

6. S. L. Miller, "A Production of Amino Acids under Possible Primitive Earth Conditions." *Science*, Vol. *117*, 1953, pp. 528–529.

7. S. L. Miller, "Production of Some Organic Compounds under Possible Primitive Earth Conditions." *Journal of the American Chemical Society*, Vol. 77, 1955, pp. 2351–2361. S. L. Miller, "The Mechanism of Synthesis of Amino Acids by Electric Discharges." *Biochimica et Biophysica Acta*, Vol. *23*, 1957, pp. 480–489. S. L. Miller, "The Formation of Organic Compounds on the Primitive Earth." *Annals of the New York Academy of Sciences*, Vol. *69*, 1957, pp. 260–274; also in *Proceedings of the First International Symposium on the Origin of Life on the Earth*. Edited by A. I. Oparin et al., Pergamon Press, New York, 1959, pp. 123–135.

8. D. Ring, Y. Wolman, N. Friedmann, and S. L. Miller, "Prebiotic Synthesis of Hydrophobic and Protein Amino Acids." *Proceedings of the National Academy of Sciences* (U.S.), Vol. *69*, 1972, pp. 765–768. Y. Wolman, W. J. Haverland, and S. L. Miller, "Nonprotein Amino Acids from Spark Discharges and Their Comparison with the Murchison Meteorite Amino Acids." *Proceedings of the National Academy of Sciences* (U.S.), Vol. *69*, 1972, pp. 809–811.

9. K. Kvenvolden et al., "Evidence of Extraterrestrial Amino Acids and Hydrocarbons in the Murchison Meteorite." *Nature*, Vol. *228*, 1970, pp. 923–926.

10. K. A. Kvenvolden, J. G. Lawless, and C. Ponnamperuma, "Nonprotein Amino Acids in the Murchison Meteorite." *Proceedings of the National Academy of Sciences* (U.S.), Vol. *68*, 1971, pp. 486–490.

David Hubel

11 Neurobiology: A Science in Need of a Copernicus

The object of neurobiology is to understand the nervous system. For man, this amounts to asking what happens in our heads when we think, act, perceive, learn, or dream. In this essay I shall try to assess our present state of knowledge in neurobiology, in order to ask, at the end, whether anything like a Copernican revolution has occurred, is taking place, or is to be expected.

The two great interlocking branches of neurobiology are neuroanatomy and neurophysiology. Anatomy seeks to describe how the brain is put together, and physiology asks how the various parts work together. Though traditionally the two fields have tended to pursue separate courses, usually housed in separate departments at universities, in fact they are very much interdependent. Most modern neuroanatomists are not content with a simple description of spatial relationships for their own sake but soon go on to ask what the structures are for. A precise drawing of a watch, a printing press, or a television set has little interest in itself, especially if we do not know that the purpose is to tell time, print books, or entertain. Physiology, on the other hand, is simply impossible without anatomy: no one can hope to learn how a watch or printing press works without knowing where the gears or springs are.

In their development, both neuroanatomy and neurophysiology have had to wait until the physical sciences could provide them with several necessary techniques. The nerve cell, or neurone, which is the unit or building block out of which the brain is constructed, is too small to see with the unaided eye, except as a mere speck, and far too small for its signals to be recorded with ordinary wires. Hence to advance beyond the most rudimentary stages anatomy required the microscope, first the light and then the electron microscope, and physiology required the microelectrode. The crowning achievements of the neuroanatomists of the past century have been the recognition that the nerve cell is the basic unit of nervous tissue, and the discovery that nerve cells are interconnected with a high degree of order and specificity. The physiologist, using the microelectrode, has made a strong beginning by learning, in electrical and chemical terms, how the neurone transmits its messages. These two sets of accomplishments by no means tell us how the brain works, but they provide an absolutely essential groundwork.

Department of Neurobiology, Harvard Medical School, Boston, Massachusetts 02115.

The Nerve Cell

Why was it so hard to establish, in the first place, that the nerve cell is the basic unit of nervous tissue? Undoubtedly the main obstacles were the minute dimensions, the fantastic shapes, and the enormous variety of shapes of these cells, plus the fact that the branches of nearby cells are largely intermingled. The word "cell" conjures up in the minds of most students of Biology 101 a shape like that of a brick or a jelly bean, but a neurone may look like an oak tree or a petunia, and it may have processes ten or twenty micrometers in diameter and several yards long. (Even if America does finally adopt the metric system, "yards" somehow seems more picturesque than meters, in describing the distance from a giraffe's head to its big toe.)

To see these nerve cells requires not only a microscope but also a stain to contrast them with their surroundings. Here one comes up against a major difficulty: the cells are normally packed together so intimately that in any one region the branching systems of hundreds of them intertwine in the densest of thickets. Adjacent branches of different cells are separated by films of fluid only about 0.02 micrometer wide, so that virtually all the space is taken up by cells and their processes. Hence when all the cells in any one region are stained, the product, under an optical microscope, is simply a dense and useless smear. The most important single advance, after the microscope itself, came around 1875 when the Italian anatomist Camillo Golgi hit upon an amazingly useful technique. He discovered a method by which, seemingly at random, only a very small proportion of cells are stained at one time, but those few cells are stained in their entirety. Instead of a hopeless morass, a good Golgi stain shows only a few cells, each complete with all of its branches. By looking at many slices of brain stained with this technique, the anatomist can gradually build up an inventory of the various kinds of cells in the tissue he is studying. To this day no one knows how or why Golgi's method works—what makes one cell in a hundred stain completely while the others are quite unaffected.

Golgi's own contributions to neuroanatomy were not otherwise earth-shaking, but his Spanish contemporary, Ramón y Cajal, devoted a lifetime to applying the new method to virtually every part of the nervous system. His gigantic *Histologie du Système Nerveux,* first published in 1911, is recognized as the most important single work in neurobiology up to the present.[1]

Our present picture of a typical nerve cell, due largely to Cajal, is illustrated in Figure 11.1. It consists of a cell body, from which emanate one axon and a number of fiberlike branches called dendrites (Greek *dendron:* tree). The axon, which is often very long, may itself give off branches near its beginning, close to the cell body, and it often branches extensively near its end. Nerve cells vary widely in their general form, as already mentioned, and tend to fall into a large number of classes, of which one is illustrated in Figure 11.1 and several more in Figure 11.2. Two nerve cells in the same class are never identical but are generally about as similar as two oak trees. On a very rough guess the number of classes of cells in the mammalian nervous system might amount to a few hundred, and one class may contain anything from thousands to hundreds of millions of representatives.

One can say, roughly, that the cell body serves for the general upkeep of the cell, the dendrites and cell body receive incoming signals from numbers of other nerve cells, the axon transports a new set of signals to the axon terminals, and the terminals distribute the information to another set of nerve cells.

The transfer of information from one nerve cell to another occurs at specialized contacts between axon terminals (and occasionally dendrites), as suppliers of information, and the dendrites and cell body (and occasionally axon terminals), as recipients. Around Cajal's time a controversy was raging over the question of whether nerve cells were completely separate entities, or whether the axons and dendrites of different cells were joined, with protoplasmic continuity, to form a network or syncytium. The question was an important one, because protoplasmic continuity between cells would mean that the signals generated by one cell would be able simply to pass to an adjoining cell without interruption. If there were no continuity, there must exist a special process for generating signals anew in each cell. Using Golgi's method, Cajal saw large numbers of individual completely stained cells but never anything to suggest a syncytium. Although this evidence was not quite conclusive, it for practical purposes ended the controversy in favor of the first alternative, that nerve cells were quite separate. In recent decades the few remaining holdouts have finally been convinced by the evidence of the electron microscope.

The points of contact between the axon terminals and the dendrites or cell bodies, where signals from one cell exert their influence on the next one, are called *synapses*. Cajal's first achievement, then, was to establish the no-

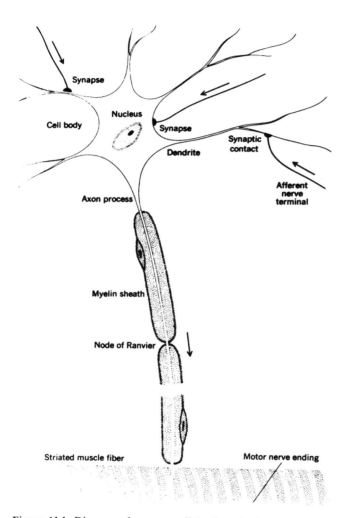

Figure 11.1. Diagram of a nerve cell in the spinal cord, showing dendrites, cell body, axon, and axon terminals.

The terminals of this nerve cell end on muscle fibers. Axon terminals of other cells are shown ending on the dendrites and cell body. From *Nerve, Muscle and Synapse,* by Bernard Katz. Copyright 1966 by McGraw-Hill Book Company, New York. Used with permission of McGraw-Hill Book Company.

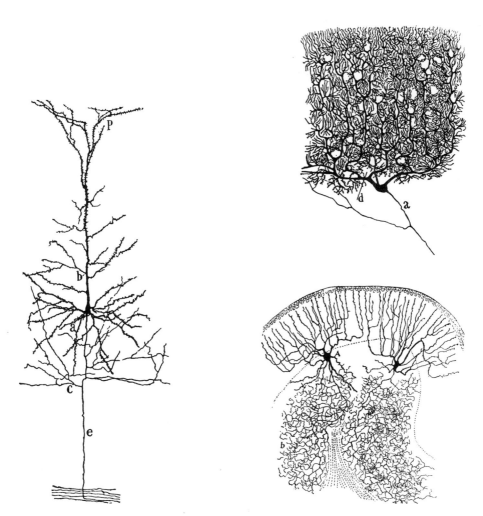

Figure 11.2. Drawings, by Cajal, of several kinds of nerve cells, to illustrate the extreme diversification. (S. Ramón y Cajal, *Histologie du Système Nerveux de l'Homme et des Vertébrés,* Consejo Superior de Investigaciones Científicas, Instituto Ramón y Cajal, Madrid, 1972.)

(a) Pyramidal cell in the cerebral cortex. The distance from C to P is in the order of 1 mm. e is the axon, c represents some of the axon collateral branches. a, b, and P are dendrites.

(b) Cell in the cerebellum. d are dendrites, a is the axon.

(c) Two cells in cerebellum. A is a cell body, b are the axon terminals. The terminals are probably associated with many synapses!

tion of a nervous system containing separate well-defined nerve cells communicating with each other at synapses. Most nerve cells receive signals from hundreds of other cells, and transmit signals to hundreds of other cells.

Cajal's second contribution was possibly even more important. He compiled massive evidence to show that the interconnections between nerve cells, while incredibly complex, are also highly structured and specific: in fact they are not only not random, as has sometimes been supposed, but are the very antithesis of what the word implies. He considered structure after structure, spinal cord, cerebral cortex, cerebellar cortex, hippocampus, optic tectum, vestibular nucleus, habenular nucleus, retina, medial geniculate, lateral geniculate, and so on and on through a list of dozens and dozens of tissues. For each he described exhaustively the architecture, identifying and classifying the various cells and showing, as far as his methods would permit, how they are interconnected. From his time on it was clear that to understand the brain the biologist would not only have to learn how different subdivisions are built up but would also have to discover their purposes and find out in detail how they work as individuals and as groups. Before this could be done, it would be necessary to learn how a single nerve cell generates its signals and conducts and transmits them to the next. Cajal perhaps never explicitly formulated the problem of understanding the nervous system in these terms, but one can hardly survey his work without taking that message from it.

The Physiology of Nerve Cells

The problem of how a single nerve cell generates electrical signals and conveys information to other cells is now reasonably well understood. This has been the great achievement of neurophysiologists over the past three or four decades. The work was done by many individuals: Loevy, Dale, Hodgkin, Huxley, Katz, Eccles, and Kuffler are some of the main contributors. One of the surprising results has been a realization that nerve cells, despite their variety of sizes and shapes, all use the same two kinds of electrical signals, which we term "graded potentials" and "action potentials." In the space available here it is possible to give only a very rough impression of these signals, but a reading of books like those of Hodgkin[2] or Katz[3] will convince the student that both processes are now understood in a rigorous and highly satisfying way.

The entire nerve cell, with its cell body, its long axon, and its multiple branching dendrites, is polarized so that the inside is about 70 millivolts negative to the outside. This is immediately apparent if one plunges a sharply pointed glass capillary tube filled with salt solution into the cell: if the microelectrode is connected to a voltage measuring device, an abrupt negative deflection is seen as soon as the fine tip penetrates the cell membrane. Two properties of the membrane are responsible for this difference in potential. First, the membrane actively transports ions, extruding sodium and bringing in potassium, so that the concentrations of the ions inside the cell are very different from those outside. Second, the ease with which ions flow through the membrane is very different for different ions. How the membrane extrudes or brings in ions selectively and why it is selectively permeable to different ions is still not clear, in molecular terms, but given these two properties, the membrane potential follows according to well-known electrochemical rules.

The nerve cell membrane, then, has a potential across it everywhere, at the cell body, the dendrites, and the axon. It is *changes* in this outside-to-inside "resting potential" that constitute the electrical signals of nerves. If something happens suddenly at any point along the nerve membrane to change the transmembrane voltage, the change will tend to spread quickly in all directions along the membrane, dying out as it spreads. A few millimeters away there is likely to be no detectable signal left. This is the first kind of electrical signal, which is the "graded potential" mentioned earlier. Its properties—for example, the rate of spread and of attenuation—can be predicted from simple physics, provided one knows a few things like the electrical capacity and resistance of the membrane. Its main function is to convey signals for very short distances.

The second type of signal is used especially to convey information over longer distances. If the membrane potential is decreased—if the membrane is *depolarized*—by a sufficient and rather critical amount, from the resting level of 70 millivolts to about 50 millivolts, a sudden and dramatic change occurs. The normal barriers to the flow of certain ions are temporarily removed, and as a result of the local flow of ions, the membrane potential first becomes reversed, reaching about 50 millivolts positive inside, and then reverses itself again, restoring the normal resting potential. All of this happens within a period of about 1 millisecond. Meanwhile the immediately adjacent region of nerve has been brought to its critical level, because the reversal of potential produces a powerful graded signal (of the kind just described) which spreads in all directions. A reversal thus occurs in the next piece of membrane, which causes a spread to the next, and so on. The result is a rapid spread of the process of transient polarity reversal along the nerve fiber. This type of signal, which travels the entire length of the nerve without attenuation, is called a *nerve impulse*.

In a typical nerve fiber in a nerve bundle in our leg, the impulse travels at a rate of about 10 meters a second. After an impulse has gone by, the nerve fiber recovers its ability to be stimulated again in about 1 millisecond, so that in many nerves the signals can be produced at rates of up to 1000/second.

All signaling in the nervous system involving distances of about 1 mm or more occurs in the form of impulses. Regardless of the type of nerve fiber we study, whether it is involved in vision, movement, or thought, the impulse signals are all virtually identical. This fact, which gradually became evident during the first half of this century, we owe largely to the work of the English neurophysiologist E. D. Adrian. It is a very important and rather surprising result: one might have thought that a single nerve fiber would have been designed to conduct several kinds of signals of different sizes or shapes. Instead there is only one kind of long-distance signal. What varies, in a given nerve fiber, is the *number* of impulses in a second.

When an impulse reaches the terminations of the axon, the cells next in line are influenced in such a way that their likelihood of producing impulses is modified. What happens at most synapses in vertebrates is unexpected, and was accepted by physiologists only slowly and reluctantly. Now, however, it is generally regarded as proved that when the impulse reaches the terminals a chemical *transmitter substance* is released from the presynaptic terminals. This chemical diffuses across the narrow space separating the two cells and affects the membrane on the far side of the synapse (the postsynaptic membrane) in one of two ways. In some synapses, which are termed "excitatory," it leads to a lowering of the transmembrane potential (a depolarization), with the result that the postsynaptic cell tends to produce impulses at a more rapid rate. In other synapses, termed "inhibitory," the effect of the transmitter is to stabilize the membrane potential, making it harder for excitatory synapses to produce depolarization. The result is to prevent impulses from arising or to reduce their rate of occurrence. Whether a given synapse is excitatory or inhibitory depends on what chemical transmitter the presynaptic cell makes and on the chemistry of the postsynaptic cell's membrane. Almost every nerve cell receives input from many terminals, usually many hundreds and often thousands, some excitatory and others inhibitory. At any instant some of the inputs will be active and others quiescent, and it is the sum of the excitatory and inhibitory effects that determines whether or not the nerve cell fires, and if it fires, how rapidly the impulses occur.

To sum up: the signals that travel along a nerve fiber are all-or-none impulses. They propagate at rates of many meters/second and can recur at rates of a few to many hundreds/second. For any given cell the direction of impulse propagation is always the same: with a few exceptions an impulse proceeds from its origin near the cell body or dendrites, along the axon, to the axon terminals. How rapidly a nerve fires depends on the net effects of excitatory and inhibitory influences at any instant. These influences are exerted by other axons terminating upon the cell body and dendrites at synapses.

The nerve cell, then, is much more than a device to send signals from one place to another. Each nerve cell constantly evaluates all of the signals reaching it from other nerves and expresses the result in the form of its own rate of signaling.

In any one nerve the most interesting question, perhaps, is what the signals mean in biological terms. For a nerve fiber in the touch-sensation system, more or fewer impulses per second can signal that a certain discrete region of skin on the body is being touched more or less vigorously. For a nerve fiber going to a muscle cell that bends the big toe, a greater number of impulses will lead to a stronger contraction. But whatever the signals of the particular nerve mean, the important thing is the rate of impulses. Perhaps surprisingly, there is no indication that the *pattern* of impulses is of any great importance. While one could have imagined a system in which evenly time-spaced impulses in a visual nerve, for example, might signal red, and irregularly spaced impulses green, this seems not to occur. Instead, it is the particular nerves that fire, and how fast they fire, that determines what color is seen and how vividly.

The whole segment of neurophysiology that deals with the generation and transmission of nerve signals forms a field of very active investigation at present. The chemistry of synaptic transmission is a particularly exciting area, with about half a dozen transmitter substances already identified and the methods by which nerves make, release, and destroy them already fairly well known. Many of the most important unanswered questions have to do with the chemistry of the nerve cell membrane, for we still do not know exactly how, in molecular terms, ions are transported across membranes or how the permeability to ions is influenced by potential changes or by transmitter substances. But in those areas progress is rapid, and in a decade or so the main activities of individual nerve cells should be known in great detail.

Structure of the Nervous System

Given a reasonable start at understanding the physiology of individual cells in the nervous system, we are in the position of someone who knows something about the physics of resisters and condensers and transisters, and who now looks inside a television set. To understand how the machine as a whole works is quite hopeless unless one knows how its elements are wired together and also has at least some idea of the purpose of the machine. The first step, then, is to learn how the various larger subunits of the brain are interconnected, and how each subunit is built up out of nerve cells. The second step is to try to understand how the cells interact, and to learn the significance of the messages they carry.

A rough caricature of a nervous system is shown in Figure 11.3. At the left, or input end, of the diagram we have groups of receptors. These modified nerve cells are specialized to transform various specific kinds of energy from the outside world into electrical signals. Some receptors respond to light, others to chemicals (taste and smell), and others to mechanical deformation (touch, hearing). The receptors make contact with other nerve cells, which in turn contact others, and so on. At each step along the way, from left to right in Figure 11.3, a number of axons converge on a single cell, and each axon may break up into many subdivisions so as to supply a number of cells. Each cell integrates the excitatory or inhibitory impulses converging upon it from other, lower-order cells. After a number of steps, sooner or later, the nerve axons terminate on muscle cells, which constitute the output of the nervous system. So we have an input, which is our only way of knowing about the outside world, an output, our way of influencing the outside world, and

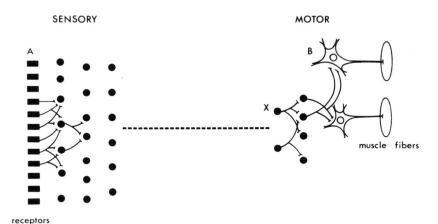

Figure 11.3. Schematic diagram of a central nervous system.

The rectangles to the left represent receptors, such as somatosensory (touch, pressure, and so on), auditory, visual. Successive stages of cells are shown to the right as circles. For simplicity, dendrites are omitted, and axons are represented as subdividing and terminating on the cells. At the right-hand end of the diagram cells B (here shown with their dendrites) have their axons terminating on muscles. Cell x is referred to in the text.

what is between, which must include perception, emotions, thought, and presumably, in ourselves though not in jellyfish or dogs, the soul.

The diagram, it need hardly be said, is oversimplified. While the main flow of traffic is left to right, there frequently exist connections between the cells of any given stage, and often there are connections in the reverse direction, from output toward input, just as there is feedback in many electronic circuits. There is not just one path from input to output, but many: there are many different arrays of receptors, specialized for light (the rods and cones of the retina), for sound (the hair cells of the inner ear), and so on, and countless different pathways and detours, rather than just one. The number of synapses between receptors and muscles may be very many or it may be only two or three. When the number is small, the circuit is usually termed a "reflex": the constriction of the pupil in response to a light shining on the retina represents a reflex involving perhaps four or five synapses. We owe to the great neurophysiologist Sherrington (1857–1952) the idea that the nervous system is built upon a foundation of reflexes of various degrees of complexity.

In thinking about a diagram such as that shown in Figure 11.3 one must be sure to remember that any given synapse is of one of two types, excitatory or inhibitory. If both excitatory and inhibitory influences impinge upon a cell at any given moment, the result may be a complete cancellation of effects.

Performance of Assemblies of Nerve Cells

Physiologists already have some idea of the kinds of operations that are performed by the nervous system close to the input end and at the very output. At the input, the nervous system is apparently chiefly preoccupied with extracting from the outside world information that is biologically interesting. Even the receptors themselves, for example, generally respond best at the very onset or cessation of a stimulus such as pressure on the skin. Obviously we need, in the world of touch, mainly to know about *changes;* no one wants or needs to be reminded 18 hours a day that his shoes are on. (A colleague to whom I gave this essay for his criticism asked whether I slept only six hours a night—or slept with my shoes on.)

In the visual system it is contrasts and movements that are important, and much of the circuitry of the first two or three stages is devoted to enhancing, in the images on the retina, the effects of contrast and of movement. At subsequent stages of the visual system the behavior of cells becomes more complex but is always orderly, and it fortunately makes sense in terms of perception. Knowing how cells in the cerebral cortex, four to six or more synapses removed from the receptors, respond to lights and patterns falling on the retina has fully confirmed Cajal's observation that the nervous system is wired in an amazingly intricate, precise, orderly way. At these levels one may find a cell that does not respond at all when the retina is completely and diffusely lit up, presumably because of mutual cancellation of excitatory and inhibitory effects, but fires vigorously if a line 1 mm long and 0.1 mm wide is moved across a tiny confined area in a certain orientation (for example, vertical or oblique), at a certain rate, and in a certain direction (up versus down). Given the already well-described building blocks of the nervous system—nerve cells, excitatory and inhibitory synapses—it is easy to think up circuits in which, in five or six stages, one might produce a cell that fires specifically to a line with a vertical orientation, but not to a line oriented in an 11 o'clock direction. It is the job of the neurophysiologist to determine just what circuits in our brains are responsible for these kinds of responses. It is known that such cells are present in newborn animals that have never used their eyes, showing that a substantial part of our nervous system is wired according to instructions that are genetically transmitted.

Although the visual system is now one of the best understood parts of the brain, we are still far from knowing how objects are perceived or recognized. Nevertheless, the amount that has been learned, over the few years that microelectrodes have been available, does suggest that a part of the brain such as the cerebral cortex is capable of being understood in relatively simple terms.

At the output end of the nervous system what counts are not the contractions of individual muscles but movements involving the coordinated contraction and relaxation of many muscles. To take a simple example: in making a fist or in grasping an object, we contract our fingers but must at the same time exert some extension at our wrist, otherwise that joint would also be flexed by the finger-flexor tendons. (One should perhaps remind the nonbiologist that the main muscles that flex the fingers are not in the hand but in the forearm!) This counteracting extension force at the wrist is exerted automatically, with no thought, as one can easily verify by making a fist and feeling the extensor muscles contract on the hairy side of one's forearm. Evidently, then, the order to "make a fist" involves the firing of cells (x) several stages from the right-hand end of Figure 11.3, cells whose axons are distributed to the nerve cells (B) that supply all the muscles involved in fist making. Other movements will bring in other circuits, perhaps involving the same muscles but in different combinations. These circuits like those involved in vision, are to some extent "built in," since a newborn baby can grasp an object elegantly, with just the right amount of wrist extension to correct the finger flexion.

It is easy, then, to visualize a few of the kinds of functions that may be performed by the nervous system, especially close to the sensory or to the motor ends. It is in between that our knowledge of function is most lacking.

At present, mainly because of Cajal, much is known about the wiring diagrams of a few parts of the brain. Even in the most thoroughly studied regions, however, our knowledge is far from complete. Cajal's work consisted mainly in describing the cells in different parts of the brain and classifying them by their shapes. To learn how these cells are wired to each other is not possible by the Golgi method alone: it requires the combined use of a number of different methods, including the Golgi stain and the electron microscope, which was not available to Cajal. Just to work out in a rough, non-detailed fashion the connections in a single structure, such as a particular part of the cerebral cortex or the cerebellum, may take one or two neuroanatomists five or ten years. Neuroanatomists, a special breed of people, often compulsive and occasionally semiparanoid, number only a few score in the entire world. Since the brain consists of hundreds of different structures each with its own special architecture, it is easy to see that an understanding just of the wiring of the brain is still many years away.

To know the connections of a structure within the brain is, however, a very different thing from understanding the structure's physiology. In the case of the cerebellum, for example, not only is the wiring known in some detail, but it is now fairly clear which synapses are excitatory and which inhibitory. For a few synapses in the cerebellum the transmitters have even been identified chemically with a moderate degree of confidence. Yet how the cerebellum works is known only in the vaguest of terms. It surely has to do with regulation of movements and muscle tone and balance, but how these functions are carried out by this magnificently patterned, orderly, and fantastically complex piece of machinery is quite obscure.

The cerebellum is admittedly a difficult place to work, because it lies at the watershed between sensory and motor processes, in the part of Figure 11.3 that I have left blank, like darkest Africa a century ago. The kinds of inputs it receives—the significance of the impulses that come to the cerebellum from the cerebral cortex, the spinal cord, and so on—are only imperfectly known, and the neural structures to which it sends its output, and which in turn are ultimately connected to the muscles, are also poorly understood. For similar reasons most parts of the brain are, at present, still relatively unexplored and only dimly understood. This is not necessarily because of technical obstacles but more because the time and effort have not yet been expended. Admittedly, new and revolutionary methods are badly needed. To give only one example, there is now no known way of studying the signals of single cells in a human without actually opening the skull on an operating table, something that is not only cumbersome but also ethically unacceptable. If we ever hope to understand something like speech, which is peculiar to man, we will have to find a way of recording from single neurons without opening the skull.

Would a knowledge of the wiring and an understanding of the moment-to-moment workings of nerve cells in the brain really represent the ultimate and only goal of neurobiology? Surely not, for certain large aspects are left quite untouched if our horizons are so limited. At present we have very little knowledge of the mechanisms that underlie memory. Neurobiology, perhaps more than most sciences, is subject to fads and fashions that sometimes amount almost to a derailment of rational thought. In one of these recent fashions, an idea was advanced that memories might be recorded in the form of very large molecules, the information being coded as a sequence of smaller building-block molecules in the same way that genetic information is coded. Few people familiar with neurobiology, especially with the highly patterned specificity of connections in the brain, took this idea seriously, but much time was nevertheless consumed in many laboratories teaching animals tasks, grinding up their brains, injecting extracts into the abdomens of other animals, testing the animals, and of course finding "statistically significant" improvements in ability to learn the same tasks or else significant differences in the chemistry of the brains of the trained animals. Today this fashion is dying out, but I refer to it only to illustrate the point that neurobiology has not always advanced or even held still on all fronts; sometimes there are actual small revolutions in a counterclockwise direction. Most neurobiologists today would agree that memory must ultimately be explained in terms of alterations in connections, for example by an increase in the efficiency of some synapses, possibly at the expense of others. Where and how this occurs, if it does, is at present completely unknown.

A second challenge of very large dimensions and of considerable excitement at present is to understand how the nervous system develops. How are the inherited macromolecular sequences of the genes translated into the specific kinds of neural patterns that Cajal described? How do the million nerve fibers of the optic nerve grow back, prior to birth, and link up with the correct cells in the brain? There are still no final answers to these questions, but the next twenty years of work by anatomists, chemists, and physiologists should be fascinating ones.

Where, in all this, is the Copernican revolution? It could be argued that so far in neurobiology there has been nothing quite analogous to such a revolution. Progress in understanding the brain has tended to be slow, even though technical advances have produced a marked acceleration in the last few decades. But there have been no abrupt upheavals to compare with those of Kepler or Newton or Einstein in physics, or of Darwin or Watson and Crick in biology. The only jump in knowledge that comes close is the contribution of Cajal, but this does not seem to me to be of quite the same order as the work of a Darwin or a Pasteur.

If a Copernican revolution has not yet occurred in neurobiology, can we expect one in the near future? Given the conceptual importance of understanding the brain, and the degree to which such an understanding could change our ideas of ourselves and the world around us, it is indeed hard to doubt that a really profound advance in this field would constitute a Copernican revolution. Each of the past revolutions has had the property of bringing some very fundamental aspect of man's study of nature into the realm of rational and experimental analysis, and away from the supernatural. If Copernicus pointed out that the earth is not the center of the universe, if Darwin showed that man is related to other living creatures, if Watson and Crick showed that biological inheritance can be explained in physical and chemical terms, then science seems mainly to be left with the question of the mind, whether or not it is something more than a machine of great complexity. One, and perhaps the only, important remaining branch of science in which a serious investigator can speak about the supernatural without necessarily being laughed at is that dealing with the brain and the mind. In *Leben des Galilei* Brecht writes:

Sechzehnhundertzehn, zehnter Januar:

Galileo Galilei, sah, daß kein Himmel war.

What I am asking is whether, fifteen to thirty years from now, some writer will be able to say:

Neunzehnhundertneunzig, zehnter Februar:

—— —— fand, daß keine Seele war.

Of course I am not trying to imply that at some particular moment in the future a discovery is likely to be made which completely explains the brain and so eliminates the soul; heaven and hell were certainly not eliminated in a day, by a single person.

How long it will be before we will be able to say that the brain—or the mind—is in broad outline understood is of course anyone's guess. As late as 1950, however, anyone who had predicted that in ten years the main processes of life and reproduction would be unraveled, except for details, would at the very least have been called optimistic. I think it will take longer to understand the brain, simply because it is such a many-faceted thing—a box brimful of ingenious solutions to a huge number of problems. How the brain untangles an auditory time sequence may be utterly different from the way in which it scans a visual scene. One may never actually solve all the separate individual problems that the brain presents, any more than one can hope to go into the physiologies of all of the species of life that inhabit the earth, or just as it seems unlikely that men will ever know the precise configuration of every protein in the human body. The hope, however, is that as each region of the brain is looked at in turn we will become more and more convinced that the functions are orderly and capable of being understood in terms of physics and chemistry, without appealing to supernatural processes. There will of course be major individual milestones, since there may well be some single mechanism by which memory works, or some one process to explain how nerve fibers find their proper destinations. Whether each of these will be advances of Copernican proportions is difficult to predict.

As we learn more about the brain, the effects of that knowledge on other fields of inquiry will be profound. The branches of philosophy concerned with such subjects as the nature of the mind and of perception will, in a sense, be superseded, as will the parts of psychology that seek to obtain the answers by indirect means. The entire field of education will be affected if we ever succeed in obtaining a scientific understanding of the mechanisms underlying memory.

My answer, then, is that a revolution of Copernican proportions has not yet occurred in neurobiology, and will perhaps not occur, at least in a single stroke. If there is one, it may be gradual, taking place over many decades. When it is over, we will know whether or not the brain is capable of understanding the brain. If the answer is yes, then we will be on the way to a far deeper understanding of ourselves.

References

I. Ramón y Cajal, *Histologie du Système Nerveux de l'Homme et des Vertébrés*. Consejo Superior de Investigaciones Cientificas, Instituto Ramón y Cajal, Madrid, 1972.

2. A. L. Hodgkin, *The Conduction of the Nerve Impulse*. Charles C Thomas, Springfield, Ill., 1964.

3. B. Katz, *Nerve, Muscle, and Synapse*. McGraw-Hill Book Company, New York, 1966.

Part III

Chemistry and Physics

Summary

This is a difficult part of the volume because it deals with simple questions—in particular, "What is everything made of, and how does it move?" We all know that the simplest questions have the most perplexing answers.

Hughes's article asks "How does the earth's atmosphere move?" Of the four articles in this part of the book it is the easiest to approach, partly because it deals with visible things (clouds sailing in the sky, for example) and partly because it invokes the familiar long-established theme of Newtonian mechanics. The classical theory of fluid dynamics, based on Newtonian mechanics, describes how air (a fluid) flows in the atmosphere, driving along clouds and rain and determining the weather. Hughes describes how, starting from a disastrous storm that wrecked the French Fleet at Balaklava, meteorologists first collected data upon the weather, then learned the fluid-dynamical mechanism behind it, and finally reached a growing competence at weather forecasting. Never content, man is now trying to control the weather. It remains to be seen if he will succeed at that, and what social effects could result. Certainly the lawyers would not be idle: Who stole the rain from my crops, or who ruined my sunshine cruise?

Seaborg's article gives the chemist's answer to "What is everything made of?" Everything, says the chemist, is made of atoms, an atom resembles (to put it naively) a tiny solar system with a central nucleus (like the sun) surrounded by a swarm of orbiting electrons (like planets); the outermost electrons of one atom can "hold hands" with the outermost electrons of another atom to form assemblies of atoms called molecules; there are only about 100 different sorts of atoms, but they can assemble in many many different patterns, so yielding the huge variety of different substances in the world; and chemistry is the study of the possible ways of assembling atoms to form molecules. From the chemical point of view, what distinguishes one sort of atom from another sort is the number of electrons it has. In naturally occurring substances, this number can range between 1 and 92, with just 4 of the intervening numbers missing. Recently, however, man has contrived to fabricate artificially these 4 absentees as well as some new sorts of atoms, with between 93 and 105 electrons. Seaborg tells the story of how these new sorts were first visualized in terms of theory and then realized in the laboratory. He was a principal participant in many of these discoveries; and his account conveys a vivid sense of the excitement of this scientific chase.

Weisskopf's article is a physicist's answer to "What is everything made of?" The physicist wants to know how an atom "works," what determines the shapes and sizes of the orbits of its electrons, what forces hold the nucleus together, and so on; and his answers involve two concepts in particular: quantum theory and energy. Energy is a rather difficult concept to grasp; but the basic idea behind quantum theory is quite easy. Bankers know that money comes in little chunks called cents: if you are rich enough, you can have any whole number of cents; but you cannot have fractions of a cent. To a banker, the whole numbers are the natural numbers, and one cent is the so-called *quantum* of money—the smallest indivisible unit of cash. So much for the quantum theory of money: let us look next at energy. Energy is the capacity to do work (physical work). It is a kind of convertible currency for physical action. Consider a pile driver with the hammer poised at the top of its run. In this position, elevated as it is in the earth's gravitational field, the hammer possesses *potential energy*, a credit balance, as it were, for the physical action that will occur when the hammer is released. On release, the hammer falls and gathers speed. A massive moving object has *kinetic energy*: indeed, as the hammer falls, its kinetic energy increases at the expense of its potential energy. Finally, the hammer strikes the pile and both come to rest. Where has the energy gone? Into *heat energy*: the hammer is heated, and so is the top of the pile, and some energy is dissipated through frictional heat as the pile jars the earth. There are a few other forms (for example, chemical, elastic, electromagnetic, and so on) in which energy may appear; but, in all forms, energy is the manifestation of or potentiality for physical action, and physics studies the conversion of energy from one form to another, just as finance deals with the conversion of one form of money into another. The quantum theory of physics recognizes that energy and several other physical quantities come in little chunks: there is a smallest possible chunk of energy called a quantum of action, and Nature abhors fractions of a quantum just as much as your bank abhors fractions of a cent.

Sachs's article is (among other things) a mathematical physicist's answer to "How do things move?" We all have an intuitive feeling for what motion is: it has something to do with "space" and "time"; an automobile covers such and such a distance in such and such a time, at a certain number of "miles per hour." And these intuitions seem to serve us quite well for the ordinary movements we notice in everyday life. However, at very high speeds (near the speed of light) some peculiar effects arise that defy our normal intuitions. To "explain" these peculiarities, the theory of relativity regards space and time, not as two distinct concepts as our intuitions would persuade us but as a single unified concept called "space-time." The gravitation of heavy objects (like the sun) "bends" space-time: or that, at least, is a picturesque way of speaking about the physical application of a branch of mathematics called differential geometry, which the relativist employs. It may seem a mysterious and unreal theory, but the physical effects it predicts are real enough. After the theory had predicted that rays of light would be bent by the sun, astronomers looked for this effect and (lo and behold!) detected it. We find it hard to subdue our intuitions and "get a feel" for space-time, but the authors of the essays in this chapter have taken great pains to present their topics via phrases and analogies, which the layman can at least reach after. In here using ordinary language (instead of the technical language of theoretical physics) to describe either the submicroscopic or the pancosmic, to transcend the five senses, they are attempting something very difficult and attempting it with their hands voluntarily tied. The lay reader should not limply surrender but should respond with his own reciprocal effort, reading, skipping, turning back a page or two, and rereading until at last the flavor of the subject permeates. Thus he will learn what it entails to vanquish the preconceptions of human intuition, in which conquests lie Copernican revolutions.

All language is treacherous and beguiling, especially the language of mathematics. This time, the stumbling block is not to the layman but to the professional scientists. If the scientist constructs mathematical theories as models of the real world, as mirrors of Nature, how can he know that he has portrayed a faithful image of reality? By the interaction of theory and experiment? By forcing his theory to predict, and challenging his experiment to contradict or verify? Yes, in large part: but theory is very powerful, and unless he is careful he will find that his theory has told him just what he may or may not observe, has indeed dictated the experiment. Heisenberg's principle of uncertainty, described in Weisskopf's article, was a radical overthrow of Victorian preconceptions about scientific determinism; yet in its turn it has become by its acceptance a contemporary preconception, not necessarily wrong because it is the current scientific dogma, but at least in peril of becoming a principle of certainty. There is a passage in Eddington's *Philosophy of Natural Science,* wherein he points to the dangerous influence of theory on experiment, asking whether Rutherford discovered the atom or made it—that is to say, conducted an experiment so conditioned by the theory that it could not but verify. You will recall, he says, Procrustes who entertained travelers at dinner and then put them to bed, stretching them out on the rack if they were too short for his bed or chopping them down to size if they were too tall. But, he adds, you may not know the sequel: that the following morning he measured them up and wrote a learned paper for the Anthropological Society of Attica on the uniformity of stature of travelers in Greece. Mathematics, despite appearances, is in a certain sense a simple subject; and there is a Procrustean temptation to cast theoretical physics and chemistry in the same simple mold. Mathematical theories sometimes rest on very simple foundations, ideas like isotropy (that is, things are the same whichever way you look) and symmetry. The relativist rests a good deal of his case on isotropies or invariances in space-time, and the theoretical physicist studies the symmetries of fundamental particles. All that ties in with the Copernican theme of a simplicity "pleasing to the mind." But does it please Nature—How far is reality really like that? That is a question of knowing *how* we know.

Glenn T. Seaborg

12 From Mendeleev to Mendelevium—and Beyond

As the title suggests, this essay begins with the great Russian chemist, Dmitri Ivanovich Mendeleev—actually, with his periodic table and its development up to a time just before the Second World War. This marked the beginning of revolutionary changes and extensions in his periodic table which are traced here. For more than half of a century following Mendeleev's monumental formulation of the periodic table in 1869, mankind was generally satisfied with the concept that the upper limit of the chemical elements would be set by uranium. The development of the understanding and experimental means of nuclear transmutation in the 1930s removed this conceptual barrier. The ensuing evolution of the whole field of the transuranium elements, continuing with ever-increasing excitement and vigor today in the search for superheavy elements, constitutes a real Copernican revolution. Here we have found, and are continuing to find, the drama attendant with the replacement of previously held concepts by new ones with important and far-reaching consequences.

Mendeleev's Revolution
The story begins, of course, on March 6, 1869, when Mendeleev and his associate, Nikolai Menshutkin, presented a paper to the Russian Chemical Society in St. Petersburg (now Leningrad) which postulated that the elements showed a periodicity in their chemical properties when they were arranged in the order of their atomic weights. This in itself was not novel. Several chemists in other countries had observed some kind of orderliness in the elements then known, the most prominent being the German, J. W. Döbereiner and his triads (1829), the Frenchman, A. E. B. de Chancourtois and his "telluric screw" (1862), and the Englishman, J. A. R. Newlands and his "law of octaves" (1864). It is not generally known that some American chemists of the nineteenth century also proposed various forms of periodic classifications, including O. W. Gibbs in 1845 and J. P. Cooke, Jr., in 1854. The only real challenge to the generally accepted validity of Mendeleev's originality, however, has come from the work of Lothar Meyer in Germany, who in 1870 produced independently a generalization almost identical to that of Mendeleev. The reason for the general acceptance of Mendeleev's preeminence is straightforward: not only did he show

Department of Chemistry, Lawrence Berkeley Laboratory, University of California, Berkeley, California 94720.

that periodicity existed in the properties of the elements then known, but he had the courage and the vision to state that his method of classification constituted a fundamental law of Nature, and that where there appeared to be deficiencies in his periodic table, they were due to gross errors in measurement of atomic weights or simply to the fact that certain elements had not yet been discovered. Indeed, Mendeleev's claim to priority in the discovery of the periodic system was not completely accepted until his predictions of missing elements were proved by experimental evidence in later years.

In Table 12.1, I show Mendeleev's periodic table as published in 1871, which incorporates improvements made in the original version of 1869. It should be noted in particular that the table predicts the existence and properties of the elements with atomic weights of 44, 68, and 72. These correspond to scandium, gallium, and germanium, as we know them, and the three elements were actually found in Nature during the period from 1875 to 1886. Many other experimental proofs of Mendeleev's "law" were made in the years that followed.

As time progressed, adjustments had to be made to the periodic table to accommodate the rapidly expanding knowledge of the properties of the elements and their atomic and nuclear structure. Also, additional elements were discovered during the late nineteenth century and the first part of the twentieth century which required some reconstruction of the Mendeleev periodic system. The most significant changes were the addition of another vertical row, or group, of elements now known as the noble gases, and the substitution of a series of elements—the rare earths, or lanthanides—in the place of a single element.

By the end of the first decade of the twentieth century the total number of elements had increased to 85, and soon thereafter the concept of atomic number, as the fundamental basis for the ordering of the elements in the periodic table, was established. During the next twenty-five years, three more elements were discovered, leaving below uranium, element 92, those having atomic numbers 43, 61, 85, and 87 as the missing elements. The understanding of atomic and nuclear structure was such by this time that it was considered probable that these elements would all be radioactive and would have to be prepared by artificial methods. As expected, these four were found to be radioactive and, for all practical purposes, must

Table 12.1. 1871 Periodic Table of Mendeleev, Including Changes to Original 1869 Table (D. Mendelejeff, *Ann.*, Supple. VIII, 1871, p. 133).

[31]	Группа I	Группа II	Группа III	Группа IV	Группа V	Группа VI	Группа VII	Группа VIII. Переход к группе I
Типические элементы · H = 1	Li = 7	Be = 9,4	B = 11	C = 12	N = 14	O = 16	F = 19	
Ряд 1-й — Первый период	Na = 23	Mg = 24	Al = 27,3	Si = 28	P = 31	S = 32	Cl = 35,5	
— 2-й	K = 39	Ca = 40	— = 44	Ti = 50?	V = 51	Cr = 52	Mn = 55	Fe = 56, Co = 59, Ni = 59, Cu = 63
— 3-й — Второй период	(Cu = 63)	Zn = 65	— = 68	— = 72	As = 75	Se = 78	Br = 80	
— 4-й	Rb = 85	Sr = 87	(?Yt = 88?)	Zr = 90	Nb = 94	Mo = 96	— = 100	Ru = 104, Rh = 104, Pd = 104, Ag = 108
— 5-й — Третий период	(Ag = 108)	Cd = 112	In = 113	Sn = 118	Sb = 122	Te = 128?	J = 127	
— 6-й	Cs = 133	Ba = 137	— = 137	Ce = 138?	—	—	—	
— 7-й — Четвертый период	—	—	—	—	—	—	—	
— 8-й	—	—	—	—	Ta = 182	W = 184	—	Os = 199?, Ir = 198?, Pt = 197?, Au = 197
— 9-й — Пятый период	(Au = 197)	Hg = 200	Tl = 204	Pb = 207	Bi = 208	—	—	
— 10-й	—	—	—	Th = 232	—	Ur = 240	—	
Высшая солевая окись	R^2O	R^2O^2 или RO	R^2O^3	R^2O^4 или RO^2	R^2O^5	R^2O^6 или RO^3	R^2O^7	R^2O^8 или RO^4
Высшее водородное соединение			$(RH^5?)$	RH^4	RH^3	RH^2	RH	

be synthesized. A typical periodic table of this time is shown in Table 12.2.

The first artificial element to be identified was technetium with atomic number 43. It was discovered by C. Perrier and E. Segrè in Italy in 1937. Molybdenum was bombarded with deuterons produced in a cyclotron at the University of California at Berkeley and shipped to Italy. The forms of element 43 discovered were radioactive, and the amounts produced were submicroscopic in quantity. The chemistry of technetium conformed with what might be expected from its position in the periodic table.

Francium with atomic number 87 was discovered in France in 1939 by M. Perey. It was found to exist in very short-lived radioactive form in very small quantities among the radioactive heavy elements that are found in Nature, that is, in uranium-containing ores such as pitchblende, carnotite, and so on. The most practical way of making this element is through the neutron bombardment of radium; in this sense, it may be classified as a synthetic element.

Element 85, astatine, was discovered by D. R. Corson, K. R. MacKenzie, and Segrè at the University of California in 1940. The element was synthesized by bombarding bismuth with accelerated helium ions. Its chemical behavior is that of a halogen, but, as expected, its properties are somewhat more metallic in character than those of the other halogens.

Promethium, atomic number 61, the last of the four missing elements, was found among the fission products of plutonium-manufacturing reactors. It is a member of the rare-earth group of elements, all members of which are very similar in chemical properties and can be separated from each other only with some difficulty. Nevertheless, it was positively identified in 1945 by J. A. Marinsky, L. E. Glendenin, and C. D. Coryell, working at Oak Ridge in the wartime Plutonium Project of the Manhattan District.

These four radioactive elements were previously and erroneously identified as masurium (Ma), atomic number 43; illinium (Il), atomic number 61; alabamine (Ab), atomic number 85; and virginium (Vi), atomic number 87. These names sometimes appear in the periodic tables of the time.

Increased Understanding

During the first quarter of the twentieth century, the studies of such great scientists as J. J. Thomson, E. Rutherford, N. Bohr, and H. G. J. Moseley

Table 12.2. Periodic Table before Second World War. (Predicted elements have their atomic numbers in parentheses.)

1 H																		2 He
3 Li	4 Be											5 B	6 C	7 N	8 O	9 F		10 Ne
11 Na	12 Mg											13 Al	14 Si	15 P	16 S	17 Cl		18 Ar
19 K	20 Ca	21 Sc	22 Ti	23 V	24 Cr	25 Mn	26 Fe	27 Co	28 Ni	29 Cu	30 Zn	31 Ga	32 Ge	33 As	34 Se	35 Br		36 Kr
37 Rb	38 Sr	39 Y	40 Zr	41 Nb	42 Mo	(43)	44 Ru	45 Rh	46 Pd	47 Ag	48 Cd	49 In	50 Sn	51 Sb	52 Te	53 I		54 Xe
55 Cs	56 Ba	57–71 La–Lu	72 Hf	73 Ta	74 W	75 Re	76 Os	77 Ir	78 Pt	79 Au	80 Hg	81 Tl	82 Pb	83 Bi	84 Po	(85)		86 Rn
(87)	88 Ra	89 Ac	90 Th	91 Pa	92 U	(93)	(94)	(95)	(96)	(97)	(98)	(99)	(100)					

57 La	58 Ce	59 Pr	60 Nd	(61)	62 Sm	63 Eu	64 Gd	65 Tb	66 Dy	67 Ho	68 Er	69 Tm	70 Yb	71 Lu

on the structure of the atom and its nucleus provided the final argument to support the framework of the periodic system as formulated by Mendeleev. These pioneer scientists, and others, developed the concepts of atomic number and electronic structure to the point where these could be correlated with each chemical element's position in the periodic table. Based on the groundwork of these investigators and the subsequent development of quantum theory and quantum mechanics and further experimental work, our present-day understanding of atomic and nuclear structure evolved. We now know that the atom consists of a positively charged nucleus consisting of protons (fundamental particles carrying one unit of positive charge with a mass about 1850 times that of the electron) and neutrons (fundamental particles, electrically neutral, with about the same atomic mass as protons); this nucleus is surrounded by electrons, whose characteristics are defined by quantum numbers and which are more loosely referred to as occurring as groups in shells and subshells. The atomic number (Z) of an element is determined by the number of protons in the nucleus of the atom which is equal to the number of surrounding electrons in the neutral atom. Isotopes of an element are essentially chemically identical forms of nuclides (atomic species), each containing the same number of protons. The isotopes of an element differ in the number of neutrons (N) and thus also differ in the mass number (A), the integral sum of nucleons (the collective name for protons and neutrons) in the nucleus.

As additional chemical elements were discovered and the principles of atomic and nuclear structure became generally understood, the periodic table evolved into a generally accepted form in which the atomic number determined the position of the element; it was bounded at the upper end by the heaviest known element, uranium (atomic number 92). Many attempts were made to discover naturally occurring elements heavier than uranium, all of which were unsuccessful. Although uranium gradually disappears due to radioactive decay, its rate of decay is slow; its half-life[1] is comparable to the age of the earth which means that much of our original inheritance of uranium is still with us. However, the increasing charge on the nucleus for the elements beyond uranium leads to increased rates of radioactive decay, that is, half-lives that are short compared to the age of the earth. Thus our original heritage of such elements has long since disappeared, except for extremely small, essentially negligible, quantities of one or two of them.

During the 1930s a number of attempts were made to predict an extension of the periodic table to include these elements which had not yet been discovered and which were not bounded by other known elements in the periodic table—that is, elements beyond uranium, the "transuranium elements." The typical periodic table before the Second World War placed them in the manner shown in Table 12.2. Thus, it can be seen that the first transuranium element, with the atomic number 93 (element 93), was predicted to be a chemical counterpart or homologue of rhenium; element 94 a homologue of osmium; element 95 a homologue of iridium; 96 a homologue of platinum, and so forth. This is the way the periodic table was uniformly published before World War II, and it should be noted that here thorium and protactinium and uranium appeared as chemical homologues of (and hence under) hafnium and tantalum and tungsten.

There were, however, a number of suggestions that somewhere in the region of uranium there should be another "inner transition series" as in the rare-earth elements. In the rare-earth elements the electrons successively enter an inner electron subshell that has room for exactly 14 electrons of a special type. These 14 elements, with atomic numbers 58 to 71, all have chemical properties similar to those of lanthanum and should all fit in the periodic table in the same single space as lanthanum (atomic number 57), and they are given the alternate group name "lanthanide" elements; for convenience they are usually listed, along with lanthanum, in a row below the main body of the periodic table. These considerations, therefore, suggested that there should be a second, heavier "rare-earthlike" series of chemical elements in the region of uranium in which a second inner electron subshell with room for 14 electrons would be filled.

One of the first periodic tables to reflect these ideas and to place properly the first 14-member rare-earth series of elements was that corresponding to the ideas of Niels Bohr in 1923 (Table 12.3). Bohr suggested that the second 14-member rare-earthlike series of elements would begin to be filled at element 95 and would be completed at element 108, as shown in Table 12.3. The 14 elements 95–108 would thus have chemical properties similar to those of eka-osmium (atomic number 94). I use the "eka" nomenclature of Mendeleev to denote an undiscovered element, in a column of the periodic table, which should have properties similar to the preceding members of the column.

Although Bohr's ideas were, in a theoretical sense, near the mark, they

Table 12.3. Periodic Table of 1923 Corresponding to the Ideas of Niels Bohr Showing Predicted Elements beyond Uranium and with Transition Series Beginning with Element 94 (Niels Bohr, "The Structure of the Atom," *Nature*, Vol. *112*, July 7, 1923, p. 30, as it appeared in L. L. Quill, *Chem. Rev.*, Vol. *23*, 1938, p. 87).

																H 1	He 2
Li 3	Be 4											B 5	C 6	N 7	O 8	F 9	Ne 10
Na 11	Mg 12											Al 13	Si 14	P 15	S 16	Cl 17	A 18
K 19	Ca 20	Sc 21	Ti 22	V 23	Cr 24	Mn 25	Fe 26	Co 27	Ni 28	Cu 29	Zn 30	Ga 31	Ge 32	As 33	Se 34	Br 35	Kr 36
Rb 37	Sr 38	Y 39	Zr 40	Nb 41	Mo 42	— 43	Ru 44	Rh 45	Pd 46	Ag 47	Cd 48	In 49	Sn 50	Sb 51	Te 52	I 53	X 54
Cs 55	Ba 56	*	— 72	Ta 73	W 74	— 75	Os 76	Ir 77	Pt 78	Au 79	Hg 80	Tl 81	Pb 82	Bi 83	Po 84	— 85	Em 86
— 87	Ra 88	Ac 89	Th 90	Pa 91	U 92	E-Re 93 [X]	109	110	111	112	113	114	115	116	117	118	

*Rare Earths:	La 57	Ce 58	Pr 59	Nd 60	— 61	Sm 62	Eu 63	Gd 64	Tb 65	Ds 66	Ho 67	Er 68	Tm 69	Yb 70	Cp 71

[X] Possible Transition Group	E-Os 94	95	96	97	98	99	100	101	102	103	104	105	106	107	108

did not lead to correct predictions for the chemical properties of the transuranium elements, nor was his placement of the transuranium elements in the periodic table generally adopted. Periodic classifications of the type illustrated by Table 12.2 were most commonly used for predicting chemical properties of the transuranium elements. As we shall see, this placement of these elements was wrong and led, therefore, to erroneous predictions.

Since the heaviest naturally occurring element is uranium (except for the extremely small quantities of one or two transuranium elements), the practical consideration of the missing transuranium elements awaited the discovery and development of nuclear transmutation processes that could be applied to their synthesis. Although the alchemist dreamed of ways to transmute one element into another by means of chemical reactions and made claims of success in effecting such transformations, actual success was not achieved before the twentieth century and then not by the use of chemical reactions. This success depended on the methods of nuclear physics which made possible the transmutation of the nucleus of an atom to the nucleus of another atom, a process requiring much larger energies than those employed in the unsuccessful chemical experiments of the alchemists. The first nuclear transmutation experiment was performed in 1919 by Rutherford, who bombarded nitrogen gas with the natural emanations of radioactive substances to convert nitrogen nuclei into oxygen nuclei. The next great step forward was achieved in 1931 by J. D. Cockcroft and E. T. S. Walton when they used their "electrostatic accelerator" to produce protons that were sufficiently energetic to initiate nuclear reactions, the first to be brought about by man-made bombarding particles. This was followed by the development of other more powerful accelerators, such as the "cyclotron" of E. O. Lawrence. These accelerators could produce positively charged bombarding projectiles of sufficient energy so that they could cause transmutations throughout the entire range of chemical elements by overcoming the repulsive force from the positive charge on the target nuclei. The discovery in 1932 by Chadwick of the neutron, observed as a product of the bombardment of light elements by radiation from natural radioactive sources, extended greatly the possibilities of nuclear transmutation because this neutral particle could enter the nucleus without being subject to a repulsive force from the nuclear charge; furthermore, the development of the nuclear chain reactor, following the discovery of nuclear fission in uranium, provided a source of neutrons of tremendous intensity.

The discovery and development of nuclear transmutation constituted, in itself, a Copernican revolution, but this is not a major theme of this essay; our more limited concern here is with the application of nuclear transmutation to the synthesis of the man-made transuranium elements and, less directly, with its application to the nuclear fission of uranium and the transuranium elements.

The first attempts to produce elements beyond uranium were made in Italy by E. Fermi, E. Segrè, and co-workers, who bombarded uranium with neutrons in 1934. They found a number of radioactive products. The radioactive products of the neutron bombardment of uranium were the object of chemical investigations during the following years by O. Hahn, L. Meitner, and F. S. Strassmann, in Germany, and by numerous other scientists. On the basis of misleading tracer chemical studies over a period of several years, some of these radioactivities did indeed seem to exhibit chemical properties that might be expected for transuranium elements with atomic numbers such as 94 or 96; that is, properties similar to those of elements such as osmium and platinum listed directly above elements 94 and 96 in the periodic table of that time (see Table 12.2). A great breakthrough in scientific thought finally came in late 1938 when Hahn and Strassmann showed that this interpretation was not correct. Their meticulous chemical experiments revealed that these products of the uranium neutron bombardments actually were radioactive isotopes of lighter elements and were thus fission product elements such as barium, lanthanum, iodine, tellurium, or molybdenum, formed by the "fission" or "splitting" of uranium into nearly equal parts. It was soon demonstrated that a large amount of energy is released in this reaction. This, the discovery of the nuclear fission of uranium (soon shown to be due to the rare isotope of mass number 235, known as uranium-235), was a monumental discovery with tremendous import for the future of the entire world. The subsequent unleashing of the tremendous energy residing in the nucleus of the atom, in the form of atom bombs or, more hopefully, as abundant electrical power, served to make uranium the key element of the "nuclear age."

Actual transuranium elements were discovered soon after this. Figure 12.1 depicts the rate of discovery of the thirteen presently known transuranium elements, and their various isotopes, in the years since the first discovery in 1940. The following sections describe the discovery of several of the key elements in this region. At the present time, research on the transuranium elements is taking place in a large number of laboratories in many countries throughout the world.

The First Transuranium Elements, Neptunium and Plutonium
The discovery of nuclear fission was soon followed by another great advance, the discovery of the first real transuranium element, with the atomic number 93, in the spring of 1940. E. M. McMillan, working at the University of California at Berkeley, was trying to measure the energies of the two main recoiling fragments from the neutron-induced fission of uranium. He placed a thin layer of uranium oxide on a thin support, and next to this he stacked very thin paper sheets to stop and collect the high-energy fission fragments from uranium. The paper he used was ordinary cigarette paper, the kind used by people who roll their own cigarettes. In the course of his

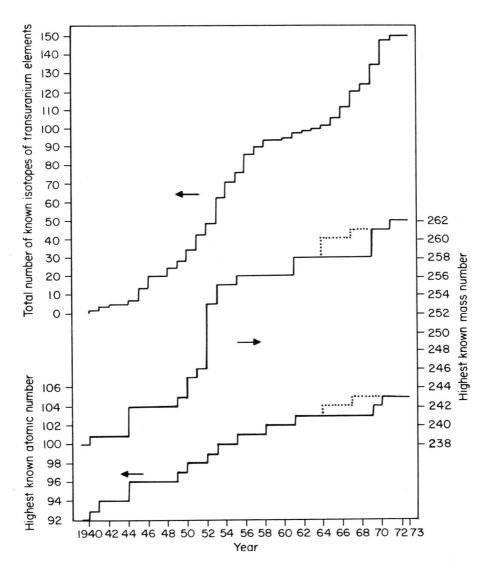

Figure 12.1. Rate of discovery of transuranium elements. (The dotted lines refer to the Soviet claims of the discovery of elements 104 and 105.)

studies, he found that there was another radioactive product of the reactions induced by neutrons, a beta particle (that is, electron) emitter with a half-life ($T_{\frac{1}{2}}$) of 2.3 days, which had low energy and hence did not recoil sufficiently to escape from the thin layer of uranium undergoing fission, as do the high-energy fission products. He suspected that this was a product formed by simple neutron capture in the abundant, relatively nonfissionable isotope[2] of uranium, $^{238}_{92}U$, also written more simply as uranium-238. McMillan and P. H. Abelson, who joined him in this research, were able to show, on the basis of their chemical work, that this product is an isotope of element 93, $^{239}_{93}Np$ (or neptunium-239), formed by simple neutron capture in $^{238}_{92}U$ followed by beta particle (β^-) decay of the previously identified $^{239}_{92}U$. The loss of a negative charge by $^{239}_{92}U$, through beta particle (that is, electron) decay, led to the daughter $^{239}_{93}Np$ representing an increase of one on the positive charge of the nucleus.

The 60-inch cyclotron of the Crocker Radiation Laboratory, on the Berkeley campus of the University of California, was used to furnish the neutrons, produced by the bombardment of light elements by accelerated charged particles. This cyclotron was used subsequently in the discovery of five additional transuranium elements (plutonium, curium, berkelium, californium, and mendelevium). It has since been moved to the Davis campus of the university and is being used, after renovation and modernization, for further research in nuclear chemistry and physics.

The experimental investigation of neptunium by McMillan and Abelson showed that it resembles uranium, not rhenium, in its chemical properties. This was the first recognized experimental evidence that an inner electron subshell is filled in the transuranium region of elements and that the assumed placement of element 93, as a homologue of rhenium, as shown in Table 12.2 was wrong.

The discoverers of element 93, as was their right, suggested the name "neptunium" and symbol "Np." It was named after the planet Neptune, the first planet beyond Uranus, for which uranium was named.

Plutonium (atomic number 94) was the second transuranium element to be discovered. By bombarding uranium with deuterons in the 60-inch cyclotron at Berkeley, E. M. McMillan, J. W. Kennedy, A. C. Wahl, and the author, in late 1940, succeeded in preparing a new isotope, neptunium-238, which decays to plutonium-238, an isotope with a half-life of about 90 years.

The first bombardment of uranium oxide with 16-MeV deuterons was performed on December 14, 1940. Alpha particle radioactivity (that is, the emission of helium ions) was found to grow into the chemically separated element 93 fraction during the following weeks. This alpha particle activity was chemically separated from the neighboring elements, especially elements 90 to 93 inclusive, in experiments performed during the next two months. These chemical experiments, which constituted the positive identification of element 94, showed that this element resembles uranium and neptunium but not osmium in its chemical properties; the periodic table of that time (Table 12.2) had suggested that element 94 should resemble osmium.

Element 94 was finally christened "plutonium" in March of 1942 and given the symbol "Pu." Plutonium was named after the planet Pluto, following the pattern used in naming neptunium. Pluto is the second and last known planet beyond Uranus.

Because of its ability to undergo fission and thereby serve as a source of nuclear energy similar to uranium-235, the plutonium isotope with the mass number 239 is the one of major importance. The search for this isotope, as a decay product of neptunium-239, was being conducted by J. W. Kennedy, E. Segrè, A. C. Wahl, and the author, simultaneously with the above-described experiments leading to the discovery of plutonium. The isotope plutonium-239 was identified and its great potential as a nuclear energy source was determined during the spring of 1941. This established plutonium as by far the most important of the transuranium elements and as an element destined to play an extraordinary role in the future of mankind. Plutonium-239, with a half-life of about 24,000 years, was produced by the beta particle decay of neptunium-239, which in turn was produced from uranium-238 by neutrons from the 60-inch cyclotron in the Crocker Laboratory.

In August 1942, B. B. Cunningham and L. B. Werner, working at the wartime Metallurgical Laboratory at the University of Chicago on the ultramicrochemical scale of investigation, succeeded in isolating about a microgram of plutonium-239 which had been prepared by cyclotron irradiations. Thus, plutonium was the first man-made element to be obtained in visible quantity. The first weighing of this man-made element, using a sample weighing 2.77 micrograms, was made by these investigators on September 10,

1942. These were dramatic events in the lives of those of us who were privileged to witness them.

Origin of the Actinide Concept

As indicated earlier, the tracer chemical experiments with neptunium and plutonium showed that their chemical properties were much like those of uranium and not at all like those of rhenium and osmium. This indicated that an inner electron shell was being filled in this region, but it was not clear which element constituted the beginning of this transition series.

For a few years following this, uranium, neptunium, and plutonium were considered to be sort of "cousins" in the periodic table, but the family relationship was not clear. It was thought that elements 95 and 96 should be very much like them in their chemical properties. Thus it was thought that these elements formed a "uranide" (chemically similar to uranium) group.

The periodic table of 1944, a portion of which is shown in Table 12.4, therefore implied that the chemical properties of elements 95 and 96 should be nearly identical with those of neptunium and plutonium. These assumptions proved to be wrong, and the initial experiments directed toward the discovery of elements 95 and 96 based on this hypothesis failed. The undiscovered elements 95 and 96 apparently refused to fit the pattern indicated by the periodic table of 1944.

Then, in 1944, the author conceived the idea that perhaps all the known elements heavier than actinium were misplaced on the periodic table. The theory advanced was that these elements heavier than actinium might constitute a second series that was completely analogous to the series of "rare-

Table 12.4. Portion of Periodic Table of 1944 Showing Predicted Transuranium Element Positions.

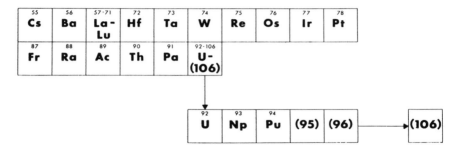

earth" or "lanthanide" elements. The lanthanides are chemically very similar to each other and, as already indicated, are usually listed in a separate row below the main part of the periodic table. This would mean that all these heavier elements really belong with actinium—directly after radium in the periodic table—just as the known "lanthanides" fit in with lanthanum between barium and hafnium. Elements 95 and 96, although similar in some of their chemical properties to uranium, neptunium and plutonium, would not be as closely related as implied by the "uranide" hypothesis.

The new concept meant that elements 95 and 96 should have some properties in common with actinium and some in common with their rare-earth "sisters," europium and gadolinium. When experiments were designed according to this new concept, elements 95 and 96 were soon discovered; that is, they were synthesized and chemically identified.

The revised periodic table, then, first published in 1945 (Table 12.5), listed the heavier elements as such a second "rare-earth" series. These heaviest elements (including the, at that time, undiscovered elements up through element 103), with the name "actinide" elements, were paired off with those in the already-known lanthanide rare-earth series.

Since the 14 elements beyond actinium (through lawrencium, element 103) belong to the actinide group, the elements thorium, protactinium, and uranium were removed from the positions they occupied in the periodic table before World War II and placed in this second "rare-earth" family. Elements 104, 105, and 106 will take over the positions previously held by thorium, protactinium, and uranium. Thus, we have the interesting result that the newcomers affected the face of the periodic table, and a preconceived notion had to be changed after many years even though the table seemed to have assumed its final form.

This concept was the key to the discovery of elements 95 and 96 (americium and curium); in addition, the recognition that these elements were members of an actinide transition series was absolutely essential to the discovery of the elements beyond curium through element 102 (nobelium). Because the chemical behavior of the actinides is similar and analogous to the lanthanides, which are difficult to separate from one another, separation of various transuranium elements from each other by ordinary chemical means is also difficult. It was, therefore, fortunate that, concurrent with the early work on the transuranium elements, techniques for chemical separation

Table 12.5. Periodic Table Showing Heavy Elements as Members of an Actinide Series. (Arrangement by Glenn T. Seaborg, "The Chemical and Radioactive Properties of the Heavy Elements," *Chemical and Engineering News*, Vol. 23, December 10, 1945, pp. 2190–2193.)

																			1 H 1.008	2 He 4.003
3 Li 6.940	4 Be 9.02												5 B 10.82	6 C 12.010	7 N 14.008	8 O 16.000	9 F 19.00	10 Ne 20.183	
11 Na 22.997	12 Mg 24.32												13 Al 26.97	14 Si 28.06	15 P 30.98	16 S 32.06	17 Cl 35.457	18 A 39.944	
19 K 39.096	20 Ca 40.08	21 Sc 45.10	22 Ti 47.90	23 V 50.95	24 Cr 52.01	25 Mn 54.93	26 Fe 55.85	27 Co 58.94	28 Ni 58.69	29 Cu 63.57	30 Zn 65.38		31 Ga 69.72	32 Ge 72.60	33 As 74.91	34 Se 78.96	35 Br 79.916	36 Kr 83.7	
37 Rb 85.48	38 Sr 87.63	39 Y 88.92	40 Zr 91.22	41 Cb 92.91	42 Mo 95.95	43	44 Ru 101.7	45 Rh 102.91	46 Pd 106.7	47 Ag 107.880	48 Cd 112.41		49 In 114.76	50 Sn 118.70	51 Sb 121.76	52 Te 127.61	53 I 126.92	54 Xe 131.3	
55 Cs 132.91	56 Ba 137.36	57 La–Lu 138.92 series	72 Hf 178.6	73 Ta 180.88	74 W 183.92	75 Re 186.31	76 Os 190.2	77 Ir 193.1	78 Pt 195.23	79 Au 197.2	80 Hg 200.61		81 Tl 204.39	82 Pb 207.21	83 Bi 209.00	84 Po	85	86 Rn 222	
87	88 Ra	89 Ac series	90 Th	91 Pa	92 U	93 Np	94 Pu	95	96										

LANTHANIDE SERIES

57 La 138.92	58 Ce 140.13	59 Pr 140.92	60 Nd 144.27	61	62 Sm 150.43	63 Eu 152.0	64 Gd 156.9	65 Tb 159.2	66 Dy 162.44	67 Ho 163.5	68 Er 167.2	69 Tm 169.4	70 Yb 173.04	71 Lu 174.99

ACTINIDE SERIES

89 Ac	90 Th 232.12	91 Pa 231	92 U 238.07	93 Np 237	94 Pu	95	96							

of lanthanide elements were being developed, and these could be used for the chemical identification of the actinides as they were being discovered through nuclear synthesis.

As a result of this new understanding the elements americium and curium (numbers 95 and 96) were synthesized and identified in 1944 and 1945, berkelium and californium (numbers 97 and 98) in 1949 and 1950, einsteinium and fermium (numbers 99 and 100) in 1952 and 1953.

Mendelevium

Mendelevium (atomic number 101) is the first element to be synthesized and discovered on a one-atom-at-a-time basis. The fundamental methods worked out for this successful experiment have served as the basis for the production and discovery of the elements heavier than mendelevium and will doubtless continue to serve as a basis for many of the experiments designed to discover still further transuranium elements. Thus, this is a pivotal element in the transuranium region.

The experiment was conducted in 1955 by A. Ghiorso, B. G. Harvey, G. R. Choppin, S. G. Thompson, and the author, at the University of California Radiation Laboratory in Berkeley. The investigators thought the techniques had advanced to a point where it might be possible to identify the undiscovered element with the atomic number 101 through the transmutation of a target that contained an unweighable and invisible amount of material (which had never been attempted before). The plan of attack involved use of the Berkeley 60-inch cyclotron to bombard with helium ions the maximum available quantity of einsteinium-253 utilizing the reaction:

$$^{253}_{99}\text{Es} + {}^{4}_{2}\text{He} \rightarrow {}^{256}_{101}\text{Md} + {}^{1}_{0}\text{n}.$$

Following this plan the first step was to gather all the einsteinium-253 that was available. This turned out to be no more than 1 billion atoms, or less than 1 million millionth of a gram.

Calculations were then made to ascertain the possible yield of the new element. These indicated the production of approximately one atom in each experiment! The calculation was correct; that is, the yield and the half-life of the product had been estimated approximately correctly. Only about one atom per experiment was produced, and still it was possible to identify the new element chemically and announce its discovery to the world!

The name mendelevium was suggested, in recognition of the pioneering role of Dmitri Mendeleev, who was the first to use the periodic system of the elements to predict the chemical properties of undiscovered elements, a principle that has been the key to the discovery of nearly all of the transuranium elements.

At this point, nuclear science had proceeded from Mendeleev to mendelevium and was now ready to go beyond.

Nobelium and Lawrencium

The next element, nobelium (atomic number 102), was also produced and identified one atom at a time by expanding upon the new techniques conceived for the production and identification of mendelevium. The difficulties that had to be surmounted here were even greater than in the case of mendelevium. In this case, the half-life of the product nucleus was so short that a direct chemical identification was not possible at the time of its discovery. Consequently, a technique involving the chemical identification of a longer-lived daughter isotope was used. This ingenious method, introduced by Albert Ghiorso, again formed the basis for subsequent experiments that have identified additional transuranium isotopes and elements.

In 1958, A. Ghiorso, T. Sikkeland, J. R. Walton, and the author, at the University of California Radiation Laboratory, reported the identification of the isotope of element 102 with the mass number 254, produced through the bombardment of curium-246 (and curium-244) with carbon-12 ions accelerated in the then-new Heavy Ion Linear Accelerator (HILAC). Any bombarding projectile heavier than helium, such as carbon-12, is known as a "heavy ion."

A. Ghiorso, T. Sikkeland, A. E. Larsh, and R. M. Latimer succeeded in the spring of 1961, following three years of preparatory work, in producing an isotope that they identified as having the atomic number 103. They employed boron as heavy ions in the HILAC, and a mixture of californium isotopes as the target. With lawrencium we reach the last element of the actinide series of elements.

Chemical Properties of Actinide Elements

The studies of the chemical properties of the actinide elements have been made with macroscopic (that is, weighable) quantities for all up to and including einsteinium (atomic number 99). Those beyond this can be studied only with amounts so small as to be unweighable, relying on their radioactive properties to work with "tracer" quantities. The chemistry of the heaviest of these has been studied using "one-atom-at-a-time" techniques. The actinide elements neptunium through einsteinium are being synthesized in increasingly large quantities through neutron irradiation processes in high neutron flux nuclear reactors in several countries.

The extensive study of the chemical properties of all the actinide elements has confirmed their similarity to the lanthanide elements and the appropriateness of fitting them into Mendeleev's periodic table as shown in Table 12.6 (which will be explained in the next to the last section).

Transactinide Elements

With element 104 we enter the relatively unexplored region of the periodic system as the first member of what we call the "transactinide" elements— that is, all elements beyond lawrencium, the last member of the actinide series. Once we have established the location of the fourteen elements in which the second inner eletron subshell is filled, it is possible to locate the positions of elements 104 to 121 and in the style of Mendeleev to predict their chemical properties with varying degrees of detail and reliability by comparing them with their homologues in the periodic table.

Consultation of Table 12.6 shows that element 104 should be a homologue of hafnium, 105 a homologue of tantalum, and so forth, until element 118, a noble gas homologous with radon, is reached. If we use the notation of Mendeleev, we may call element 104 "eka-hafnium," element 105 "eka-tantalum," and so on.

Elements 104 and 105

Elements 104 and 105 have been synthesized by heavy ion bombardments and identified using "one-atom-at-a-time" techniques. These are the heaviest known elements at the time of writing this essay.

There are competing claims for the discovery of, and hence competing suggestions for the names of, these elements. The two groups of investigators are Ghiorso, M. Nurmia, J. Harris, K. Eskola and P. Eskola, working at the Lawrence Berkeley Laboratory, and G. N. Flerov, Y. Oganesyan, V. A. Druin, and co-workers at the Laboratory for Nuclear Reactions at Dubna in the Soviet Union. The former group found alpha particle emitting isotopes, and the latter found isotopes decaying by spontaneous fission, a pro-

Table 12.6. Modern Periodic Table Showing Predicted Locations of Transuranium Elements.

H 1																	He 2
Li 3	Be 4											B 5	C 6	N 7	O 8	F 9	Ne 10
Na 11	Mg 12											Al 13	Si 14	P 15	S 16	Cl 17	Ar 18
K 19	Ca 20	Sc 21	Ti 22	V 23	Cr 24	Mn 25	Fe 26	Co 27	Ni 28	Cu 29	Zn 30	Ga 31	Ge 32	As 33	Se 34	Br 35	Kr 36
Rb 37	Sr 38	Y 39	Zr 40	Nb 41	Mo 42	Tc 43	Ru 44	Rh 45	Pd 46	Ag 47	Cd 48	In 49	Sn 50	Sb 51	Te 52	I 53	Xe 54
Cs 55	Ba 56	La 57	Hf 72	Ta 73	W 74	Re 75	Os 76	Ir 77	Pt 78	Au 79	Hg 80	Tl 81	Pb 82	Bi 83	Po 84	At 85	Rn 86
Fr 87	Ra 88	Ac 89	Rf 104	Ha 105	(106)	(107)	(108)	(109)	(110)	(111)	(112)	(113)	(114)	(115)	(116)	(117)	(118)
(119)	(120)	(121)	(154)	(155)	(156)	(157)	(158)	(159)	(160)	(161)	(162)	(163)	(164)	(165)	(166)	(167)	(168)

LANTHANIDES

Ce 58	Pr 59	Nd 60	Pm 61	Sm 62	Eu 63	Gd 64	Tb 65	Dy 66	Ho 67	Er 68	Tm 69	Yb 70	Lu 71

ACTINIDES

Th 90	Pa 91	U 92	Np 93	Pu 94	Am 95	Cm 96	Bk 97	Cf 98	Es 99	Fm 100	Md 101	No 102	Lr 103

SUPER-ACTINIDES

(122)	(123)	(124)		(153)

cess that does not characterize an isotope with a clearly identifying energy as does alpha decay.

The Berkeley research group has proposed that element 104 be named rutherfordium (symbol Rf) after Lord Ernest Rutherford, the British nuclear physicist, while the Dubna group has proposed kurchatovium (symbol Ku) after the Soviet nuclear physicist, Igor Kurchatov.

Ghiorso and co-workers suggest that element 105 be named hahnium (symbol Ha), after Otto Hahn, the discoverer of nuclear fission in uranium, while Flerov and co-workers suggest nielsbohrium (symbol Ns), after Niels Bohr, who played the key role in developing the early theory of electron structure of atoms.

Beyond Element 105

Our discussion has now carried us to the outer boundary of the known elements, namely, element 105—four elements beyond mendelevium. These last four elements were synthesized and identified on a one-atom-at-a-time basis, using techniques that were conceived in connection with experiments that led to the discovery of mendelevium. On the basis of the simplest projections, it is expected that the half-lives of the elements beyond element 105 will become shorter and shorter as the atomic number is increased, and this is true even for the isotopes with the longest half-life for each element. This is illustrated by Figure 12.2 in which the half-lives of the longest-lived isotopes (for alpha and spontaneous fission decay) of transuranium elements on a logarithmic scale are plotted against the increasing atomic number. Thus, if this rate of decrease could be extrapolated to ever-increasing atomic numbers, we would expect half-lives of the order of 10^{-10} second for the longest-lived isotope of element 110, and 10^{-20} second for element 115, and so forth, with decay by spontaneous fission becoming of dominating importance beginning with element 106. This would present somewhat dismal future prospects for heavier transuranium elements, but fortunately other factors have entered the picture in recent years. These have led to an increased optimism concerning the prospects for the synthesis and identification of elements well beyond the observed upper limit of the periodic table

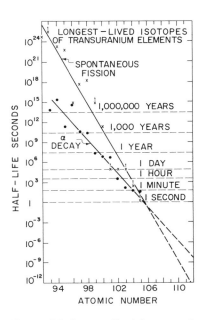

Figure 12.2. Longest-lived isotopes of transuranium elements.

—elements that have come to be referred to as "superheavy" elements. The following sections will deal with these exciting prospects, which represent another breakthrough in the overthrow of preconceived notions.

Superheavy Elements

In order to set forth the considerations giving rise to the possibility of identifying elements far beyond the region of atomic number 100, we must briefly review the theories of nuclear structure (as distinct from electronic structure) that have led to a recent optimism about superheavy elements. A prime requirement for the existence of a chemical element is the stability of its nucleus. The story began about twenty-five years ago, when Maria Goeppert Mayer at the University of Chicago and O. P. L. Haxel, J. H. D. Jensen, and H. E. Suess of the University of Heidelberg began to develop a "shell" model of the nucleus that consisted of particles moving in a field of nuclear force. The collection of particles (neutrons and protons) was shown to be particularly stable when the nucleus contained a "magic" number, or closed shell, of neutrons or protons. The stable structure could be regarded as shells, or spherical orbits, whose capacity for nuclear particles is filled; it is analogous to the filled electron shells of the noble gases. Magic numbers or closed shells of neutrons (N) or protons (Z) are generally recognized as being 2, 8, 20, 28, 50, and 82 in the elements below uranium in the periodic table. The magic number $N = 126$ is also significant in this region, as can be seen in the special stability of lead 208 ($^{208}_{82}$ Pb, or $Z = 82$ and $N = 126$), which has a doubly magic nucleus. The shell theory has evolved through many stages, to the point where the energy of single nucleons (protons or neutrons) in a deformed (nonspherical) nuclear field can be calculated as a function of distortion by using "Nilsson orbitals," a method developed by the Swedish physicist Sven Gösta Nilsson.

Complicated theoretical calculations, based on such filled shell (magic number) and other nuclear stability considerations, have led to extrapolations to the far transuranium region. These suggest the existence of closed nucleon shells at $Z = 114$ and $N = 184$ that exhibit great resistance to decay by spontaneous fission, the main cause of instability for the heaviest elements. Earlier considerations had suggested a closed shell at $Z = 126$, by analogy to the known shell at $N = 126$, but due to the coulomb field of repulsion this shell is predicted to be weak in the proton case. Although $Z = 164$ may represent an additional hypothetical point of stability, as suggested by the calculations of W. Greiner and co-workers, this possibility has not been investigated in much detail. It also corresponds to a much more extensive extrapolation and any predictions are therefore much more uncertain.

Table 12.7. Closed Proton (Z) and Neutron (N) Shells with Closed Electron (Noble Gas) Shells for Comparison. (Predicted shells shown in parentheses.)

Z	N	e^-
2 (He)	2 (^{4}He)	2 (He)
8 (O)	8 (^{16}O)	10 (Ne)
20 (Ca)	20 (^{40}Ca)	18 (Ar)
28 (Ni)	28 (^{56}Ni)	36 (Kr)
50 (Sn)	50 (^{88}Sr)	54 (Xe)
82 (Pb)	82 (^{140}Ce)	86 (Rn)
– – – –	126 (^{208}Pb)	– – – –
(114)	– – – – – –	(118)
(126)	(184)	
		(168)
(164)	(318)	

Table 12.7 illustrates the known closed proton (Z) and neutron (N) shells and the predicted closed nuclear shells (shown in parentheses) that might be important in stabilizing the superheavy elements. The indicated shells at $Z = 164$ and $N = 318$ are included for completeness, although there appears to be little hope of ever observing these on earth; it is, of course, conceivable that nuclei containing such extremely large numbers of protons or neutrons may someday be observed by identification of their characteristic radiation emitting from stars. Included by way of analogy are the long-known closed electron shells observed in the buildup of the electronic structure of atoms. These correspond to the noble gases, and the extra stability of these closed shells is reflected in the relative chemical nonreactivity of these elements. The predicted (in parentheses) closed electronic structures at $Z = 118$ and $Z = 168$ will be further discussed in the section on chemical properties of superheavy elements.

Enhancing the prospects for the actual synthesis and identification of superheavy nuclei is the fact that the calculations show the doubly magic nucleus $^{298}114$ not to be the single long-lived specimen but to be merely the center of a rather large "island of stability" in a "sea of spontaneous fission." In Figure 12.3 nuclear stability is depicted in a scheme that shows regions of known or predicted stability as land masses in a sea of instability representing forms of decay. The grid lines show "magic" numbers of protons or neutrons giving rise to exceptional stability. The doubly magic region at 82 protons and 126 neutrons is shown by a mountain; a predicted doubly magic but less stable region at 114 protons and 184 neutrons, by a hill at the island of stability. The ridges depict areas of enhanced stability due to a single magic number. The submerged ridges show isotopes that are unstable but more stable than nearby ones because of increased stability afforded by closed shells of nucleons in the nucleus.

Stability against decay by spontaneous fission is highest near the center of the island, and instability increases as the edges of the region are reached. The doubly magic nucleus $^{298}114$, for example, is predicted to have a very long half-life for decay by the spontaneous fission process, a substantially shorter half-life for decay by the emission of alpha particles, and it is predicted to be stable against beta particle decay. Adjacent odd A or odd Z nuclei may be even longer lived. Half-lives as short as nanoseconds (billionths of a second) are discernible with the instruments now used in the detection of new elements, so that there is quite a margin for error in employing the predictions. These indicate that maximum stability should be found at 110 protons and 184 neutrons ($^{294}110$), taking into account decay by both spontaneous fission and alpha-particle emission, and this nucleus is possibly stable against decay by beta particle emission.

If the half-life of a superheavy nucleus should be as long as a few times 100 million years, this would be long enough to allow the isotope to survive and still be present on the earth (as in the case of uranium-235, which has a half-life of 700 million years), provided that it was initially present as a result of the cosmic nuclear reactions that led to the creation of the solar system. Although so long a half-life is unlikely, adventurous scientists have pursued this remote possibility. Consultation of the periodic table (Table 12.6) will show that element 110, for example, is a homologue of platinum and should have chemical properties similar to those of that precious metal. Therefore, searches have been made for element 110, and also for its neighboring elements, including element 114, in naturally occurring platinum, lead, and other ores by workers in a number of laboratories.

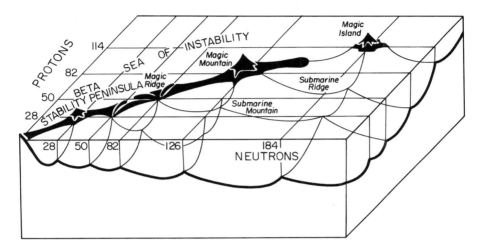

Figure 12.3. Allegorical representation of island of stability (Magic Island).

Every attempt to find evidence, direct or indirect, of the superheavy elements in nature has not been conclusive enough to give a clear-cut answer. Because of the physical limitations inherent in any experimental technique, it is not possible to say that the superheavy elements do not exist in Nature, but such existence appears unlikely. The results of such searches establish that the concentration, if they are present, is extremely small—for example, much less than one part in a million million parts of ore. Searches have also been made in cosmic rays, meteorites, and moon rocks, with generally negative results except for some indirect evidence of former presence in meteorites during the early history of the meteorite's life. The postulated current synthesis of a broad range of chemical elements, possibly including superheavy elements, in stars might enhance the prospects for finding even shorter-lived (half-life much less than 100 million years) superheavy elements in cosmic rays; elements as heavy as uranium have apparently been found in cosmic rays emanating from such stars.

Synthesizing Superheavy Elements
It appears that the superheavy elements will be observed on earth only as the result of their creation by man on earth. There appear to be nuclear synthesis reactions that might make this possible, and these should be

effected by bombarding target nuclei with sufficiently energetic projectiles consisting of heavy ions, possibly very heavy ions. Unfortunately, the yield of the desired product nuclei is predicted to be very small because the overwhelming proportion of the nuclear reactions lead to fission rather than to the desired synthesis of superheavy nuclei through amalgamation of the heavy ion projectile and the target nucleus.

The accelerators required to effect such nuclear reactions are complicated machines. Each projectile heavy ion must be multiply charged to a high degree by the removal of orbital electrons so that it can be accelerated in a machine of reasonable size, and it is very difficult to obtain intense beams of such ions. And each such ion must be accelerated to very high energies in order that it can overcome the repulsion of its positive charge by the positive charge of the target nucleus (that is, overcome the so-called coulomb repulsion) which is necessary in order to allow the amalgamation of the projectile with the target nucleus. So-called heavy ion accelerators that meet these criteria to varying degrees are in operation or under construction in various parts of the world. Two of these that have special capabilities are the SuperHILAC (the rebuilt HILAC) situated at the Lawrence Berkeley Laboratory and the tandem cyclotron combination at the Dubna Laboratory. In addition, powerful machines are under construction, notably the UNILAC heavy ion accelerator of the Gesellschaft für Schwerionenforschung (GSI) at Darmstadt, Germany. Scientists now working with or planning to work with these machines will be devoting a tremendous effort toward the synthesis and identification of the superheavy elements.

An inherent difficulty in the synthesis of the superheavy nuclei situated near the center of the island of stability is the simultaneous requirement that there be a sufficient number of neutrons as well as protons in the product nucleus. The desired product nuclei have a larger ratio of neutrons to protons than the constituent projectiles and target nuclei, and hence somewhat unusual nuclear reactions are required. There is no way of knowing prior to the actual experimental attempts which nuclear reactions will be most effective.

The identification of the presumed small yield of superheavy elements after their synthesis will require extraordinary ingenuity by the participating scientists. One method will be to steer the charged superheavy product nuclei, individually projected from the target as the result of the reaction by which they are produced, through a series of electric and magnetic fields interspersed with counters and detectors in order to determine the atomic and mass number. Another method will be to subject the recoil superheavy products, after catching them on some kind of a collector, to chemical separation procedures based on chemical properties such as those predicted in the following section. It is, of course, possible that even if the superheavy elements are relatively stable toward radioactive decay it will be impossible to detect them because there exist no nuclear production reactions with sufficient yields of these desired products. There are a number of aspects of the dynamics of the reaction of the heavy incident nuclear projectile with the heavy target nucleus which are not yet understood and which might present serious or fatal impediments to the production of superheavy nuclei; for example, the coulomb repulsive force exerted on a heavy projectile may be too large relative to the nuclear attractive force to allow the desired amalgamation reaction to occur and nuclear matter may be too viscous to allow the required ready amalgamation of the incident heavy ion with the heavy target nucleus. We might be faced with the tantalizing and philosophical dilemma of feeling certain that the island of stability is there in principle, but man lacks the ingenuity to reach it or Nature has imposed insuperable roadblocks to make arrival at the island impossible.

Chemical Properties of Superheavy Elements
So far we have dwelt largely on the nuclear constitution and stability of the superheavy elements. If the nucleus is sufficiently stable, we can then and only then begin to consider experiments concerned with the chemical properties of these elements. (The upper limit so far as stability of electronic structure is concerned seems to be at $Z = 170$ or above.) The chemical properties of an element depend almost entirely on the arrangement of electrons outside the nucleus, and these properties are closely related to the predicted positions in the periodic table.

Straightforward predictions suggest that elements 104 to 118 should be located by a simple extension of the seventh row (period) of the periodic table as shown in Table 12.6. This is much the same as was done for elements 93 and following just before the Second World War, with the important difference that now element 107 (rather than element 93) is correctly placed as eka-rhenium (that is, is placed directly below rhenium); the difference,

107 minus 93 equals 14, is accounted for by the 14-member heavy rare-earth, or actinide, series of elements in which an inner electron subshell is filled in a sequential manner. Then it is obvious that elements 119 and 120 should start the eighth row at the extreme left under the elements francium (Fr) and radium (Ra); that is, elements 119 and 120 should be eka-francium and eka-radium, respectively.

Although it probably brings us beyond the expected "island of stability," it is instructive to consider the predicted positions of even heavier elements. Quantum theory suggests that there should be, at about this point, a new type of inner subshell with 18 places for electrons, in addition to the expected third 14-member electron subshell. Table 12.6 is constructed upon the assumption that both of these subshells are filled at this point, resulting in a composite 32-member (18 plus 14) inner transition series following element 121 and thus ending at element 153. The author has suggested "superactinide" elements as the name for this series of elements and has suggested that they be placed in the periodic table as shown in Table 12.6. Then it is a simple matter to suggest that elements 154 to 168 should be located as shown; here element 168 would be a noble gas (or perhaps a noble liquid since its boiling point is predicted to be above room temperature).

Modern, high-speed computers have made it possible to calculate the electronic structures of elements up to element 168. These calculations agree, in general, with the foregoing considerations but suggest that the actual picture is somewhat more complicated. However, for the present purpose, Table 12.6 gives an overall view that is very helpful in understanding the whole picture in an oversimplified manner.

In particular, Table 12.6 enables us to predict some chemical properties of the elements in the "island of stability," the region of especial interest. Thus, element by element, number 110 falls under platinum (Pt), 111 under gold (Au), 114 under lead (Pb), and so on; this suggests that these pairs will resemble each other in their chemical properties.

If the superheavy elements can be produced and are as stable as predictions indicate, proof of their existence will be aided by the use of chemical techniques as well as methods of nuclear detection. The chemical separations must be designed on the basis of some guidance as to the nature of the element sought, and so detailed predictions of the chemistry of the new elements may be imperative. Independently of these possibly practical considerations, the prediction of the chemical properties of these elements represents an intellectual exercise of substantial dimensions replete with satisfying and challenging aspects.

Epilogue

Thus we have started with the periodic table of Mendeleev, described its development up to the time just before the Second World War and from there proceeded to mendelevium, and beyond. The beyond includes future chapters yet to be written. In the course of these developments, we have encountered a number of Copernican revolutions, including perhaps some semi- or maybe micro-Copernican revolutions.

Mendeleev's periodic table represented a breakthrough in thinking. Then the next preconceived notion to go was the general feeling that the periodic table should end at uranium. Now we have elements beyond uranium, like plutonium, which are destined to play a tremendous role in the future of mankind. The next dogma to be overcome was the habit of thinking that the beginning transuranium elements should fit into the periodic table as homologues of the platinum metals. The discovery of nuclear fission in uranium helped in the overthrow of this concept, and the discovery of the first transuranium element clinched it. But even this new concept of the position of the early transuranium elements in the periodic table was soon proved to be wrong, leading to the actinide concept. This made it possible to discover additional transuranium elements and to place not only all the early transuranium elements but elements far beyond this into proper places in a new projected periodic table. Following this, in recent years the preconceived notion of a relatively near upper limit to the periodic table—due to the rapidly decreasing half-lives, especially for spontaneous fission, as the atomic number increases—has been overthrown. The theoretical prediction of new closed nuclear shells has opened the possibility for superheavy elements. The intense experimental activity in many laboratories throughout the world should reveal whether we can take this exciting new forward step.

Notes

1. The half-life of a radioactive isotope is the time during which half of any initial quantity undergoes radioactive decay.

2. In this nomenclature the mass number (sum of neutrons and protons, or A) is given at upper left and the atomic number (protons, or Z) is given at the lower left of the chemical symbol.

Rainer K. Sachs

13 Relativity

Introduction

For more than two centuries prior to 1900 scientists' view of Nature was dominated by the ideas of Newtonian physics. The exception to this rule is Maxwell's theory of electricity, magnetism, and light. Although it was not clearly recognized at the time, Maxwell's theory does not fit very well into Newtonian physics. In this essay, Maxwell's theory is counted as part of special relativity; logically, though not historically, it belongs there.

Newtonian physics makes certain basic assumptions: that space and time are two distinct entities; that each is absolute—given a priori; that lengths and time durations are independent of the observer; that electricity and magnetism are distinct phenomena; that mass is conserved; and so on. Some of these assumptions, particularly those about space and time, seem obvious from our everyday experience. And the success of Newtonian physics made its assumptions seem well grounded. Relativity denies the foregoing assumptions and some similar ones; it relegates them to the role of preconceptions that must be abandoned.

Special relativity was introduced around 1905 by Lorentz, Einstein, Minkowski, Poincaré, and others to analyze high-speed phenomena. Today it has been checked (within its own asserted domain of validity) by literally billions of observations. Based on a conceptual revolution, it led to scientific and technological revolutions. The conceptual revolution involved unifying Newtonian concepts: space and time were unified as "space-time"; electricity and magnetism became two aspects of one entity, "electromagnetism"; it was realized that energy conservation and momentum conservation are both just partial statements of a single law, "energy-momentum" conservation; and so forth. The new concepts led to revolutionary scientific discoveries. For example, quantum theory (discussed in Weisskopf's essay) was also introduced in the twentieth century. By combining special relativity with quantum theory it was possible to predict correctly that positrons (antielectrons) must exist. Finally, the technological revolution grew from the scientific one. Its most arresting result has been nuclear weapons. Not all aspects of these revolutions can properly be called Copernican. In this essay we shall focus

Departments of Mathematics and of Physics, University of California, Berkeley, California 94720.
The author gratefully acknowledges the hospitality of Cambridge University and grants from the SRC and NSF during part of the time this essay was being written.

on the unifying conceptual revolution, which is Copernican in a very vivid and contemporary way.

To a mathematician, unification means simplification. In fact (non-quantum) special relativity is mathematically simpler than Newtonian theory. Nonetheless the concepts of special relativity are hard, precisely because of their Copernican character. Our physical intuition is often still stuck in the older Newtonian mold; our wrong preconceptions are so strong they constantly try to sneak back into the arguments. I think that when we read of how hard it was for Copernicus to abandon geocentric ideas, it does not really sink in. How can we put ourselves in his shoes at a time when only a fool can believe the earth is the center of the universe? With special relativity the difficulty is often the opposite. As long as we stick to mere formalism everything goes smoothly. But even today when we try to get a "feel" for what's really happening, we sometimes become frustrated and confused, symptoms of a Newtonian preconception sabotaging the qualitative argument.

There are also some historical similarities between special relativity and Copernicus's heliocentric theory. In 1887, Michelson and Morley measured how fast light travels from one point to another on earth. They found that the earth's motion around the sun does not affect the measured speed of light in the way that was expected from Newtonian physics. This experiment, subsequent confirmations of it, and several variations on it led to a period of confusion. Many rather arbitrary theories were suggested. Lorentz, commenting on these theories in 1904, might almost have been discussing the Ptolemaic system of epicycles:[1]

Poincaré has objected . . . that, in order to explain Michelson's negative result, the introduction of a new hypothesis has been required, and that the same necessity may occur each time new facts will be brought to light. Surely this course of inventing special hypotheses for each new experimental result is somewhat artificial.

A few years later, most specialists had been convinced by special relativity, but scientists in other areas and laymen remained skeptical. In a 1908 talk, Minkowski stressed the same themes—novelty, experimental verifiability, simplicity, and incompatibility with long-held beliefs—that characterize Copernicus's work:

The views of space and time which I wish to lay before you have sprung from the soil of experimental physics, and therein lies their strength. They are radical. Henceforth space by itself, and time by itself, are doomed to fade away into mere shadows, and only a kind of union of the two will preserve an independent reality . . . it is only in four dimensions that the relations here taken under consideration reveal their inner being in full simplicity . . . on a three dimensional space forced on us *a priori* they cast only a very complicated projection . . . there will be ample suggestions for experimental verifications . . . which will suffice to conciliate even those to whom the abandonment of old-established views is unsympathetic or painful.

The second section of this essay discusses special relativity, with the emphasis on how painful the abandonment of preconceptions can be.

General relativity was introduced by Einstein about ten years after special relativity. It integrates special relativity and Newtonian gravitational theory into a new, coherent scheme. The empirical foundations of general relativity are still somewhat shaky, but most physicists accept it. It has become very fashionable recently, primarily because of the discovery of pulsars. It is believed that a pulsar is a neutron star—a star about as massive as the sun with a diameter of perhaps 10 miles.[2] Newtonian gravitational theory cannot be used for analyzing such a dense star; general relativity (or one of its rivals) is indispensable.

Like special relativity, general relativity involves a conceptual revolution. One must somehow imagine that space-time, instead of being given a priori, can be distorted by the matter in it. Gravity and space-time are unified into two aspects of a single entity, distorted space-time. The section on general relativity discusses a few general relativistic phenomena, again with emphasis on preconceptions.

This essay in no sense attempts to argue that relativity must be correct. Moreover, the reader who wants a broad survey of relativity, a summary of the empirical evidence for it, or a balanced account of its applications must go to one of the longer treatments.[3] I shall not even be able to offer a convincing reason why our Newtonian prejudices are not realized in nature. Perhaps there is some underlying theme of "less is more": subtleties are required to ensure mathematical simplicity and a unity in diversity; subtleties duly occur.

Special Relativity

One way to describe the light from your lamp is as a beam of particles, called photons. The analysis of light and photons has played a central role in relativity from the start, so we will often discuss photons. The first subsection analyzes the speed of light and the preconception that speeds add "normally." The second subsection discusses the rest mass of photons, the energy of photons, and some preconceptions about mass. The third subsection is a digression on a recent discovery in cosmology which has a slightly counterrevolutionary implication. The last subsection analyzes relativity of simultaneity and the most stubborn of our wrong preconceptions, the simultaneity is absolute.

The Speed of Light

By comparatively inaccurate direct measurements, such as the Michelson-Morley experiment already discussed, and highly accurate indirect ones it has been found that the speed of light is absolute, is universal. Imagine a photon coming to us from a star. Suppose that on the way the photon passes a nonrotating rocket that is not disturbed by any external influences. If an observer on the rocket carefully measures the photon speed he will find a definite number c, about 1 billion feet per second, no matter how his rocket is moving. The observer uses the rocket as a frame of reference; he uses meter sticks, clocks, and so forth fixed to the rocket. Since the rocket is not rotating and is undisturbed, it is said to determine an "inertial" frame of reference. A second nonrotating, undisturbed rocket moving with respect to the first would determine a second inertial frame of reference. The speed of light is called absolute or universal because it is the same in all inertial frames of reference. That may sound reasonable; but accepting it involves dropping the very vivid preconception that speeds add in the usual way. To see what is involved, let us think of a more familiar situation, replacing photons by cars.

Imagine you are sitting on the grass strip in the middle of a North-South superhighway. Suppose you see blue cars, representing the photons of blue light, going both ways. You measure the speed of the northbound cars; each one is going at exactly 50 mph, representing the speed of light c. Each southbound car is also going 50. That in itself is a little odd—what if they want to stop?—but not yet painful.

Your friend comes along the grass strip, driving a mower at a steady 20 mph North. He stops and you ask him whether he found the pattern of car speeds peculiar.

"Yes."

"Aha!" you say, "You found the northbound cars were overhauling you at 30 mph and the southbound ones were meeting you at 70, didn't you?"

"No. The northbound ones were overhauling at 50; I guess their ground speed must have been 70. The southbound ones were meeting me at 50;[4] their ground speed must have been 30."

"Nonsense."

"Hop on and you'll see." You get on. The mower speeds up to 20 again. Without looking at the road or grass you measure the overhauling speed and the meeting speed. Both are 50. Why? Because car speed is universal. The mower stops and you measure again. 50 again, both ways. On the ground you found a basic law of Nature: the speed of cars is 50 mph. On the mower your friend has found exactly the same basic law. The two of you have found a metalaw: the basic laws of Nature are the same in all inertial frames of reference, even in inertial frames moving relative to each other like the mower and the grass. That all inertial frames are equal is one underlying idea of special relativity.

Perhaps you have trouble accepting the result that no matter how fast you go a photon that overhauls you always overhauls you at speed c; Copernicus also had problems. The peculiar behavior of photon speeds is related to the unification of space and time into space-time. Time intervals alone or distances alone do not have absolute numerical values, though a unified entity, the "distance-time" (technically called an "interval") does. Now a speed is a distance divided by a time. When its numerator and denominator become fouled up, the speed gets fouled up in such a way that speeds no longer add the way we expect. The inventors of special relativity were somehow able to overcome their preconceptions and to turn the argument around. Accepting the universality of the speed of light, they guessed that space and time must be unified.

Rest Mass and Energy

In Newtonian physics it is usually taken for granted that every particle has a positive mass and that total mass is conserved (that is, that mass can neither be created nor destroyed). In a certain sense special relativity denies both preconceptions.

In special relativity every isolated particle is assigned a rest mass. The rest mass is just a number, independent of the speed of the particle and absolute in the sense that every careful observer will assign the same rest mass, no matter how the observer is moving. Roughly, the rest mass is the amount of material in the particle. Somewhat more precisely, rest mass is often measured by letting the particle bang into a known particle. In a low-speed situation the more the known particle recoils, the bigger the unknown rest mass. But, in contrast to the Newtonian assertion, there are particles of zero rest mass. In fact, photons have zero rest mass. Despite having zero rest mass, a photon has nonzero energy and momentum; photons can cause recoils in collisions, but of course not in a low-speed situation because they always go at the speed of light.

Again in contrast to Newtonian theory, special relativity makes no claim that rest mass is conserved. For example, a pion is a particle with a nonzero rest mass. Often a pion decays ("explodes"), and the only fragments are photons, each of which has zero rest mass.

By way of partial compensation, special relativity demands that the law of energy conservation be modified to include what is called rest-mass energy. Imagine a particle of positive rest mass m at rest in the laboratory. Nineteenth-century Newtonian physics would say that the particle contributes nothing to the total energy measured in the laboratory. Special relativity assigns a rest-mass energy mc^2 to the particle. Thus, if a pion at rest decays, it cannot simply disappear without descendants. It must make fragments whose total energy is mc^2.

If the fragments are photons, one says, somewhat loosely, that rest mass has been converted to energy. More accurately, one says rest-mass energy has been converted to photon energy. A nuclear power plant works by converting rest-mass energy into other forms of energy. In that case, much less than 1 percent of the total rest-mass energy is actually converted;[5] but, roughly speaking, c is so large that a little bit of m makes a lot of mc^2.

To discuss what happens when rest-mass energy goes into photon energy, we must say a bit more about photons. Visible light, radio waves, x rays, and so on are all beams of photons. What distinguishes the various kinds of photons? Their frequency. Frequency is important in various ways. First, for visible photons, different frequencies mean different colors. A low-frequency visible photon gives red, a higher-frequency visible photon gives blue, at higher frequencies still one has violet. Second, frequency can be grasped intuitively by imagining that you are on an ocean shore and waves are coming in. If you count the number of wave crests that hit the shore each hour, you get the wave frequency f in wave crests per hour. It takes quantum theory to explain why a photon, a particle, has a wavelike quantity such as frequency associated with it. Finally, the frequency is important because quantum theory requires that the energy of a photon with frequency f is hf, where h is a universal constant, called Planck's constant.

Is the frequency of a photon an absolute quantity, independent of reference frame, like the speed c or the rest mass 0 of the photon? No. You can see roughly why if you imagine that at the ocean you set out in a motorboat from shore. You will hear the waves slap on the boat; the faster you go, the more waves slap per hour; that is to say, running into the wave gives a higher frequency. Photons are similar. Going in the same direction as a photon and forcing it to overhaul you give a lower frequency, a lower energy, and for blue photons a shift toward the red (even though the speed c is not changed). Running head on at the photon gives a higher frequency, a higher energy, and for blue photons a shift toward the violet. For example, recall that on the grass of the superhighway you saw all the cars as blue, representing photons of blue light. What would your northbound friend on the tractor see? The northbound cars would be red, because of the low frequency; and the southbound cars would be violet, corresponding to a high frequency.

This dependence of observed frequencies on the relative motion of source and observer is known as the "Doppler shift." As discussed in Part I, the Doppler shift can often be used to estimate how fast a distant celestial body is running away from us or toward us.

The Microwave Photons[6]

A recent discovery in cosmology has clarified and slightly modified the statement, made in the first subsection, that the laws of nature are the same in all inertial frames of reference. Nowadays one should add a minor qualification. After all, we live in a particular universe, not in the set of all universes consistent with the laws of nature. Consequently, there may be an inertial frame of reference which is "more equal" than the others because our particular universe looks especially simple in the preferred reference frame. In fact, such seems to be the case.

It has been found that there are microwave photons, somewhat similar to the photons that carry signals to your television set, which come to us from deep in the universe. Oddly enough, we observe the same pattern of microwave photons no matter which way we look. For example, to within the accuracy of our measurements, frequency is the same in all directions. One says the microwave radiation is "isotropic," that is, is direction independent. The isotropy, and some other features of the radiation, seem to indicate a deep simplicity of the universe as a whole, so cosmologists are quite excited by the discovery. What concerns us here is a less profound question, the implication of the isotropy for special relativity.

We see the microwave photons at the same average frequency in all directions, just as on the superhighway you saw all the cars as blue. As discussed in the previous subsection, a very high-speed rocket would not see such a simple pattern. High-frequency photons would be seen coming toward the nose, low-frequency photons would be seen coming toward the tail. The rocket would be like your friend on the tractor who saw red northbound and violet southbound cars, a less symmetric pattern. Thus the earth, or probably more accurately the center of mass of our local group of galaxies, defines a preferred reference frame. One might say it is "absolutely at rest" or "at rest with respect to the microwave photons."

To have a preferred reference frame is counter to the spirit of special relativity. Has the discovery of the microwave photons led to a kind of Ptolemaic counterrevolution? No. The amount of energy carried in by the microwave photons is very small. For the vast majority of purposes, they can be wholly neglected. To the extent that they can be neglected, special relativity remains valid. All that has happened is that the domain of validity of special relativity has been defined a little more precisely than before.

Relativity of Simultaneity

Returning to our theme of preconceptions, we shall discuss relativity of simultaneity. Mathematically one of the simplest effects, it is intuitively perhaps the hardest of all because the preconception that simultaneity is absolute is so stubborn.

Sometimes a star suddenly increases its output of light a millionfold or more. Then one speaks of a nova. The onset of a nova will here serve as an example of an "event"; an event is sometimes defined as "something which happens at one place at one time, like an explosion." For our purposes it will not matter that a star is an extended body; we shall think of a nova as a point light source that instantaneously gets much brighter. Another example of an event is the execution of Charles I, which took place January 30, 1649, on the earth—one point at one time for our purposes.

Now suppose an astronomer were asked whether a particular nova occurred before, simultaneous with, or after the execution. A nineteenth-century astronomer, believing in the absoluteness of simultaneity, would have seen nothing wrong with the question as such. He might have pointed out that one must be careful: none of our measurements are infinitely accurate, the light from the nova takes some time to reach us, and so on. But he would assume that if one sufficiently careful observer concludes "before" all other sufficiently careful observers must agree. Special relativity denies this preconception. The question "before, simultaneous with, or after?" need not have any absolute, reference frame independent answer at all. An ideally careful observer using instruments fixed on the earth might conclude "before" even though a similar observer in a high-speed rocket concludes "after."

Such ambiguities occur only when nothing can get from either event to the other at a speed less than or equal to the speed of light. For example, suppose a student throws a tomato (event 1) and the tomato hits a speaker (event 2). Then something—the tomato itself, for example—was able to get from event 1 to event 2 at a speed less than 9 billion feet per second. Here all careful observers will agree that the throw was earlier than the splash.

Even when there is an ambiguity about "earlier than, simultaneous with, or later than," all observers can still agree on statements about the unified entity, space-time. We may not be able to say a nova was absolutely earlier than King Charles's execution. But we can be sure the nova there and the execution here were separated in space-time; the "interval" between them is nonzero.

General Relativity

According to general relativity, the gravitational effects of a massive body can be described as a distortion, or curvature, of space-time. This idea was suggested by a fact known to Galileo and later indirectly verified to very high accuracy: different bodies dropped from rest in the same gravitational field fall at the same rate. Now if gravitational effects do not depend on the particular moving body, what do such effects depend on? Einstein suggested regarding gravity as a property of the one thing common to all bodies—the space-time in which they move.

To think general relativistically one must overcome various preconceptions. First, just as in special relativity, one must consider space-time rather than space separately and time separately. But space-time is already rather hard to think about and distorted space-time somewhat harder still. So we will here talk about a special case: static (that is, intrinsically time-independent) gravitational fields. In discussing static gravitational fields, though not in general, it is proper to think about space separately and time separately. That will simplify our discussion.

Another preconception to be overcome is that space must be flat. Curved space is discussed in the next subsection. Harder is overcoming the preconception that time cannot be distorted. We have seen in the previous subsection that concepts of time, such as simultaneity, may depend on the observer's state of motion. In the second subsection here we analyze a situation where the way time flows depends instead on location; the effect is due to gravity and is thus a specifically general relativistic one. The final subsection on the Pound-Rebka experiment analyzes the same effect in more realistic terms.

The Distortion of Space

There is a simple way to check whether space is distorted. We pick any three points and stretch strings very tautly between them to get a triangle. Then we measure the angles. If the sum of the angles does not come out 180° for all triangles, space is distorted. General relativity predicts that if we were able to conduct this experiment with a triangle just circumscribing the sun there would be a deviation of about one part in a million from 180°.

Actual observations use planets and photons. In these observations the distortion of space and the distortion of time, to be discussed later, are both relevant. An analogy that (unfairly) ignores this complication may help clarify roughly how the curvature of space can be related to gravity.

Imagine a big rubber sheet. In the analogy you are not supposed to think about the direction perpendicular to the sheet, just about the directions within the sheet. The two-dimensional sheet represents all of three-dimensional space. For a situation where gravity is negligible, one pictures the sheet as flat and horizontal. Then a rolling marble on the sheet (representing a photon) will just keep going.

Now suppose instead that a heavy ball, representing the sun, is placed on the sheet. There will be a distortion near the ball. The fact that there is a distortion can be noticed by measurements within the sheet. For example, suppose we choose three points on the sheet, join them by the shortest lines that lie wholly within the sheet, and measure the angles. In general, we will not get 180° for the sum. Now again imagine a marble rolled on the sheet, coming close to the ball. Its path will be disturbed. One can notice that the marble has changed direction by comparing its initial direction, far away from the ball where the distortion is negligible, to its final direction.

The analogy suggests that one could get a photon to go in a circle. For the sun and probably even for a neutron star this is false—the curvature just is not big enough. But it is likely (not certain) that there are "black holes." A black hole is a theoretical prediction of the ultimate state of a collapsing heavy star. The star presumably goes into a state so dense that nothing can escape far from its surface, not even photons. From far outside, the black hole thus looks black. But as the star collapses it leaves its gravitational distortion behind in the outside world, rather like the grin of the Cheshire cat. General relativity predicts that a photon could go in a circle in the gravitational field.

The Distortion of Time
The preconception that time cannot be distorted is far more stubborn than the preconception that space is flat. This subsection gives an example of time distortion, with the numerical values exaggerated for vividness; the next subsection discusses the actual numbers.

Imagine a ski lift, consisting of a cable and many seats. The seats are attached firmly to the cable in an "endless chain": they go to the top, are carried around a big pulley, come back down, are carried around another big pulley, and start back up. Suppose the lift is run, night and day, at a steady pace. Now suppose the attendant at the bottom finds 1000 seats pass each hour and the attendant at the top finds 999 seats[7] pass each hour, what do they conclude?

First they look to see whether some clown in the middle of the slope is removing one upbound seat each hour and restoring it on the way down. No culprit is found (in any case the gap would be noticed). Then they trade watches, but they still find 1000 seats per hour at the bottom and 999 at the top. Now they look for subtler explanations. Perhaps a cold wind at the top

slows down the top watch. To check this they use various watches of different construction. Surely a cold wind will affect differently constructed
watches in different ways. Always they find the same ratio, 999:1000.

Now what? They finally decide that there is some unknown "thing," perhaps at the top or perhaps at the bottom, which affects all good watches
equally. But time itself is really just whatever is common to all good
watches. So they are in effect deciding that the unknown thing is a location-
dependent distortion of time.

According to general relativity, the situation is the following. The earth's
gravity distorts time. The distortion depends on height above the earth. Far
away from the earth the distortion is negligible, so one may regard the
nearby distortion as a distortion of time "relative to time at infinity." It is
the different distortions at different heights which lead to the unexpected
behavior of the ski lift. The unknown thing our attendants guessed at is
essentially the gravitational potential (gravitational potential energy per
unit mass for a small body in the gravitational field of the earth).

The Pound-Rebka Experiment

Pound and Rebka, at Harvard, measured an effect like that discussed in the
previous section. In oversimplified terms, their experiment consisted of
shooting a photon from the bottom of a 60-foot tower to the top. They compared the photon frequency at the top with that at the bottom and found
a slight difference.

To estimate the frequency difference they obtained, we can argue as follows.[8] The climbing photon loses ("kinetic") energy as it struggles up
against the earth's gravitational field. But by the relation energy $= hf$, given
earlier and discussed more extensively on page 303, a loss in energy gives a
loss in frequency, so the top frequency is lower. How much lower? The
photon's loss in kinetic energy is balanced by its gain in gravitational potential energy. That gain is about $hfgH/c^2$, where H is the 60 feet the photon
climbs and g is the acceleration due to gravity at the earth's surface—about
32 feet per second squared. If you plug in the numbers you will get about

$$\frac{\text{Frequency difference}}{\text{Frequency}} = gH/c^2 = 0.000000000000002,$$

not a very big difference. But Pound and Rebka were able to measure the
difference and obtained the predicted value to good accuracy.

How is the photon frequency related to the ski lift? After all, as discussed on page 302, a photon frequency is interpretable as a certain number of wave crests per unit time. In the argument there is no essential difference between wave crests per hour and seats per hour: both are frequencies. How is general relativity relevant? In two ways. First, it, or one of its competitors, must be used to make the argument systematic. Second, the predicted value does not really depend on the fact that we are using photons; note for example that Planck's constant h does not enter into the final result. When we calculate "any old thing per unit time," it is believed to be the "unit time" that is different (distorted) at the top compared to the bottom.

Conclusion

The earth is nearly rigid. It rotates so slowly that the rotation speed at the surface is much less than the speed of light. Its gravity is very small compared to that of a neutron star. It seems our imagination is crippled by these rather accidental features. If we lived near a very wobbly, rapidly rotating neutron star, the ideas of relativity might seem quite natural to us. Like Copernicus's heliocentrism, Einstein's relativity asserts that earthbound ideas obscure the actual simplicity of the world.

Notes

1. The Ptolemaic system is outlined in Neyman's introductory essay, Chapter 1. The two quotes given here are from H. A. Lorentz et al. *The Principle of Relativity*. W. Perret and G. B. Jeffery translators, Methuen, London, 1923.

2. Neutron stars are discussed in Chapter 5, the essay by G. and M. Burbidge.

3. Claude Kacser, *Introduction to the Special Theory of Relativity*. Prentice-Hall, Englewood Cliffs, N.J., 1967, is excellent. Intended for liberal arts students, it requires only a good knowledge of high school algebra; but it is far more precise and systematic than the treatment here. It has a short treatment of general relativity, including the Pound-Rebka experiment. An outstanding modern graduate physics text on relativity is by Steven Weinberg, *Gravitation and Cosmology*. John Wiley & Sons, New York, 1972.

4. For actual cars, this number would be about 69.99999999999999, almost the 70 that you anticipated.

5. The main limit is a quantum theoretical conservation law called baryon number conservation.

6. Microwave photons are also discussed in Chapter 5, the essay by G. and M. Burbidge.

7. For a ski lift of reasonable height, on earth, 999.99999999999 would be more realistic.

8. The argument uses both Newtonian gravitational theory and special relativity. These theories, in their standard form, do not fit well together, so there are ambiguities. For example, should one write "kinetic" energy of a photon $= hf$ or write kinetic energy plus gravitational energy of a photon $= hf$? We use the first, which gives the correct answer to high accuracy. A more systematic argument requires using a theory, such as general relativity, which includes both Newtonian gravitational theory and special relativity in a comprehensive framework.

Victor F. Weisskopf

14 The Impact of Quantum Theory on Modern Physics

Il est à peine nécessaire de faire remarquer combien la théorie des Quanta s'écarte de
tout ce qu'on avait imaginé jusqu' ici; ce serait là, sans aucun doute, la plus grande
révolution et la plus profonde que la philosophie naturelle ait subie depuis Newton.

—Henri Poincaré (1912)

Physics before 1900

The discovery of the quantum of action by Max Planck at the turn of our
century was the beginning of one of the most fruitful and also most revolutionary developments in science. Rarely have our views regarding the basis
of the properties and behavior of matter been changed and expanded as
profoundly as in the three decades following Planck's discovery. It is useful
to sketch the situation of physics before the year 1900 in order to evaluate
the significance of the new insights gained by the quantum ideas.

Physics in the second half of the nineteenth century was dominated by
two important developments. One was the success of the atomic view of
matter; the other was the growing insight into the nature of electric and
magnetic phenomena.

The view that matter is an aggregate of atoms or molecules (the molecules
being made up of atoms) is a very old one. It received its scientific basis
from the investigations of Dalton at the beginning of the nineteenth century. The language of chemistry was based upon the concepts of atoms and
molecules.

The atoms were considered as unchangeable. Different atoms combine to
form a great variety of chemical compounds. These compounds are usually
made up of molecules, each molecule of a given chemical compound is a
certain well-defined combination of atoms. Chemistry deals with the way
atoms join to molecules, with the "bonds" that keep them bound within the
molecule, and with the methods and processes that lead to changes and new
combinations in the structure of molecules. Chemistry and physics were
rather separate sciences at that time; chemical and physical properties of
materials were kept rather separate. The former referred to the characterization and transformation of chemical compounds, the latter referred to the
quantitative description of material properties in bulk, such as density,
elasticity, conductivity, viscosity, surface tension, and so on. The concept of

Department of Physics, Massachusetts Institute of Technology, Cambridge, Massachusetts
02139.

the atom or molecule did not play an essential role in the language of physics.

The first important use of the molecular concept in physics was the kinetic theory of gases, an idea conceived in the nineteenth century, which was brought into a quantitative form mainly by Clausius and Maxwell. The properties of dilute gases were successfully described by assuming that the gas consisted of molecules whose dimensions are small compared to their average distance in the gas. They move randomly in straight lines until they collide with one another and with the walls. The higher the temperature of the gas, the faster do the molecules move. The kinetic energy of the molecules is proportional to the temperature. Loschmidt determined the size of these molecules to be of the order of 10^{-8} cm. Many properties of gases could be explained by this theory. The pressure of a gas on the wall is the result of the impacts of the molecules; this gives the right dependence of the pressure on the temperature. The viscosity and the thermal conductivity of gases are explained satisfactorily by this picture. The significance of that theory and its subsequent development by Boltzmann and Gibbs are twofold: First, it gave great support to the atomic hypothesis and, second, it put the idea of heat as atomic random motion on a firm quantitative basis. Thermodynamics became a branch of mechanics.

The other important development in the nineteenth century concerned the electromagnetic phenomena. Within a short time interval—roughly between 1834 and 1865—the connection between electricity and magnetism was discovered, the concept of the electromagnetic field was conceived, its laws and its interaction with currents and charges formulated and brought into a symmetric and elegant form by Maxwell in his famous equations. He recognized that light is an electromagnetic wave. Optics has become a branch of electricity. This extraordinarily rapid growth in the understanding of electric and magnetic phenomena is paralleled only by the rapid development of quantum mechanics in our century. The electromagnetic nature of light and the kinetic nature of heat were the two great milestones in the advance of physical knowledge in the nineteenth century.

Toward the end of the century certain problems and difficulties appeared within the framework of these ideas, which prepared the ground for the subsequent revolution. These difficulties can be divided into internal and external ones, the former being logical inconsistencies within the theories, the latter being the results of new experiments which seem to contradict the theories. One of the most important internal problems of electromagnetism was connected with the electromagnetic effects of moving charges, currents, and light sources. These effects seem to distinguish certain inertial frames of reference and, therefore, would allow the determination of an absolute rest system in space. For example, the observed light velocity should depend upon the motion of the observer. Einstein's theory of relativity—a child of the twentieth century—has solved this problem once and for all, by one of the most ingenious ideas ever conceived in science. It was not electrodynamics but mechanics and kinematics that needed a new formulation. It led to a new conceptual framework for the unification of mechanics, electrodynamics, and gravity, which brought with it a new conception of space and time. This framework of ideas can be considered, in some ways, as the crown and synthesis of nineteenth-century physics, rather than a break with the classical tradition. In contrast, quantum theory was such a break; it was a step into a world of phenomena that did not fit into the web of ideas of nineteenth-century physics.

The internal problems of kinetic gas theory are more directly connected with the quantum revolution.

Heat energy is energy of motion. It was L. Boltzmann who studied carefully how the energy of heat at a given temperature T is distributed over the different ways in which a mechanical system can move. In a gas, for example, the molecules travel randomly in all directions; they also can rotate around their axis, the atoms within the molecule can vibrate against one another. Each of these different modes of motion is called "a degree of freedom" of the molecule. Boltzmann and his contemporaries found that the heat energy should distribute itself evenly over all these different modes of motion. In the average each degree of freedom should get the same heat energy. This is called the "equipartition theorem." In fact, the energy that each degree of freedom should get is proportional to the temperature; let us call it ε. Boltzmann derived the famous relation $\varepsilon = \frac{1}{2} kT$, where k is the so-called Boltzmann constant. For example, one may conclude that in an ideal gas a molecule has 5 degrees of freedom, since it can move in three dimensions and can rotate around an axis whose direction is determined by two angular coordinates. One then would conclude that the heat energy per molecule is five times the unit ε. This result was borne out by experiments at ordinary temperatures. However, the whole idea of 5 degrees of freedom per molecule is most questionable. Why is there no vibrational motion? (Indeed, gases of some heavier molecules show a heat energy larger than 5ε.) There is a contradiction in principle here which we will call the "Boltzmann paradox." Any piece of matter of any size must have an undetermined number of degrees of freedom. It may undergo deformations of all kinds, it may have an internal structure that can be set in motion. Not only can the molecule rotate and vibrate, but its constituents can move in various ways against each other. The electrons within the atoms have their own motion that must be counted as additional degrees of freedom, the atomic nuclei have an internal structure that can be set in motion, and so on. The equipartition theorem should be valid for any degree of freedom. Even if the molecules were very hard to deform, the energy taken by them would still be ε per degree of freedom; hence, in the classical theory, the specific heat is undetermined; the concept is self-contradictory since, logically, the number of degrees of freedom of any finite chunk of matter is infinite, whatever its size.

The same paradox appears even more forcefully in the classical theory of a radiation field at a given temperature. Here the number of degrees of freedom is infinite (an infinite span of frequencies is available). Since every material at finite temperature emits some radiation, it would follow from the theory that it is impossible to establish a thermal equilibrium between a material and its radiation in a cavity inside or outside the material. Energy would flow into the insatiable radiation field with its finite number of degrees of freedom. The classical equipartition theorem destroys the theory upon which it is based.

We now come to the external difficulties, coming from experimental results that contradicted the accepted theories. There were difficulties at low temperatures. The heat energy of gases and solids at low temperatures (usually below 100 °Kelvin) is much less than the expected value of ε, multiplied with the number of degrees of freedom. This is in contradiction to the equipartition theorem. At low temperature, ε is very small; it seemed that matter does not like to accept small amounts of heat energy.

Let us consider electric and optical phenomena. Toward the end of the nineteenth century the electron was discovered as the carrier of currents in metals and as an important part in the structure of atoms. Electrons within the atoms were recognized to be responsible for the radiation effects of matter. It was observed that light is emitted by matter with well-defined frequencies that are characteristic of the material (the "spectrum" of the material). It was natural to assume, therefore, that electrons are elastically bound in atoms and perform oscillatory vibrations. Elastic vibrations do have characteristic frequencies, which could be the ones that are typical for each material. But again the equipartition theorem causes some trouble. It predicts a radiative behavior of matter in plain contradiction with the facts. The theorem would require that the electrons should oscillate with all the available frequencies, at any temperature, only weaker at low temperatures and stronger at higher ones. In fact, however, matter emits lower frequencies at low temperatures and higher ones at more elevated temperatures. When heated, a piece of matter glows red, then yellow, and finally bluish white. It was not immediately recognized that the classical theories do not account for this trivial daily experience. As often in the history of science, the relations of simple and most generally known facts to a theoretical edifice of ideas are slow in getting recognized.

The final deadly blow against a classical view of atomic structure came as late as in 1911 when Rutherford published his findings of large-angle scattering of α-particles by thin metal foils. He correctly interpreted this observation as proving that most of the mass and the positive charge of the atom is concentrated in a small nucleus. Hence he concluded that the electrons in the atom are *not* elastically bound but must move around the nucleus like planets around the sun. With this observation, any hope was lost for an understanding of atomic dynamics based upon classical mechanics and electrodynamics. The planetary model of an atom would not emit light with selected definite frequencies. What is even worse, it would not exist for any length of time: the electrons would quickly fall into the nucleus because of their energy loss through radiation. A simple calculation shows that an orbit of atomic size (10^{-8} cm) would not last longer than 10^{-9} sec. A planetary system of a nucleus and electrons would never exhibit the typical characteristic properties of size, shape, symmetry, and stability against collisions that atoms manifestly exhibit; moreover, it would quickly collapse into an inert electrically neutral body consisting of a nucleus with its electrons sitting at its surface.

Toward the end of the nineteenth century many other observations were made which pointed toward the impossibility of a classical explanation of atomic structure. None were as conclusive and direct as Rutherford's discovery. It took many decades and a continuous barrage of a growing array of facts before it became clearer to the physicists that the nature of the atoms cannot be explained by the concepts of classical physics. The most elementary observations that every chemist makes in his daily work are at variance with these classical principles. He observes his atoms and molecules exhibiting properties characteristic to the species, all members of one kind, completely identical with one another. This identity of properties and behavior survives collisions, tight packing in liquids or solids, and subsequent evaporation; it can be recovered after spark discharges, ionization, and any other violent interference. When the original conditions are restored, the atom or molecule regenerates itself with identical properties and no trace of the previous history is left. These are features that are utterly alien to the framework of ideas in classical physics. In particular, an atom consisting of electrons circling around the nucleus, like the planets around the sun, would never exhibit the stability and the regenerative ability according to the laws of classical mechanics.

The discovery of radioactivity at the end of the nineteenth century shook the physicists even more than the complex web of observations concerning the chemical, electric, and optical properties of the atom. Here they observed particles and radiations exhibiting energies that exceed by factors of the order of a million the energies one has associated with inner-atomic processes. The law of the conservation of energy was put in doubt, there was a heightened feeling of mystery surrounding the structure of atoms. They did not know that they were looking at phenomena pointing to the next level of insight, into the internal structure of the nucleus. Nevertheless, these phenomena did provide the sharp tools that were necessary to penetrate into the details of atomic structure. Rutherford made use of radioactively emitted charged particles when he discovered the existence of the atomic nucleus.

Basic Discoveries Leading to the Quantum

Around the turn of the century, a number of discoveries were made and new ideas conceived which proved to be the basis of the subsequent great revolution in physics.

The first one was an exact measurement, by Rubens and Kurlbaum, of the radiation emitted by a material kept at a fixed temperature ("black radiation"). At low frequencies the observed radiation was what one would expect from Boltzmann's equipartition theorem, but it broke down completely at high frequencies. In the last year of the nineteenth century, Max Planck wrote his historic paper in which he showed that the correct radiation is obtained by making a seemingly abstruse assumption: the energy that the atom emits or absorbs in form of radiation, must be "quantized"; it can assume only values that are an integer multiple of a quantity $h\omega$ (ω being the frequency of the radiation). Here h is the famous constant of Planck; it has the dimension of an action and therefore is referred to as the "quantum of action."

The next important step was the discovery of the existence of light quanta. In 1905 Einstein was able to infer it from the behavior of the blackbody radiation at high frequencies. He concluded from this rather indirect evidence that the energy of monochromatic light must always be an integer multiple of $h\omega$. (Planck assumed only that the energy absorbed or emitted by an atom is quantized.) A direct evidence of what Einstein suggested was found much later (1915) in the photoelectric effect, when it was observed that the energy of electrons ejected from a metal by monochromatic light is independent of the intensity and always equal to the energy carried by the light quantum that ejected the electron.

Another important discovery came from a detailed study of the spectrum of atomic radiation. It is the combination principle of Rydberg (1900) and Ritz (1908). The values of the many characteristic frequencies emitted (or absorbed) by a given atom were found to be the differences between a number of so-called term values T_i:

$$\omega_{ik} = T_i - T_k.$$

These term values are characteristic for each atom. Such a situation runs counter to any model of oscillatory vibrations in the atom.

Then, there were the experiments of Franck and Hertz (1914), in which the energy transferred to an atom by a beam of electrons was measured. They concluded from these experiments (not immediately but in the course of several years of experimentation and interpretation) that energy can be supplied to an atom only in definite characteristic amounts. How great was the surprise, when it turned out that these amounts are equal to hT_i, where T_i are the above-mentioned term values and h is again Planck's constant! The energy of atoms seems to be quantized in the sense that it can take on only certain definite values, namely the term values hT_i. These are the same term values that determine the characteristic frequencies of the atom by the above-mentioned combination principle! They form a series of energy values that represent the typical "quantum states" of the atom. (See Figure 14.1.) For the simplest atom—the hydrogen atom—Balmer (1884) constructed a, formula that represented very well its spectral lines. From that formula follows a simple expression for the term values of hydrogen:

$$T_i = -\frac{R}{n^2},\tag{14.1}$$

where n is any positive integer number and R is the Rydberg constant $R = 3.290 \times 10^{-15}$ sec^{-1}. In other atoms the term values do not follow such simple and regular patterns. However, they have one property in common with Equation 14.1: they become denser with increasing values of T_i and reach a limit (T_∞).

The existence of the light quantum offers an immediate logical bond between the combination principle and the Franck-Hertz discovery of the

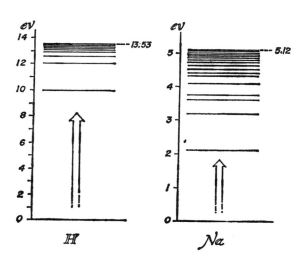

Figure 14.1. Energy of quantum states of hydrogen (H) and sodium (Na).

quantization of atomic energies: light emission or absorption by an atom is a transition from one energy state hT_i to another hT_k; the emitted or absorbed frequency is determined by the conservation of energy, since the energy of the light quantum must correspond to the energy difference of the two atomic states.

Today it is hard to realize the utter absurdity of such ideas in the framework of classical physics. It was inconceivable that certain ways of internal atomic motion are allowed and others not, and that atoms perform *instant* jumps from one state to another (*natura non fecit saltus*), producing a *continuous* electromagnetic wave.

In spite of its conceptual difficulties, the idea of characteristic atomic energy levels hT_i already contains the main principle of quantum theory and offers an immediate explanation to many outstanding problems. It provides an explanation for the stability of atoms against influences from the outside, disturbances and collisions. There is definite difference in energy between the lowest level and the next higher level. Usually the atom finds itself in the lowest level: the ground level or ground state. The atom cannot accept energies less than the energy difference $hT_2 - hT_1$ between the ground state and the next higher level. As long as the effects from the environment are less energetic than that amount, the atom will remain impervious to any outside influence.

The failure of the equipartition theorem also is easily understood in these terms. If the average thermal energy per atom is much less than the difference $(hT_2 - hT_1)$, the atom cannot participate in the thermal energy exchanges; it remains unchanged in its ground state. Under these conditions the internal atomic degrees of freedom cannot be excited, and this is why they do not count in the equipartition of thermal energy.

Early Quantum Theory

In the first two decades of the twentieth century, the physicists faced two fundamental paradoxes: the existence of light quanta and the existence of discrete atomic energy levels. The first is totally at variance with the laws of classical electrodynamics and the well-known observations of interference of light which seem to prove that light is a continuous wave motion; the second is at variance with the laws of mechanics, in particular with what one would expect of a planetary system consisting of electrons and a nucleus as the atom was found to be. The first successful attempt to a systematic approach of formulation, if not understanding, was Niels Bohr's famous contribution in 1913. He was able to reproduce the energy levels of the hydrogen atom by postulating that the angular momentum of the atomic electron must be a multiple of Planck's constant h. This condition, indeed, leads to the expression (14.1) and determines the value of the Rydberg constant as $R = \frac{1}{2}(me^4/\hbar^2)$, in excellent agreement with the observation. It also fixes the size of the orbits; in particular, the radius of the ground state orbit becomes equal to the "Bohr-radius" $a_B = \hbar^2/me^2$. In these two expressions, m is the mass of the electron and e is its charge; $\hbar$ is Planck's constant divided by 2π. By one simple but still mysterious assumption, Bohr established the energy and size scale of atomic phenomena. Chemistry began to be a branch of physics.

In the subsequent years Bohr's assumption was generalized, mainly by A. Sommerfeld and his school, so that it can be applied to other mechanical systems, not only the hydrogen atom. Bohr has recognized a most important principle that follows from all these quantum rules: In the limit of high energy levels, the classical and the quantum predictions become identical. This is the famous correspondence principle; a quantum system that is excited to a relatively high energy level loses its typical quantum effects, its behavior resembles in the limit what one would expect from a classical system.

The Wave Nature of Particles

In the third decade of this century the great paradox of quantum theory was solved and a new rational description of the dynamical behavior of small systems was found in terms of the new quantum mechanics. It was discovered by a unique effort of a small group of physicists in a momentous development that took place essentially in the years 1924 to 1928. The leading spirits were Niels Bohr, Werner Heisenberg, Erwin Schrödinger, Max Born, Wolfgang Pauli, Louis DeBroglie, and Paul Dirac. Never before have so few done so much in such a short time.

It is not appropriate to describe here in any systematic way how the ideas of quantum mechanics were conceived. We will try to sketch the development of its most important ideas without following the historical course of events. One may consider the discovery of the particle-wave duality as the first step toward quantum mechanics: not only were light waves found to exhibit particle properties, but particle beams were found to exhibit wave properties. Here a fundamental discovery was made—the wave nature of particles. The result is bewildering and highly unexpected. Many experiments had to be performed before the physicists were really convinced that the wave effects were not caused by some other phenomenon. All these experiments, however, only made it more and more clear that waves play a part in the motion of electrons and also of other atomic particles such as protons. Physicists can be proud of the fact that these facts were arrived at first by theoretical reasoning before the direct experimental verification. Einstein concluded the existence of light quanta in 1905, but the photo effect was not measured reliably until 1915. DeBroglie (1923) and Schrödinger (1926) predicted the existence of particle waves several years before its experimental demonstration by Davisson and Germer in 1927.

The experiments made it possible to measure the wavelength of this mysterious "electron wave." The wavelength depends upon the speed of the electron: the higher the speed, the smaller the wavelength; for electrons with an energy of a few electron volts the wavelength is of the size of the atoms. It is a very small wavelength indeed, and this is why the wave nature of electron beams is not easy to detect. In most practical applications of electron beams, such as television tubes, the wave nature plays no role whatsoever.

The dual nature of electrons as particles and waves contains the clue to the riddle of atomic structure. The unexpected properties of the atomic electrons are directly connected with their wave nature. In order to understand this connection, one must realize the peculiar behavior of waves when they are confined to a limited region. It is well known that confined waves —a violin string, sound in an organ pipe, vibrations of a membrane, electromagnetic waves in a cavity—exhibit special waveforms with characteristic frequencies; they are the proper vibrations of the system. Schrödinger considered an electron bound by the atomic nucleus as a case in question; the electron wave is kept confined in the neighborhood of the nucleus by the attractive force between the opposite charges of electron and nucleus. He calculated the shapes, the frequencies, and the characteristic patterns that should develop when an electron wave is confined by a nucleus. It is a straightforward problem of dynamics of confined waves, once the relation between the wavelength of the electron wave and the kinetic energy of the electron is known. The ensuing relation that determines the electron wave is called the "Schrödinger equation." The result was a series of distinct vibrations, each of them with a characteristic pattern (see Figure 14.2). The frequencies of the electron-wave vibrations corresponded exactly to the Balmer formula (Equation 14.1) and gave the correct energies of the observed quantum states. The spatial extensions of these vibrations—the sizes of the wave functions—are of the order of the Bohr radius; their shapes and patterns are characteristic for the physical properties of the hydrogen atom.

The wave nature of the electron immediately "explains" the fact that the electron can assume only certain well-defined states of motion in the atom. This result is of fundamental significance. A connection was found between the wave nature of the electron and the existence of discrete atomic states. The properties of the atoms, which seemed so strange and incomprehensible on the basis of the planetary model, fall into place when considered as a confined wave phenomenon. A confined electron wave assumes certain well-defined states that form a series of vibrating patterns, beginning with the simplest pattern, which vibrates with the lowest frequency, and including more complicated patterns of higher frequencies. The states are characterized by certain quantum numbers that order them according to the number of nodes and other geometrical properties of the wave.

The higher the frequency (or energy), the finer the patterns become and the closer and denser are the frequencies and energies. At high frequencies (energies), the patterns are so varied and fine grained and the energy values so close that they almost can be considered smooth and continuous. Hence the motions will be nearly those of an ordinary particle without wave properties. In the wave picture, too, the quantum phenomena cease to be important at high energies and the atom behaves as if it were an ordinary planetary system.

The hydrogen atom in its ground state vibrates in the simplest possible pattern, a spherically symmetric wave without nodes. Other atoms, however, exhibit more complex patterns even in their ground states. This is explained by an important principle which was first discovered by Wolfgang Pauli in 1925. It states that, when more than one electron is confined in an atom, not more than two electrons may assume the same pattern. Thus, added electrons will have to assume higher patterns in the scale. The ground state of

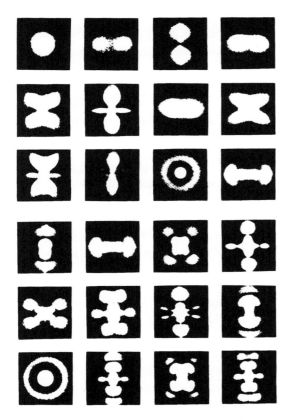

Figure 14.2. A reconstruction of characteristic wave patterns in the hydrogen atom.

a complex atom is similar to one of the excited states of a simpler atom. The pattern of the last added electron determines the configuration of the atom and determines thus the way the atoms fit together, whether they form a crystal, a liquid, or a gas. Quantity becomes quality in the atomic world; one electron more may lead to a complete change of properties.

The Complementarity between Particle and Wave Pictures
It must be realized that the wave picture of the electron is not yet the solution of the problem. Obviously, an electron cannot be a wave in the classical sense of the concept, just as a light quantum cannot be a classical particle.

The energy of a particle or a light quantum, for example, is determined by the frequency of the wave, not by its intensity; the intensity is connected with the number of particles present. Light waves and particle waves do show interference phenomena, but they transfer energy or momentum in lumps when they hit an object as if they were true particles. Nevertheless, the seeming successes of the wave picture indicate that the real solution must have some important traits in common with the wave picture. We are faced with the fundamental question: How can an electron or a photon be a particle and a wave at the same time?

The unexpected dual characteristic of matter shows that our ordinary concepts of particle or wave motion are not adequate for a description of what goes on in the atomic world. One of the features of classical physicists that must be questioned is the "divisibility" of such phenomena. This is the idea that every physical process can be thought of as consisting of a succession of particular processes. According to this idea, theoretically at least, each process can be followed step by step in time and space and the electron must be either a particle or a wave. It cannot be both at the same time because a careful tracing of the electron within the atom must decide this question and put it in either one or the other category. Here the problem of the divisibility of atomic phenomena comes in. Can one really perform this tracing? There are technical problems in the way. If one wants to "look" at the detailed structure of the orbit, he must use light waves whose wavelength is as small as an atomic orbit. Such light has quanta of an energy that would be far more than enough to tear away the electron from the atom. When it hits the electron, it will knock it out of its orbit and destroy the very object of our examination.

This situation is not peculiar to experiments when light is used to trace the electron orbit. Quite generally, all measurements which could be used for a decision between the wave and the particle nature of the electron (or the proton, or any other entity) have the same property. If one performs these measurements, the object changes its state completely in the performance itself, and the result of the measurement applies not to the original state but to the state into which the object was put by the measurement. That latter state, however, is a state of very high energy which no longer shows any wave properties.

The quantum nature, the coarseness of light or of any other means of observation, makes it impossible to decide between wave and particle. It does not allow us to subdivide the atomic orbit into a succession of partial motions, be it particle movements or wave oscillations. If we force a subdivision of the process and try to look more accurately at the wave in order to find out where the electron "really" is, we will find it there as a real particle, but we will have destroyed the subtle individuality of the quantum state. The wave nature will have disappeared, and with it all the characteristic properties of the atom.

Conversely, the same is true for light waves. If one wants to measure the electric or magnetic field step by step, as it oscillates in a light wave, one is forced to use light of such intensity—of so many quanta—that the typical light quantum effects have disappeared.

The wave nature of the electron and the particle nature of a light beam are predicated upon the indivisibility of the quantum state. The great new insight of quantum physics is the recognition that the individual quantum states form an indivisible whole, which exists only as long as it is not attacked by penetrating means of observation. In the quantum state the electron is neither a particle nor a wave in the old senses. The quantum state is the form a system assumes when it is left alone to adjust itself to the conditions at low energies. It forms a definite individual entity, whose pattern and shape correspond to a wave motion with all the peculiar properties spreading over a finite region of space.

At this stage of our discussion it will appear quite natural that predictions of atomic phenomena sometimes must remain probability statements only. The prediction of the exact spot where the electron will be found after the quantum state has been destroyed with high-energy light is a case of this kind. If the quantum state is examined with pinpointing light, the electron will be found somewhere in the region of the wave, but the exact point cannot be predicted with accuracy. Only probability statements can be made— such for example, as that the electron will be found most probably where the electron wave was most intense.

The impossibility of measuring certain quantities relating to atomic particles is the basis of the famous uncertainty principle of Heisenberg. It states, for example, that one cannot determine with full accuracy both velocity and position of an electron. Clearly, if one could, the electron would be recognized as a particle and not as a wave. The Heisenberg principle states that no measurement can be performed with sufficient accuracy to decide between the wave or the particle nature of the electron. This principle expresses a negative statement that certain measurements are impossible. This impossibility is more than a mere technical limitation that someday might be overcome by clever instrumentation. If it were possible to perform such measurements, the coexistence of wave and particle properties in a single object would collapse, since these measurements would prove one of the two alternatives to be wrong. We know from a great wealth of observations that our objects exhibit both wave and particle properties. Hence, the Heisenberg restrictions and the corresponding lack of strict causality must have a deeper root: they are a necessary corollary to the dual nature of atomic objects. If they were broken, our interpretation of the wide field of atomic phenomena would be nothing but a web of errors, and its amazing success would be based upon accidental coincidence.

Quantum mechanics has given us an unexpected but powerful answer to a great dilemma. On the one hand, atoms are the smallest parts of matter; they are supposed to be indivisible and endowed with every detailed specific property of the substance. On the other hand, atoms are known to have an internal structure; they consist of electrons and nuclei, which necessarily must perform mechanical motions not unlike the planets around the sun and therefore cannot be imagined to exhibit the required properties.

The answer lies in the discovery of the quantum states that do have the properties of indivisibility, wholeness, and specificity, but the range of this behavior is limited. Only if they are exposed to disturbances smaller than a characteristic threshold will they retain their identity and their specific properties. If they are exposed to stronger disturbances, the atoms lose their typical quantum properties and exhibit the untypical behavior expected from the mechanical properties of its internal structure.

The quantum state cannot be described in terms of a mechanical model. It has a particular way of escaping ordinary observation because of the fact that such observation necessarily will obliterate the conditions of its existence. Niels Bohr used a special term for this remarkable situation; he called it complementarity. The two descriptions of the atom—the wavelike quantum state on the one hand and the planetary model on the other—are complementary descriptions, each equally true but applicable in different situations.

Atomic phenomena present us with a much richer reality than we are accustomed to meeting in classical macroscopic physics. But the description of the atom cannot be as "detached" from the observing process as classical descriptions were. We can describe atomic reality only by telling truthfully what happens when we observe a phenomenon in different ways. The same electron can behave quite differently as we observe it in two complementary situations. These features, however, do not make electrons less real than anything else we observe in Nature. Indeed, the quantum states of the electron are the very basis of what we call reality around us.

Summary

Quantum theory has caused a revolution in our view of the material world which pervades all fields of natural science. The language in which natural phenomena are interpreted has undergone a thorough change. Concepts such as quantum state, excitation energy, transition probability, wave function, orbital, resonance, light quantum, and so forth have replaced the concepts of classical mechanics and electromagnetism.

Quantum mechanics had its most extensive success in the interpretation of atomic and molecular phenomena where the energy exchanges are below the nuclear excitation threshold. Here a singular feat of unification was achieved: The structure and dynamics, the properties and behavior of atoms, molecules, and their aggregates are understood to be based upon one simple and well-understood force: the electromagnetic interactions between electron and atomic nuclei.

The new approach to atomic dynamics led to three important insights into the character of the atomic world.

First it led to a conception of typical dimensions in size and energy which dominate atomic phenomena. Their scale and measure are understood. The sizes of the electron waves confined by the electrostatic attraction of the nucleus necessarily are of the order of the Bohr radius $a_B = \hbar^2/me^2$ and their energies are of the order of the Rydberg unit $R = me^4/\hbar^2$. Thus, atomic dimensions are reduced to basic fundamental constants. Quantum mechanics also defines a typical length and energy for nuclear phenomena. The nature of the nuclear force is not yet understood, however, and therefore cannot be expressed in terms of fundamental constants. Taking into account the empirical knowledge of the strength and character of this force, one obtains a characteristic length of several 10^{-13} cm and an energy of the order of 10^6eV as the natural scales for nuclear phenomena.

Second, quantum mechanics introduces a "morphic" trait, which previously was absent in physics but not in Nature. The atomic world abounds in characteristic shapes, typical properties, ever recurring qualities, from the identity of all atoms of a given kind to the faithful reproduction of living species. The morphic character has its root in the electron wave function. It assumes special shapes and patterns, defined by the electric field which the electron faces within the atom. These patterns are characteristic for each atomic and molecular species, and reflect the specific symmetry of the situation. They are the fundamental shapes of which all things in our environment are made. They always appear, identical and unchanged, whenever the atom finds itself under the same conditions, independent of its previous history. It is remarkable that we actually find in the world of atoms what Pythagoras and Kepler sought vainly to find in the motion of the planets. They believed that the earth and other planets move in special orbits, each unique to the planet and determined by some ultimate principle that is independent of the particular fate and the past history of our planetary system. There is no such principle in the motion of planets, but there is one that dominates the motion of particles in atoms and nuclei.

The *third* new insight of quantum mechanics concerns the nature of a quantum state. It is a well-defined indivisible entity. The results of certain measurements, however, are expressible only in terms of probability distributions. This lack of complete predictability is a consequence of the fact that the measurement interferes with the system and destroys the subtle individuality of the quantum state. This situation often is described as a breakdown of causality since a complete knowledge of the quantum state does not lead to exact predictions of the results of all measurements. Here is the important difference between the classical and the quantum mechanical description. The former considers the state of a system such that every detailed feature is amenable to an exact measurement. The latter describes the system as being in a particular quantum state of which not every measurement is determined. The detailed structure of a system is contained in the totality of all its quantum states, reflecting the fact that the measurement of a feature that is not determined in a given quantum state would involve transitions into other states. The most striking example of this situation is the radioactive decay of a nucleus in its ground state. The time of decay cannot be predicted exactly. Any attempt to pry into the internal dynamics of the nucleus in order to find out what causes an individual decay would inevitably transfer the nucleus into many of its excited states, and thus completely change its decay.

The successes of quantum mechanics are ample proof that the ideas introduced by the quantum revolution have vastly improved our understanding of the behavior of matter. A language was created for a rational description of the atomic world, and much deeper insights were gained into the ways in which Nature builds up the entities of which all matter is composed.

Our knowledge is far from complete; there are two different kinds of frontiers at which new knowledge and better understanding are sought: the "internal" and the "external" frontiers. The first kind includes the studies of the structure and behavior of complex atomic aggregates, which yet defy our understanding, in spite of the fact that the principles that govern atomic structures and interactions are well known. Examples of largely nonunderstood structures are amorphous substances, membranes, matter under highly ionized conditions (plasma physics), and complicated macromolecules that, under suitable conditions, give rise to the varied phenomena of life. There exist only rudimentary methods to deal with complex and highly organized forms of matter, where nonlinear relations between relevant magnitudes are the rule, and new kinds of superstructures appear. The situation is quite different from the behavior of well-separated atoms or simple atomic aggregates, where the quantum states are widely spaced in energy and therefore are relatively impervious to disturbances. New concepts and new ways of thinking and formulating will be necessary to get a better understanding of hierarchies of structures that are found in nature.

The "external frontiers" of physics are those where phenomena are studied which are outside the world of atoms and their aggregates, that is, outside the world which is explainable by electric interactions between changed particles. They are the frontiers of nuclear physics, of high-energy physics and of astrophysics. In these fields the principles that govern the processes are largely unknown; we do not know the nature of the nuclear forces, we do not know the basic laws that govern radioactivity, and we are deeply puzzled by the phenomena discovered at very high energies, the various new particles and antiparticles. The astronomical discoveries also present us with new phenomena whose significance and origin are not fully recognized, such as quasars, pulsars, and exploding galaxies. Some of these new discoveries may find their interpretation within the realm of quantum mechanical ideas, but some may indicate a new way of natural behavior which is beyond comprehension with present concepts.

Note

Some parts of this essay have been published in a different context in *Die Naturwissenschaften,* Vol. *60,* 1973, pp. 441–445.

J. Hughes

15   The Copernican Legacy for Meteorology

According to the glossary of the American Meteorological Society, the goals of meteorology are complete understanding, accurate predictions, and artificial control of weather. Here, the term weather designates the phenomena in our atmosphere, particularly those developing near the surface of the earth. They affect our daily lives more than any other phenomena studied by physical sciences. The purpose of this essay is to explore what relationship the science of meteorology bears to the Copernican revolution. We begin with a brief sketch of meteorology. Then we proceed to some historical details. Efforts at weather control are discussed at the end.

The weather interests of an average citizen center at what happens in the given locality on a given date. These are details of large-scale developments in the earth's atmosphere as a whole. The large-scale phenomena stem from the facts of geography and the geometry of the earth and of the solar system as explained by Copernicus. Because of the tilt of the earth's axis, any particular locality receives sometimes more and sometimes less radiation from the sun as the earth moves around it on its annual journey. At all times during this journey the equatorial regions receive more radiation than the polar regions. The result is an excess of heat at the equator and a deficit at the poles. By a variety of convective processes in the fluid envelope of the earth, heat is transferred from the equator to the poles so that the temperature contrast is reduced to about 15 percent of the mean temperature of the earth.

One would expect from this simple picture that air would rise at the equator then drift toward the poles, where it would be cooled, and then drift back again along the surface of the earth to equatorial regions. Steady northerly winds would then blow in the Northern Hemisphere and southerly winds in the Southern Hemisphere. However, such a simple wind regime does not exist. The reason is that the rotation of the earth and the tilt of its axis introduce considerable complications, the understanding of which constitutes the so-called problem of *general circulation of the atmosphere*. Early in the nineteenth century, a part of these complications was explained by Coriolis (1792–1843; an eminent teacher at the Ecole Polytechnique in Paris) by an apparent force that now bears his name. The distribution of land and water and the location of mountains also influence the movement of the air mass. The result is that in broad bands of lati-

Office of Naval Research, Atmospheric Sciences Program, Arlington, Virginia 22217.

tude, excepting the tropics and subtropics, the atmosphere moves on the average, statistically, from West to East.

That broad movement, though, includes many individual features such as storm circulations that rotate clockwise in the Northern Hemisphere and which are characterized by comparatively low atmospheric pressure, by cloudiness and precipitation; correspondingly, there are air masses of comparatively high pressure, so-called *anticyclones,* that rotate in an anticlockwise direction, generally bringing clearing and cooler or cold weather. These two types of air masses have definite life cycles and generally follow one after the other. When these two kinds of air masses are juxtaposed, the transition regions between them (from warm to cold, moist to dry, rain or snow to clear) are known as *fronts,* after the Norwegian school of meteorology.

These complicated movements of air occur in a lower stratum of the atmosphere called the *troposphere.* The troposphere contains most of the water vapor and pollution of the atmosphere. It is where the clouds form and precipitation occurs. Above the troposphere, where present-day jet aircraft fly in an attempt to get above the "weather," there is a comparatively dry air mass called the *stratosphere* whose movement, along with that of the troposphere, constitutes the general circulation.

The general circulation of the atmosphere together with that of the oceans determines the mechanism for redistributing the excess heat from the tropics to the poles and maintaining a heat balance. This balance makes life possible in the biosphere. Numerous other features of the general circulation help to redistribute heat. These are smaller-scale circulations variously called hurricanes, typhoons, or cyclones, born in the tropics at sea. These storms sometimes drift to land causing great destruction in their path, but they also replenish the water resources of the land masses. Smaller in size than the hurricane, but more violent in the force of its spinning winds, is the tornado.

An additional atmospheric driving force results from the changes that water substance undergoes in the atmosphere. The invisible water vapor rising from the ocean can rise only so far until it is cooled to a temperature where it condenses into visible water droplets and becomes a cloud. If the cloud rises enough, the water will freeze to ice. Each change, vapor to water drop or water drop to ice, liberates heat; and this heat further drives the general circulation. The clouds in turn are dissipated in rain or evaporated to appear as new clouds at other localities. Also the earth's cloud cover reflects some of the solar radiation and so regulates terrestrial temperature by limiting the amount of sunlight that reaches ground level.

These various atmospheric phenomena present many problems to puzzle the meteorologist. He does not yet fully understand the general circulation or how a hurricane or a tornado starts, or how electric charges are separated in a cloud to trigger thunderstorms. Even details of the nucleation process by which a cloud forms water drops or ice elude him (normally, clean water without "nuclei" refuses to freeze), hampering his efforts to modify weather by introduction of artificial nuclei into a cloud.

While the previous paragraphs may appear pessimistic, the progress in our understanding of the general circulation has been conspicuous. As a result, reasonably good weather forecasting is now possible. Improving prediction helps to protect us from disastrous weather, to manage our economy and plan human activity better. So much for an outline of the subject of meteorology. Now for some historical developments.

Meteorology, as a science with a mathematical background, began its existence in the mid-seventeenth century, with Pascal showing how air pressure depends upon height above sea level. Mariotte (1620–1684) and Boyle (1647–1706) lent their names to the laws of this dependence.

The first purely theoretical derivation appeared in 1823 when Laplace published the fourth volume of his *Celestial Mechanics*. The Laplace derivation contained a product term of air density (weight of a unit volume of air) and acceleration due to gravity that stems directly from Newton's second law. This is the bridge from Copernican theory to meteorology via Newtonian mechanics.

Recognition in mid-nineteenth century that local weather is imbedded in broader regional patterns of weather has characteristics of a quasi-Copernican revolution in its own right. As mentioned earlier, since time immemorial, public interest in weather centered around the happenings in particular places and at particular points in time. In particular, the early forecasters treated their problems as being of purely local character. The realization that the solution of local problems requires global information and concerted action to make this information available were Copernican in character.

The first steps in this direction seem to have occurred in the mid-eighteenth century due to Lomonosov (1711–1765) in Russia (a distinguished contributor to many sciences and to polar exploration, also founder of Moscow University now named in his honor) and to Lavoisier (1743–1797) in France (the father of modern chemistry who identified oxygen as an element). These two scientists called attention to the practical benefits that would accrue to the farmer and the seafarer from creating geographical networks of stations for simultaneous weather observations, including barometer readings. The invention of the telegraph eventually made creation of such networks practical and possible but not until a century later.

The first to organize an effective weather reporting network by telegraphy was Leverrier (1811–1877), then director of the Paris Astronomical Observatory. The reader may remember Leverrier as the astronomer who predicted the existence of the planet Neptune and calculated its future position where it was actually discovered. A particular storm served Leverrier in demonstrating the need for synoptic observations of weather that could give warning of hazardous meteorological situations. The storm in question occurred in the Black Sea on November 14, 1854, during the Crimean War when it destroyed the French fleet at Balaklava. Since the day before there was also a storm in the Mediterranean, the question arose whether the two occurrences were in reality a single storm moving across southern Europe. The French minister of war asked Leverrier to investigate the matter.

From 250 replies to his correspondence with meteorologists and astronomers all over Europe, Leverrier, with the aid of his meteorological assistant, Lias, was able to trace the path of the single storm and thus established the usefulness of simultaneous meteorological observations (Figure 15.1). Leverrier's effort was well received, and by June 1855 he had a network of thirteen telegraph weather stations reporting from various parts of France. Two years later he had extended the network internationally to stations as far away as St. Petersburg, Vienna, Lisbon, Madrid, and elsewhere. Similar networks quickly followed in other countries. During the subsequent century, but especially after the development of aviation, the networks of reporting meteorological stations increased in practically all countries by a considerable multiple. Also, the number of different observations performed and their complexity increased.

The latest developments in observational meteorology and in rapid dis-

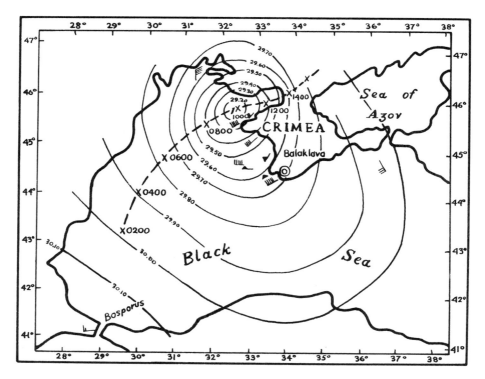

Figure 15.1. The storm that brought disaster to the French fleet and the British supply fleet at Balaklava at 1000 local time on 14 November 1854, when it was at its height.

This chart is a modern interpretation of the old weather data taken from H. Landsberg's fascinating account in the December 1954 issue of the *Scientific Monthly* of the historical events of the time and the circumstances under which the synoptic weather chart was launched. The closed lines are the lines of constant pressure in inches of mercury. The barbed arrows give the wind direction and force. The dashed line connecting the x's is the track of the storm.

semination of the data are most impressive and promise a new era in meteorological observations. In Ptolemaic fashion, the earth is now the center of its own little system of planets, the observing satellites at various distances from the earth. As illustrated in Figure 15.2, they can observe and transmit the genesis and demise of storms. They can also observe the distribution and change in cloudiness, the extent and change in ice cover, and the conditions of the space environment. Eventually, the satellites will be equipped with sensors that will give an approximate temperature profile of the atmosphere. This in turn will permit derivation of the wind regime of the upper atmosphere. We do get these data now from a balloon radio sounding network twice a day, but with a fixed time lag resulting from the time it takes to prepare a sounding balloon, more than an hour for time of ascent, and the additional time for receipt, reduction, transmission, and collation of data. When these data are eventually plotted and mapped and a forecast based on it finally issued, quite frequently, the weather will have changed.

Satellites hold out the possibility of transmitting notice of rapid atmospheric changes immediately, particularly if they can be programmed to recognize significant cloud patterns, moisture accumulations, or temperature changes. A satellite capable of "announcing" a phenomenon similar to that in Figure 15.2 would in a sense be giving a very short-term forecast of suddenly emerging meteorological events that now come to our attention only many hours later, after the meteorological data have been reported, mapped, and analyzed. Such an idea is not fanciful. Techniques of pattern recognition, computer processing, wider exploitation of the infrared and microwave spectrum, and continued improvement of satellite optics for better resolution, all combined hold promise of a better short-term forecast and more rapid updating of the general forecast.

Satellites, then, create their own kind of revolution for meteorology in a way that would probably be most pleasing to Copernicus. Wherever he is, I think he must be nodding in approval as we arrange our own little private planetary system to meet our various needs in communication, in navigation, and in total earth surveillance for better forecasting of the conditions and use of our environment.

It is important that the reader recognize that the early collections of weather data in the mid-nineteenth century, necessary as they were for establishing a scientific base for meteorology, did not imply an ability to fore-

Figure 15.2. Satellite photo of a hurricane showing spiral arms of the storm.

cast weather. In 1874 an International Subcommittee on Meteorological Telegraphy queried its membership as to whether, given the contemporary state of weather science, they had a right to issue definite prophesies or predictions of the weather, apart from their simple telegraphic reporting of the facts. The majority queried were in favor of storm warnings, and a minority wanted only factual reporting. The subcommittee recommended that storm warnings be issued for severe storms only.

Through the various government meteorological services, meteorological observations were now on an organized and systematic basis. The accumulating mass of data more than ever required a theoretical foundation. Newton (1642–1727) and Kepler (1571–1630) in dealing with Copernican theory were able to formulate laws of motion for planetary orbits. Galileo (1564–1642) derived and systemized laws of motion for earthbound objects. The laws of fluid motion which we call *hydrodynamics* that would apply to the fluid of the atmosphere were a more difficult problem. Fluids possess a property called *viscosity,* a sort of internal frictional stickiness. In the middle of the eighteenth century the French mathematician and philosopher d'Alembert (1717–1784) formulated equations of fluid motion out of which arose a famous paradox in which there is no resistance of a submerged body to fluid flow if the viscosity is neglected.

Most important to theoretical meteorology were the studies of Laplace (1749–1827) whose grand opus placing Copernican theory on a mathematical basis contained chapters on statics (the general mass of the gas at rest) and dynamics (the general mass of the gas in motion) of gases and the theory of atmospheric tides that form in the fluid of the atmosphere just as they do in the fluid of the ocean. From his work on atmospheric tides, Laplace was able to show that for explaining the general wind regime the tidal motion was inconsequential.

After the founding in France of Ecole Polytechnique in 1772 by Gaspard Monge (1746–1818), the French geometrician and Jacobin, the fundamental mechanics of theoretical meteorology continued to evolve through a line of distinguished teachers and graduates of that school. Of special importance to meteorology were Laplace, Lagrange (1736–1831), Guy Lussac (1778–1850), Navier (1785–1836), Arago (1786–1853), Babinet (1794–1872), Poisson (1781–1840), and Coriolis. In 1821, Navier began studies of fluids leading to the Navier-Stokes equations, which govern the motion of a viscous fluid.

Oberbeck used them in his study of the influence of *friction* (the drag of the lower boundary of the air mass in contact with the earth) on atmospheric dynamics.

Coriolis made a very important contribution to the fundamental equations on which our system of computed weather forecasts depends today. An artillery shell in flight in the Northern Hemisphere veers to the right of its path due to the rotation of the earth. Toward the middle of the nineteenth century Coriolis derived a quantitative expression for this deflection. Although he did not apply his theorem to atmospheric motion, he solved the general problem of acceleration of the air mass relative to the earth's motion. The Coriolis acceleration is essential to the balancing of the various forces in atmospheric dynamics. Unfortunately, his theorem was thought to be of little relevance to meteorology; so it had, in effect, to be rediscovered by Ferrel, a decade or so later. Figure 15.3 is a simplified description of the Coriolis effect using a merry-go-round to simulate the rotation of the earth.

William Ferrel (1817–1891), a Nashville, Tennessee, schoolteacher, happened upon a copy of Laplace's *Celestial Mechanics.* He was familiar with Matthew Fontaine Maury's[1] *Physical Geography of the Sea.* Aware of the mismatch of the ideas of mechanics with the empirical observations collated and mapped by Maury, he tried to build a mathematical theory of the general circulation, and arrived independently at Coriolis's deviating force of terrestrial rotation.

Ferrel made no use of the equation of state of a gas or the concept of fluid viscosity. He was not yet aware of the important equations due to Poisson and Navier, and the problem of surface friction was too much for his analysis. Though unrecognized by his contemporaries, his was a major advance in the slow but steady progress of theoretical meteorology.

Shortly after the appearance of Ferrel's *Recent Advance in Meteorology* in 1885, the studies of Helmholtz on the general circulation began to enter the literature. Helmholtz (1821–1894) was a versatile scientist of commanding stature in the nineteenth century, equally famous as a physicist and a physiologist. He is one of the founders of the law of the conservation of energy. Helmholtz introduced into hydrodynamics what is familiar to every television viewer today of the cartoon weather maps, namely, the weather front. He studied discontinuous fluid motion and the properties of so-called

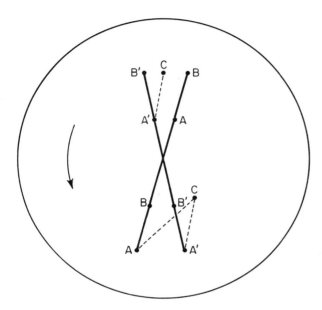

Figure 15.3. A simplified description of the Coriolis effect.

Consider a platform rotating, as shown, counterclockwise. In the lower part of the figure, point A has higher eastward velocity than point B. An object thrown and aimed along A to B carries with it A's eastward velocity and hence travels the distance AA′ eastward before it lands at C. Meanwhile, the point B with the lower velocity has traveled only to point B′. Thus, to A now at position A′ and looking at B now in position B′, the object appears to have taken the path A′C and to have been deflected to the right of its target. The actual path, of course, was A to C. Similarly in the upper part of the figure, an object thrown from A toward B carries with it a lower westward velocity and lands at point C. Again, to A now at A′ the object appears to have taken the path A′C and to have deflected to the right of the target.

free jets. Others, before him, had discovered the "fronts" (surfaces of contact of different air masses), but in a paper of 1888 he considered analytically the conditions for the existence and stability of separation surfaces in the earth's atmosphere. Helmholtz noted the possibility of wave formation on the frontal surface, the eddying of such waves, and their relation to cyclones and anticyclones.

Helmholtz had a student, Heinrich Hertz of radio fame whose work on the adiabatic curve (an adiabatic process is one in which a gas can be warmed under compression or cooled under expansion when no heat is exchanged with its environment), and whose memoir, *A Graphical Method for Determining Changes in the State of Moist Air,* introduced an important analytical tool into meteorology. This curve or chart, as the meteorologist uses it today, is essentially a plot of the temperature and water vapor against altitude as determined by a balloon sounding. Entered on the same chart is a grid of lines derived from thermodynamics and the gas laws. From the chart the meteorologist can draw conclusions about the stability of the atmosphere, estimate cloud base height, estimate thickness of fog or stratus, and much else at a given location. Figure 15.4, though not an adiabatic chart, illustrates some of the information such a chart provides.

In turn, Wilhelm Bjerknes (1862–1951), a student of Hertz, became one of the most distinguished meteorologists of the twentieth century. In a paper published in 1904, "Weather Forecasting as a Problem in Mechanics and Physics," he laid down the conditions and directions for the solution of what he considered the main problem of meteorology, namely, the forecasting of future states of the atmosphere, that is, the weather forecast. For Bjerknes this problem encompassed all the other problems of meteorology. As he saw them, the necessary and sufficient conditions for the rational solution of the forecasting problem are:

1. An accurate enough knowledge of the initial state of the atmosphere.
2. An accurate enough knowledge of the laws governing how one state of the atmosphere develops from another.

Bjerknes's knowledge of the physics of his time was comprehensive enough to marshal together a system of equations that determine future values of velocity, density, pressure, temperature, and humidity of a parcel of air. For prediction, some of his equations had to be time dependent. They were the so-called hydrodynamical equations of motion.

What Bjerknes lacked was his condition 1. Also, while his condition 2 was complete enough, the differential equations involved admitted no analytical solution, and he had no practical way of integrating them numerically. Finally, there was the bête noire of friction. Further essential progress had to await the technology of rapid data processing and a better fulfillment of Bjerknes's condition 1. The exploitation of the physical sciences by the Second World War produced these technologies.

There was one man who could not wait for the electronic computer to appear. He was L. F. Richardson (1881–1953), a British mathematical physi-

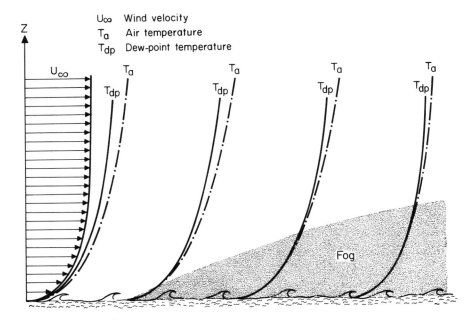

Figure 15.4. The formation of fog at sea.

This figure, through courtesy of Professor Shen Lee of the University of Missouri, shows
in a rather graphic way that when the temperature and dew point coincide, saturation
occurs, and water vapor condenses, marking, in this case, the top of a fog bank that is
building up progressively downwind from the supply of moisture. The same information
could be obtained, in principle, from the plot of a meteorological balloon sounding on
an adiabatic chart, which is a rather complicated kind of thermodynamic diagram
containing a half dozen or so different kinds of lines.

cist who during the First World War served as an ambulance driver at the
front and, in whatever free time he had, worked on the first numerical
weather forecast. From a grid of available initial values of temperature,
water vapor, and wind over Europe and England, and using the four equations of conservation of mass, momentum, water vapor, and energy, he
derived through a cycle of computations a future state of the atmosphere.
Although he used hand machine computations, Richardson's plan of calculation was similar to that used on present-day electronic computers.
Richardson repeated the cycle over and over, and his computation of a
single forecast took him the duration of the war.

The pressure changes Richardson computed were one to two orders of magnitude too high, and his experiment was regarded as a failure. The experiment was not without value to the progress of dynamic meteorology. The practicing meteorologist of the time in making his prognosis of the weather moved the pressure systems around on his map by intuition and hunch, relying on his experience gained in the analysis of many other weather maps. The whole process was highly subjective. Richardson tried to replace a subjective by an objective technique that relied purely on factual initial data and mathematical law. He may be rightfully thought of as having exercised a Copernican initiative. He tried something new.

When after the Second World War high-speed electronic computers made it feasible to repeat Richardson's attempt, a very crude theoretical model of the atmosphere was used. It eliminated certain solutions to the hydro-dynamical equations of no meteorological interest. The computed forecasts resulting were pressure changes closer to observations and encouraging enough to open not only a new era of meteorological research but also were of practical application to the daily forecast.

The foregoing theoretical developments, stemming from the work of Bjerknes, might be described as terrestrial theory of our atmosphere. No possibility of extraterrestrial factors influencing our weather was contemplated. Extraterrestrial effects on weather are a long-standing part of weather lore based mostly on superstition. There is accumulating, however, some evidence of extraterrestrial influence on hurricane and typhoon development, thunderstorm development, and precipitation. Correlations have been found between hurricane, typhoon, and thunderstorm development and lunar phase. Precipitation amounts have been found to be correlated with both lunar phase and meteor showers.

These extraterrestrial influences, if they are real, are not sufficient in themselves to be used for devising a general forecast technique, but they do appear capable of intensifying certain meteorological situations. In general, we do not take into account any of these possible influences in making our daily forecast because most of our experience leaves us with a preconceived idea that terrestrial causes are sufficient to predict weather. In deference to Copernicus, let us consider a few cosmic possibilities.

J. J. Fernandez-Partagas of the University of Miami and NOAA scientists Carpenter and Holle, examining a case file of a thousand hurricanes and

typhoons and 2400 tropical storms, found that about 20 percent more of
the hurricanes and typhoons form near new and full moon in the lunar
synodic cycle (the interval during which the moon goes through all its
phases and makes a complete revolution with respect to the earth) than near
the quarters. The case file extended over a 78-year period, showing a
stronger peak at new moon than at full moon. During the same 78-year
period, North Atlantic tropical storms that did not later become hurricanes
tended to form near the lunar quarters.

Hurricanes or typhoons form under a certain combination of meteorological circumstances that are not considered to include lunar phase. Recognizing that fact, the NOAA scientists "discourage systematic application
of these general results to a single storm, a single location, or at a single
time; the observed relationships do not invariably occur to allow such an
interpretation." Nevertheless, we are left with a clue and a legitimate hypothesis that a lunar synodic phase could influence tropical storm development further into a hurricane or typhoon.

Thunderstorm development bears a certain correlation to lunar phase.
Mae De Voe Lethbridge of Pennsylvania State University, after examination
of the thunderstorm data from 108 stations in the eastern and central United
States for the years of 1930–1933 and 1942–1965, found a peak in thunderstorm frequency two days after full moon, using full moon as the key day.
The peak is more pronounced if the key day has a declination for the moon
of 17° or more at full moon.

This case of a possible extraterrestrial influence on a thunderstorm is of
special interest to the meteorologist because to date he has not been able
to identify the particular process that the cloud uses to separate electric
charge. Lethbridge has suggested some interesting topics whose study might
lead to new correlations and clues, and we do need additional clues in
thunderstorm research.

Precipitation and lunar phase have a conspicuous correlation. Bradley
and Woodbury of NYU and Brier of NOAA used precipitation data from
1544 weather stations in the United States that operated continuously over
a period of a half century to study possible lunar influence on precipitation.
When they plotted their dates of excessive precipitation in terms of the
angular distance between moon and sun, they noted a pronounced departure
from normal expectancy. They concluded that "there is a marked tendency

for extreme precipitation in North America to be recorded near the middle of the first and third weeks of the lunar synodical month, especially on the third to fifth days after the configurations of both new and full moon." The second and fourth quarters of the lunation cycle are correspondingly deficient in heavy precipitation, the low point falling about three days previous to the date of an alignment of the earth-moon-sun system.

Perhaps one of the most fascinating possibilities of an extraterrestrial influence on our weather is that of the Australian scientist E. G. Bowen. In examining the daily rainfall records of Sydney, which were available for almost a hundred-year period, Bowen found a pronounced peak in rainfall for 12 January that was consistently high compared to the mean rainfall of the month. Data from 23 stations around New South Wales also showed a rainfall peak for 12 January. Bowen sought and analyzed data from the national weather services of other parts of the world, including New Zealand (South Island), South Africa, Rhodesia, Great Britain, Japan, the Netherlands, and the United States of America. These data, too, showed a peak in rainfall for 12 January, and in addition, rainfall peaks for 29 December, 2 and 22 January, and 1 February.

Bowen proposed that these rainfall peaks were related to meteor showers through which the earth passed in its orbit around the sun. He found a thirty-day lag in time of the earth's passage through the shower and the incidence of the rainfall peak, as for example:

Meteor Shower	Date	Date of Rainfall Peak	Time Difference Days
Bielids I	November 27–15	December 29	—
Bielids II	December 2	January 2	31
Geminids	December 13	January 12	30
Ursids	December 22	January 22	31
Quadrantids	January 3	February 1	29

The two Bielids showers of meteors require an explanation. They are associated with Biela's comet. In 1846 the originally single stream was observed to split into two parts after which there were two meteor streams. Bielids II make their appearance on a fixed date (2 December). Bielids I retrogress in time by one day in every five or six years. In 1885 they appeared on 27 November but by 1940 their appearance had shifted to 15 November.

The Bielids I were a shower of particular interest to Bowen because the meteor particles were concentrated in one part of the orbit and had an orbital period of about 6.5 years. Bowen reasoned that if his hypothesis was correct there should be a rainfall periodicity of six years or so in the late December and early January rainfall. He did find such a periodicity. He also found a similar peak for the meteor shower associated with the Giocobini-Zinner comet, which has an orbital period of 6.5 years that seems to be related to rainfall peaks on 9 November. This finding was based on about a half century of data from 48 U.S. stations.

Bowen's findings and hypothesis provoked a considerable reaction in the meteorological community. Other meteorological singularities were resurrected from various sources to show that there was nothing unusual about Bowen's. Although the idea might have been delightful to the mentality of a Copernicus, it proved to be indigestible to more than a few meteorologists.

Bowen found himself confronted with various criticism: In opposition, it was said that the rainfall variations were not real and were within the normal statistical fluctuations or could be explained in terms of well-known climatological phenomena. Various other but small data collections were presented to contradict his hypothesis. Bowen's most effective answer to his critics was the quantity of data he amassed in support of his hypothesis. He defined something he called a station year as the product of the number of rainfall stations and the length of the record in years. His argument was based on 15,000 station years; that of four of his critics altogether on 490 station years.

Presumably, observations over several forthcoming decades will resolve the dispute about the factual situation. If Bowen's hypothesis is confirmed, there will be a most interesting question to resolve: granting the temporal covariation of rainfall peaks and meteor showers, what about the causal relation? What about the detailed mechanism of the phenomenon?

The questions about the factual aspect of Bowen's findings and of the mechanism behind them bring us to the last subject to be briefly discussed in the present essay: weather modification. Here two questions are dominant: What are the facts? What is the mechanism behind them?

There are numerous aims in weather modification. Most frequently the aim is to increase precipitation; occasionally it is to decrease it. Other efforts are made to suppress hail or lightning, to attenuate hurricane winds, and to dissipate fog. Most of these attempts are based on a property of clean water, namely, that, in the absence of so-called freezing nuclei, water can be brought to a temperature below the freezing level and fail to freeze, remaining "supercooled." The introduction into supercooled water of freezing nuclei results quickly in freezing. In addition to turning water from its liquid state into solid, there is the important phenomenon of release of heat.

Quite frequently the upper part of tall clouds, whether in winter or in summer, are composed of supercooled water droplets. Most of the weather modification procedures consist in the so-called *seeding* of clouds with ice nucleating chemicals, usually with silver iodide smoke. The hope is that, by adjusting properly the dosage, the method of seeding (whether from the ground, or from aircraft flying on a preassigned level, or by rockets bursting in the clouds), all, or at least some, of the above aims of weather modification can be achieved. The essential assumption is that in the supercooled clouds there is a deficiency of freezing nuclei. In clouds or fog that are not supercooled, condensation nuclei rather than freezing nuclei attract the water vapor.

Over the last quarter of a century, many cloud seeding experiments were performed in several countries. In the United States alone a sum of about $100 million was spent on such experiments by various governmental agencies. No one has exact figures on how much has been spent on commercial cloud seeding operations (as distinct from experiments) by agricultural interests, power companies, resort owners, and other consumers of commercial weather modification services.

What are the facts? How can one establish the facts? How can one determine whether cloud seeding did in fact affect weather, for example the precipitation? The first idea was to compare the precipitation during the period with cloud seeding (perhaps one or more years) with that over a similar period of time in the past, when there was no seeding in the given locality. Also, a refinement of such direct comparison was employed, using a control area near the target of seeding operations, supposed not to be affected by seeding. However, in 1952 Brier and Enger showed that the results of such

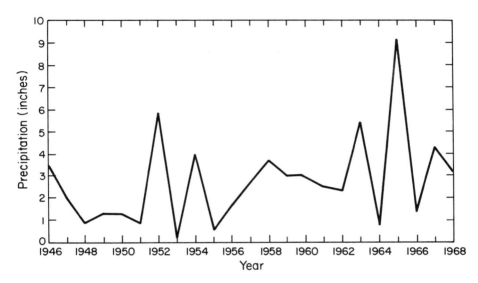

Figure 15.5. A twenty-year record of the rainfall at China Lake, California.

Annual precipitation in inches by calendar year, 1946 through 1968. From *Weather at the Naval Weapons Center*, 1946–1968, by D. L. Farnham and I. C. Vercy of the Systems Development Department.

evaluations depend upon the choice of the control area and of the historical period of reference. Figure 15.5, giving a twenty-year record of rainfall at China Lake, California, illustrates the phenomenon, familiar from biblical times, that sequences of several "fat years" are customarily followed by sequences of several "lean years" and vice versa. Thus, any claims of the effects of cloud seeding, based on comparisons with the past are countered by an irrefutable argument (generally ignored by commercial cloud seeders and by other enthusiasts), namely, that large temporal changes in precipitation occur naturally.

The answer to the question "What are the facts?" can be obtained by a method of *randomized* experiments. This method was invented by R. A. Fisher in the 1920s for agricultural trials and since spread to other domains of experimentation, such as biology and medicine. As is usually the case with a novel idea, the adoption of randomized experimental designs in many new fields has had to contend with antagonists, and weather modification has not been an exception. Currently, however, the opposition has diminished markedly. Most serious investigators now accept the need for randomization.

Briefly, the simplest randomization of an experiment with rain making consists in first, defining *experimental units*. These may be single clouds satisfying certain conditions such as minimum diameter or height or other condition. Another experimental unit that has been used in rain making is a 24-hour period, say, from noon to noon, of a day on which the morning observations show at least a specified amount of moisture in the air. For brevity, we will use this latter case in our discussion as an illustration of a randomized experimental design.

During the period of the experiment observations are made in the morning in order to determine whether the 24 hours beginning at noon will be an experimental unit. Then, in the affirmative case, a figurative coin is tossed (actually, a table of the so-called *random numbers* is used). If the coin falls heads, then the approaching experimental unit is subjected to the treatment: the cloud seeding begins as planned. If the coin falls tails, the same experimental unit is left as a control, without seeding. In either case, observations of rainfall and related atmospheric phenomena are made.

An essential point is that the determination whether the approaching 24-hour period is or is not going to be an experimental unit must be made *before* the randomized decision to seed or not. The reason for this restriction is that the decision as to whether the approaching 24-hour period is or is not suitable to be an experimental unit is always somewhat subjective and the advance knowledge whether it would be seeded may cause a subconscious bias in the decision. Cases where such bias occurred (and ruined the experiment) are on record in all domains of experimentation (see essay by W. G. Cochran, Chapter 21, Part V), including weather modification.

After a substantial number of experimental units have been accumulated, about one-half seeded and others not, the precipitation data are analyzed in order to determine whether the difference between the seeded and not seeded precipitation is *statistically significant*. We speak of statistical significance when the calculations show that such differences as have been observed (or larger) could be produced by randomization only very rarely.

What is rare or very rare is a subjective matter. Customarily, we speak of significance when the frequency of chance occurrence (denoted by P) is between 0.01 and 0.05. When it is 0.01 or less, we speak of *high significance.*

As mentioned, the seed-no seed randomization is the simplest form of experimental design. There are others, more complicated ones, and the method of calculating statistical significance depends on the method of randomization.

As of now, out of the many weather modification experiments performed only a few appear to have been strictly randomized and appropriately evaluated. The only assertion that appears safe is that, on occasion, cloud seeding has some effect on precipitation. In some cases this effect is "positive," that is, an increase in precipitation. In some other cases this effect is "negative," a loss of rainfall. In some cases these increases and decreases in precipitation were unintended, unanticipated, and on occasion quite large. Grossversuch III, an experiment of seven years' duration conducted in Switzerland to suppress hail, is an example of an experiment showing increase in rainfall and incidentally also of hail. But about an equal number of experiments show an unambiguous decrease in the precipitation. Recent examples of this class are the Whitetop project in Missouri and the Arizona experiment conducted by the University of Arizona. As carefully as these latter two experiments were planned, there are still questions about the interpretation of the negative results.

Until more is known about the rainfall mechanism in clouds, we will not know surely just how we are influencing a cloud, if at all, with artificial nucleation. There are observations in the vicinity of New Zealand to show that clouds are capable of generating their own freezing nuclei in quantities that exceed the number in the environment outside the cloud by orders of magnitude. A possible source for the abundance of these freezing nuclei are the ice whiskers (Figure 15.6) around ice crystals observed by F. K. Odencrantz at the Naval Weapons Center. As Odencrantz and his colleagues point out, the turbulence and gustiness in the cloud could fracture the hairlike ice whiskers, thus creating an abundance of ice nuclei. Another possible source is the tiny ice crystals expelled from a larger piece of ice when a cloud becomes sufficiently electrified for whatever ice crystals it contains to go into an electrical discharge called corona. When there is

Figure 15.6. Ice whiskers around an ice crystal, courtesy of Dr. Fred Odencrantz, Naval Weapons Center, China Lake, California.

a high enough electric field around an object as beneath, near, or in a thunderstorm, a glow that we call corona appears around the object. It is sometimes seen around aircraft, and old sailors used to call it St. Elmo's fire when it appeared off the masts and yardarms of their ships. This second process implied by an experiment by George Dawson at the University of Arizona would indicate an intimate connection between the state of electrification of a cloud and a natural nucleation stage and then in turn the precipitation or rain process.

To be sure, there have been enthusiastic claims of success for weather modification that may actually be successes. The analyses allegedly supporting these claims, though, are generally made by the group conducting the experiment. One would like to see some independent analyses of these claims by which to judge their validity.

The general conclusion is that, as of now, in mid-1973, it seems premature to speak of an established weather modification technology. However, the mere finding that cloud seeding did affect the rainfall in some cases, even though the nature of the effect was not infrequently unexpected, constitutes a starting point for theoretical discussion and further experiments of a revolution in atmospheric science that would surely amaze even Copernicus.

Note

1. Maury in a sense is the father of oceanography in this country. In a period just prior to the Civil War, he systematically analyzed data from the logs of sailing ships that had accumulated in the Depot of Charts and Instruments. His sailing charts were valuable time-savers for sailing vessels. The Depot of Charts and Instruments later became the Hydrographic Office and is now the Oceanographic Office.

Part IV

Quasi-Copernican Revolutions in Mathematics

Summary

The familiar branches of science, like astronomy, biology, physics, are the results of our efforts to systematize the facts observed and to explain these facts by appropriately devised mechanisms, necessarily hypothetical. All branches of mathematics stem from this endeavor, either directly or indirectly. Geometry is directly connected with our efforts to systematize, in a logically rigorous form, our perceptions of the space in which we live and which we probe. Similarly, calculus stems directly from our perceptions of speed and of acceleration. Other branches of mathematics, those indirectly connected with our studies of the real world, arise from earlier mathematical disciplines when their gradual development hits a snag. Occasionally, the unraveling of such difficulties reveals an unfounded preconception. The abandonment of this preconception creates a quasi-Copernican revolution and gives rise to a new branch of mathematics.

The first essay tells the story of the emergence of non-Euclidean geometry. The incident began in the late 1820s through independent works of János Bolyai and N. I. Lobachevski. The unfounded preconception they identified and abandoned was that Euclid's axiom on parallel lines must be true, and therefore, must be provable. In actual fact this axiom is not provable and may well be not true in the space in which we live. In fact, some astronomers work on the assumption that our space is not "Euclidean."

The second essay deals in some detail with the quasi-Copernican revolution initiated by Georg Cantor (early twentieth century) and discusses briefly another revolution now in progress, due to Kurt Gödel. The Cantor revolution abandoned the preconception, more or less, that every infinity is just like any other infinity. Contrary to this idea, Cantor showed that, roughly speaking, whatever an infinity, there is another infinity still more infinite. Cantor's work resulted in the emergence of a novel mathematical discipline, set theory, which in turn generated other branches of mathematics.

Gödel's revolution started from the abandonment of the firmly established belief that there shall be no "we shall never know" in mathematics. This doctrine was dramatically proclaimed by David Hilbert in 1900. Its essence is that every meaningful mathematical assertion must be either demonstrably true or demonstrably false. When in doubt, just work hard and, with a degree of mathematical talent and with some luck, you will find out which! Contrary to this doctrine, Gödel and his followers showed that in every well-developed mathematical discipline there must be meaningful propositions about which "we shall never know": they are "undecidable."

Undoubtedly, as in the case of Copernicus, the works of Cantor and Gödel were motivated by esthetic interests rather than by intentions to build mathematical tools for studies of Nature. Nevertheless, indirect connections with studies of Nature exist.

The third essay in this part is concerned with computers. Its contents range from mathematical symbolism, to functions of the brain, to technology. As described elsewhere in this volume (see Chapter 11 on neurobiology, Part II), thus far little is known about the miraculous functions of our brain, the functions of memory and of thought. How can one hope to build machines capable of remembering and thinking, at least within a limited domain? Yet, the digital computers now in common use do just that: they remember, they do arithmetic for us with unbelievable speeds, and also they do some other forms of "thinking."

The conceptual origin of computer technology goes back to the mid-nineteenth century and is due to the cooperation of two individuals. The moving spirit seems to have been Charles Babbage, an exceptionally brilliant holder of Newton's chair at Cambridge University, much ahead of his time but a colorful and controversial character. His principal co-worker was Augusta Ada, Countess of Lovelace, daughter of Lord Byron, the poet. The machines they attempted to build were never completely constructed. The electronic technology did not exist at the time, and the machines had to be purely mechanical, memory and all.

There is no doubt that the current omnipresence of computers is producing a revolution in our lives. But in the beginning was there a quasi-Copernican revolution involved? Possibly because of a difference in the definition, there is disagreement. The author of the essay seems to think not—the author of this summary thinks: Yes.

Eugene Lukacs

16 Non-Euclidean Geometry

Introduction
Prejudice and preconceived ideas, based on tradition and philosophical doctrines, often influence people's opinions and judgments. Examples can be found in the history of the sciences; we mention here only the opposition to the heliocentric theory in astronomy or to the theory of evolution in biology. It is somewhat surprising that similar situations occur also in the history of mathematics.

In this article we discuss the development of the theory of parallels. This mathematical problem could be solved only after deeply rooted prejudices had been overcome and the dominant philosophical ideas concerning the nature of space had been abandoned.

Euclid's *Elements*
If we study the early history of science, we note that one branch of mathematics, namely geometry, reached full maturity already during Greek antiquity. Some geometry was transmitted to Greece from Egypt and Babylon, even more important were the results obtained by Greek philosophers such as Thales, Pythagoras, Democritus, Eudoxus, and Theaetetus. But their contributions did not yet constitute a methodical presentation of the subject. The first known systematic account of geometry was presented in the famous *Elements* of Euclid.

One knows very little about the life of Euclid. The most reliable information[1] can be found in a commentary on the first book of Euclid, written by Proclus (A.D. 410–A.D. 485). Proclus mentioned that Euclid collected theorems of Eudoxus and Theaetetus in his *Elements* and that he gave correct demonstrations of propositions which had been only somewhat loosely proved by his precursors. We can infer from the statements of Proclus that Euclid lived in Alexandria during the reign of Ptolemy I around 300 B.C. His place of birth as well as the dates of his birth and of his death are unknown.

Euclid's presentation of geometry is based on strict logical deduction from a set of definitions, postulates, and common notions (axioms). This organization of the book explains why his work exerted a lasting influence on the development of mathematics.

Bowling Green State University, Bowling Green, Ohio, formerly at The Catholic University of America, Washington, D.C.

We shall be concerned only with the first book (chapter) of Euclid's *Elements* in our discussion. It starts with a list of 23 definitions, 5 postulates, and 5 common notions. Among the definitions one finds also some that are objectionable; these are not true definitions but can be considered only as somewhat vague appeals to our intuition (for instance, the definition: a point is that which has no parts). The postulates contain the fundamental geometric assumptions, the axioms (common notions) list general assumptions concerning magnitudes: for example, the statements: (a) things that are equal to the same thing are also equal to one another and (b) the whole is greater than the part are listed among the axioms.

Certain primitive notions (for example: point, line) appear in the definitions; various properties of these are listed in the postulates and axioms. Euclid and his predecessors as well as geometers of later periods believed that geometry described the physical space in which we live; hence, the theorems state relations between actual objects in real space. It was therefore very natural to appeal to intuition to justify the definitions and to consider the postulates and axioms as self-evident truths. This belief persisted through many centuries and was even strengthened by the influence of the German philosopher Immanuel Kant (1724–1804). In his philosophy the existence of synthetic judgments a priori plays a great role. Synthetic judgments a priori are apodictically true statements not derived from experience. Kant discusses the existence of synthetic judgment a priori in his famous book *Critique of Pure Reason*. He argues that the theorems of the geometry of Euclid are apodictically true and are at the same time independent of experience so that a geometry other than the Euclidean is unthinkable.

It is the purpose of this article to describe the revolutionary change in the views of mathematicians and philosophers concerning the nature of geometry. This was brought about by certain discoveries made in the first half of the nineteenth century and can be rightly compared to the Copernican revolution of the sixteenth century.

Euclid's Fifth Postulate
Euclid's fifth postulate,[2] the parallel postulate, will play an important role in our considerations. It can be stated in the following way which is somewhat different from Euclid's original formulation but is equivalent to it.

This formulation is due to J. Playfair (1748–1819). Given a line ℓ and a point P in a plane such that P does not lie on ℓ. Then there exists one and only one line through P parallel (that is, not intersecting) the line ℓ.

The fifth postulate is an assertion about the whole infinite extent of straight lines, and its validity is therefore not as obvious to our intuition as the statements of the other postulates. Even the early commentators of Euclid felt that the parallel postulate was not sufficiently evident to be accepted without proof. Proclus wrote that a number of authors had attempted to prove the fifth postulate. He mentioned in this connection Posidonius and Geminus (of the first century B.C.) and Ptolemy (second century A.D.). Proclus himself also tried to prove the parallel postulate. These attempts were continued by Arab mathematicians such as Alhazen (around A.D. 1000), Omar Khayyam (around A.D. 1100), and Nasir Eddin (1201–1274). In the twelfth and thirteenth centuries Arabian and later also Greek texts of Euclid became available again in Europe. The first criticism of the fifth postulate appeared, however, only in the sixteenth century after Proclus's commentary had been published in Basel in the year 1533. From this time on geometers worked again on the problem of proving the fifth postulate. We do not discuss here the first attempts by Italian and other European geometers directed toward this aim. These efforts, as well as the work of subsequent writers, led to the realization that the parallel postulate could be replaced by other, equivalent statements. We list a few of these. (1) A line parallel to a given line has constant distance from it. (2) Similar triangles exist which are not congruent. (3) At least one rectangle (that is, a quadrilateral with four right angles) exists. (4) A line perpendicular to one leg of an acute angle intersects the second leg of the angle. (5) The sum of the angles in any triangle equals two right angles. (6) Triangles with arbitrarily large area exist.

We still have to discuss the remarkable contribution of Gerolamo Saccheri (1667–1733), who devoted his book, *Euclides ab omni naevo vindicatus* (Euclid freed of every flaw) to the proof of the fifth postulate. He noted that the first 26 propositions of Euclid's elements were obtained without using the parallel postulate so that substantial geometric results were independent of it. Saccheri believed firmly that the parallel postulate could be proved and thought that the method of indirect proof (which occurs already in Euclid's *Elements*) might yield the desired result. The method of indirect proof proceeds as follows. One assumes tentatively that the proposition to be proved is false. Conclusions are then drawn, if one of these is absurd (contradicts some earlier proposition), then the tentative assumption is false and the original proposition is established. Saccheri starts with an isosceles birectangle. This is a quadrilateral that is constructed in the following way: (Figure 16.1) Let AB be a line segment and erect perpendiculars to AB in the points A and B and make the sides $\overline{AC}$ and $\overline{BD}$ equal to each other. It is then easy to show that the angles at C and at D are equal. This permits three possible hypotheses: these angles are either (a) right angles, or (b) acute angles, or (c) obtuse angles. Saccheri shows that if one of the assumptions (a), (b), (c) is true for a single isosceles birectangle then it is true for all such quadrilaterals. The postulate of parallels follows from (a), so that Saccheri's aim was to show that (b) and (c) lead to absurdities. Using the assumption, also made tacitly by Euclid, that the straight line is infinite, he shows that hypothesis (c) leads to a contradiction. The hypothesis (b) of the acute angle caused greater difficulties. Saccheri derived a number of consequences and arrived at the following conclusion. Let ℓ be a straight line and A be a point outside ℓ. If hypothesis (b) is true, then there exist two lines (Figure 16.2) p and p' through A which are asymptotic to ℓ; one to the right, the other to the left. The lines p and p' divide the totality of lines through A into two classes. The first class consists of the lines that intersect ℓ, the second of those that have a common perpendicular with ℓ.

At this point Saccheri committed an error. He believed that the statement just obtained led to an absurdity so that (b) was also disproved. Hence, he felt that he established the validity of (a) and therefore also of the fifth postulate.

J. H. Lambert (1728–1777) did work similar to Saccheri's. His starting point was a trirectangle, that is, a quadrilateral with three right angles (Figure 16.3). This was constructed by erecting perpendiculars a_1 and a_2 at the points A and B, respectively. Let C be a point of the perpendicular a_1 and construct a perpendicular b_1 to a_1 in C. The line b_1 intersects a_2 in a point D. There are again three possible hypotheses concerning the angle D. It is either (1) a right angle, or (2) an acute angle, or (3) an obtuse angle. Assumption (1) leads again to the parallel postulate; assumption (3) leads again to a contradiction. Lambert studied the consequences of hypothesis (2) and obtained a number of unusual results but did not arrive at a con-

Figure 16.1. Isosceles birectangle. Right angles at A and B. $a_1 = a_2$.

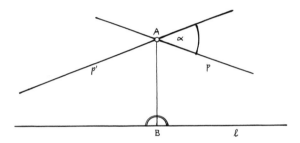

Figure 16.2. Two lines p and p' asymptotic to ℓ.

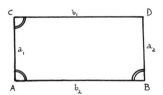

Figure 16.3. Trirectangle. Right angles at A, B, and C.

tradiction. His book *Theorie der Parallellinien* (*Theory of Parallel Lines*) was written in 1776 but was published only posthumously in 1786.

The discussion of these attempts shows that geometers tried unsuccessfully to prove Euclid's fifth postulate for about 2000 years. Deeply ingrained habits of thought proved to be an insurmountable handicap. A very radical change of attitudes was necessary to overcome these difficulties.

The Discovery of a New Geometry

The philosophy of Immanuel Kant was dominant at the beginning of the nineteenth century. As a consequence, the belief that the statements of the fundamental relations between spatial objects were synthetic judgments a priori was generally accepted. Thus Kant's authority increased the difficulty of overcoming the prejudices of the past. It took, therefore, not only mathematical insight but also courage to change the basic viewpoint. This situation was compounded by the resistance of the human mind to radical innovations. Fortunately, it happened that two scientists arose who were capable and willing to confront the philosophical prejudices and dogmatic attitudes of their predecessors and of their contemporaries. These were J. Bolyai and N. I. Lobachevski.

János Bolyai was born in the city of Kolozsvár, Hungary (now Cluj, Rumania) in Transylvania on December 15, 1802. His father was Farkas (Wolfgang) Bolyai (1775–1856). F. Bolyai was also a mathematician, he studied in Göttingen, one of his fellow students was Carl Friedrich Gauss (April 30, 1777–February 23, 1855) who became his friend and with whom he corresponded after having returned to Hungary from Germany. From this correspondence one sees that Gauss, as well as F. Bolyai, attempted to give an indirect proof of Euclid's fifth postulate. János Bolyai was taught mathematics by his father at an early age, then he studied in Marosvásárhely, Hungary (now Tirgu-Mares, Rumania) and enrolled at the Royal College for military engineers in Vienna (Austria) in 1818. In 1823 he was sent as a lieutenant to Temesvár (now Timişoara). He was subsequently transferred to various cities of the Austro-Hungarian empire and retired at his own request in 1833. From this time on he lived in Domáld and in Marosvásárhely. His interest in the parallel postulate might have been stimulated by the work of his father. First he also tried to prove the fifth postulate but became soon convinced of the futility of this attempt and came to the conclusion that he had to redirect his efforts. He started to work on the development of a new geometry, different from the one due to Euclid. On November 3, 1823 he wrote to his father:[3]

I have now resolved to publish a work on the theory of parallels as soon as I shall have put the material in order. . . . The goal is not yet reached but I have made such wonderful discoveries that I have been almost overwhelmed by them and it would be the cause of continued regret if they were lost. When you will see them you too will recognize it. In the meantime I can only say this: *I have created a new universe from nothing*.

Wolfgang Bolyai suggested that János's work be published as an appendix to his treatise, usually referred to as "Tentamen," which is the first word of the long Latin title of the book. Tentamen carries an imprimatur of 1829 but was actually published in 1832. The title of János Bolyai's work is *Appendix scientiam spatii absolute veram exhibens* and is usually referred to as the *Appendix*. Wolfgang Bolyai sent the *Appendix* to his friend Gauss first in June 1831, and since this did not reach Gauss, he sent it a second time in January 1832. He received the following reply from Gauss,[4] dated March 6, 1832:

If I commenced by saying that I am unable to praise this work you would certainly be surprised for a moment. But I cannot say otherwise. To praise it, would be to praise myself. Indeed, the whole contents of the work, the path taken by your son, the results to which he is led, coincide almost entirely with my meditations, which have occupied my mind for the last thirty or thirty-five years. So I remained quite stupefied. So far as my own work is concerned, of which up to now I have put little on paper, my intention was not to let it be published during my lifetime. . . . On the other hand it was my idea to write down all this later so that at least it should not perish with me. It is therefore a pleasant surprise for me that I am spared this trouble, and I am very glad that it is just the son of my old friend who takes precedence of me in such a remarkable manner.

From other letters that Gauss wrote as well as from notes found among his papers after his death, it is clear that Gauss had developed a new geometry.

A letter, dated 1831, shows that he was convinced that this new geometry was free of contradictions. However, Gauss refrained from publishing his results and wrote about this work only to a few friends whom he asked not to reveal this information. The reason for this strange attitude was his conviction that he would be misunderstood. He wrote to one of his correspondents that he feared "the clamour of the Boeotians." (Boeotia was a province in Greece whose inhabitants had in antiquity the reputation of being dull nitwits.) As a consequence of Gauss's reluctance to present publicly his views, two younger mathematicians, J. Bolyai and N. I. Lobachevski received rightly credit for this discovery.

Nikolai I. Lobachevski was born in Russia on October 11, 1773. He studied at the University of Kazan and became a member of the faculty of this university at an early age (Figure 16.4). He stayed there until his death on February 24, 1856. He published an article[5] "On the Principles of Geometry" in the *Kazan Messenger* in 1829. In this paper he replaced the parallel postulate by the following contrary assumption: It is possible to draw through a point C outside a line ℓ more than one line, coplanar with C and ℓ, and not intersecting ℓ. He built a consistent geometrical structure on this assumption. During the period from 1835 to 1855 Lobachevski wrote several complete presentations of his new geometry: *New Elements of Geometry* (1835–1838, in Russian), *Geometrische Untersuchungen zur Theorie der Parallellinien* (1848, in German), and his last book *Pangeometry* (1855, published in Russian and in French). Gauss read Lobachevski's German book and praised it highly. Nikolai I. Lobachevski and János Bolyai made their revolutionary discovery of the same non-Euclidean geometry almost simultaneously without knowing of each other. János Bolyai heard of Lobachevski's German book in 1848 and studied it carefully. Lobachevski probably never heard of János Bolyai's work.

The author's efforts to obtain a picture of János Bolyai resulted in a somewhat strange experience. Some books, among them an encyclopedia published in Hungary (*Új Magyar Lexikon,* Budapest 1959), contain a picture with the caption János Bolyai. But, this heading is erroneous because it is a print made after a portrait of Farkas (Wolfgang) Bolyai, the father of János Bolyai. The portrait was painted by János Szabó and is in the possession of the Hungarian Academy of Sciences. It is also reproduced, with the correct caption in G. W. Dunningham's biography of C. F. Gauss (Hafner Publishing Company, New York 1955).

In 1960 the Hungarian post office issued a stamp to commemorate the one hundredth anniversary of János Bolyai's death. This is the reproduction of a picture that the Hungarian Mathematical Society received from the director of a secondary school (called Berzsényi Dániel Gimnázium) fifteen to twenty years ago. The director of this school remarked that the picture

Figure 16.4. The Russian mathematician Nikolai Ivanovich Lobachevski (1793–1856). Permission of the Granger Collection.

Figure 16.5. The Bolyai stamp.

represented János Bolyai. However, subsequent research disproved this statement. According to a note by János Bolyai himself, two portraits representing him existed. One was in the possession of his father Farkas (Wolfgang) Bolyai; this János destroyed during one of the frequent quarrels with his father. The second was a miniature on a medallion; children played with it so much that the paint was completely rubbed off, and it was then thrown away. János Bolyai himself stated that no portrait of himself was left. He even added that this was not important since this is not the way a person's memory should be preserved. The picture on the Bolyai commemorative stamp (Figure 16.5) is therefore apocryphal, but it testifies nevertheless to the great esteem and admiration that the Hungarian nation, as well as all mathematicians, have for the work and for the accomplishments of János Bolyai.

We discuss next a few theorems of the non-Euclidean geometry of Bolyai and Lobachevski; this geometry is today called hyperbolic geometry. For the sake of simplicity we restrict ourselves to plane geometry, but we note that Bolyai and Lobachevski developed also hyperbolic trigonometry.

The assumptions of hyperbolic geometry are the usual Euclidean postulates with the sole exception of the parallel postulate, which is replaced by the following statement:

Postulate (H): Given a straight line ℓ and a point P not on ℓ, there exist at least two straight lines through P which do not intersect ℓ.

Some geometrical theorems are valid, independently of the choice of the Euclidean or hyperbolic parallel postulate. We mention a few examples: (1) Given a straight line ℓ and a point P on ℓ, one can construct a line p through P which is perpendicular to ℓ. (2) Given a straight line and a point P not on ℓ, it is possible to construct a straight line p through P which is perpendicular to ℓ. The distance from P to the point Q of intersection of p and ℓ is shorter than any segment connecting P and a point of ℓ which is different from Q. (3) It is possible to bisect line segments or angles. (4) If one side and two adjacent angles of two triangles are equal, then the triangles are congruent. (5) Vertical angles are congruent. (6) Two lines ℓ and ℓ' that form equal angles with a transversal t have a common perpendicular. (See Figure 16.6.)

Let ℓ be a straight line and let P be a point not on ℓ. (See Figure 16.7.) We draw a perpendicular from P to ℓ; it intersects ℓ in a point Q. We draw

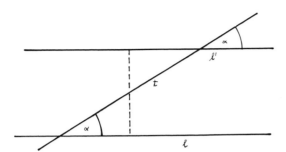

Figure 16.6. Lines forming equal angles with a transversal.

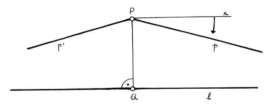

Figure 16.7. The angle of parallelism.

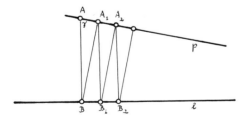

Figure 16.8. Parallel lines converge asymptotically.

a half-line x that does not intersect ℓ and rotate it so that the angle between PQ and x decreases but only as long as x does not intersect ℓ. The possibility to do this is assured by postulate (H). The half-line x will reach a limiting position p that corresponds to the least value of the angle between x and PQ. It is easily seen that p does not intersect ℓ. By symmetry we can draw on the other side a half-line p'. If p and p' would be continuations of each other, then they would form a single line that would be the unique parallel in contradiction to (H). We produce these half-lines (see Figure 16.2); they form four angles: two vertical angles that contain the segment $\overline{PQ}$ ($\overline{PQ}$ perpendicular to ℓ) and its extension and two vertical angles not containing any point of the line on which the segment $\overline{PQ}$ is located. The lines p and p' are called the parallels to ℓ; any line through the angle α does not intersect ℓ. These lines are said to be ultraparallel to ℓ. The angle between p and $\overline{PQ}$ (equals the angle between p' and $\overline{PQ}$) is called the angle of parallelism.

Let ℓ be a line and let A be a point not on ℓ. (See Figure 16.8.) Drop the perpendicular AB from A to ℓ and draw the parallel p to ℓ through A. The angle γ between BA and p is the angle of parallelism. We drop the perpendicular from B to p, it intersects p in A_1. From A_1 we draw a perpendicular to ℓ, it meets ℓ in B_1. We see that $\overline{AB} > \overline{BA_1} > \overline{A_1B_1}$ so that the distance between p and ℓ decreases as A moves to the right on p. It is possible to show that this distance tends to zero so that parallel lines converge asymptotically.

The following statement is an immediate consequence of the definition of parallelism and ultraparallelism: two lines that are perpendicular to the same line are ultraparallel.

We show next that the angle of parallelism ($\measuredangle BAp$ in Figure 16.2) decreases as the distance $\overline{AB}$ increases. Let (Figure 16.9) ℓ be a straight line and draw AB perpendicular to ℓ in A. Extend AB and let B' be a point more distant from ℓ than B. Draw the parallel p to ℓ through B and write γ for its angle of parallelism. Let p' be the parallel to ℓ through B'; we shall show that the angle between p' and $B'A$ is less than γ. We draw through B' a line q so that the angle between q and AB' equals γ. Since the lines q and p form equal angles with the transversal BB', they have (see Example 6, Figure 16.6) a common perpendicular CC' that passes through the mid-point 0 of BB'. We draw through C' a line p'' parallel to p. This line forms, by (H), an acute angle with CC'. We choose a point D in the angle between q and p'' and draw the line $B'D$. This line cannot intersect p' and does therefore not meet p. The angle $AB'D$ is smaller than γ, hence, p' and AB' enclose also an angle less than γ; that is, the angle of parallelism at B' is less than the angle of parallelism at B. Thus, the angle of parallelism γ depends on the distance $AB = h$. It can even be shown[6] that the angle of parallelism decreases as h increases and can be made arbitrarily small by selecting h sufficiently large.

We mention a surprising consequence of this property. The perpendicular to one leg of an acute angle does not intersect the other leg, provided the perpendicular is sufficiently far from the vertex. Let α be an acute angle (see Figure 16.10) between the half-lines a and a'. Select the point B on a so far away from the vertex A that the angle of parallelism corresponding to the distance AB is less than α. We erect in B a perpendicular ℓ to AB and draw the parallel to ℓ through A, the leg a' of α is ultraparallel to ℓ so that ℓ cannot intersect the leg a'.

We consider (Figure 16.11) a figure consisting of two parallel lines a and b and a transversal $\overline{AB}$. We call such a figure a singly asymptotic triangle (for short: s.a.t.) and denote it by $\Delta(A, B, a, b)$ and say that A and B are vertices of the s.a.t. It is easy to show that two s.a.t. $\Delta(A, B, a, b)$ and $\Delta(A', B', a', b')$ are congruent if $\overline{AB} = \overline{A'B'}$ and $\angle A = \angle A'$. One can also prove that the interior angle at a vertex of a s.a.t. is smaller than the exterior angle at its second vertex.

Using properties of s.a.t., one can show that Saccheri's isosceles birectangle (see Figure 16.1) has two equal acute angles opposite its right angles. This result can then be used to show that the sum of the angles of any triangle is less than two right angles.

We mention a few more rather surprising results of hyperbolic geometry:

(i) The area A of a triangle with angles α, β, γ ($\alpha + \beta + \gamma < 180°$) at its vertices is given by the formula

$$A = \left(1 - \frac{\alpha + \beta + \gamma}{180}\right) \pi,$$

where the angles α, β, γ are measured in degrees. An interesting consequence of this formula is the fact that there are no triangles of arbitrarily large size

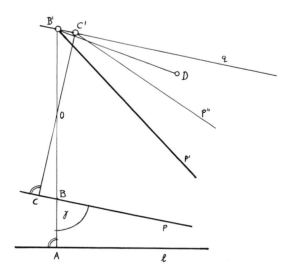

Figure 16.9. Decrease of the angle of parallelism with increasing distance.

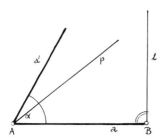

Figure 16.10. Perpendicular to one leg of an acute angle.

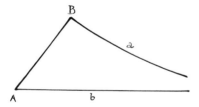

Figure 16.11. A singly asymptotic triangle.

in hyperbolic geometry. If we consider a triangle that is formed by three lines, any two of which are parallel, then $\alpha = \beta = \gamma = 0$, and the area of the triangle equals π. We see then from the formula just given that this is the greatest area a triangle can have.

(ii) Two triangles are equal if their angles are equal.

(iii) Circles can be defined in the usual way; however, the length of the circumference of a circle is not proportional to its radius but grows faster than the first power of the radius.

Lobachevski and Bolyai derived independently these and many other theorems and developed a system of plane and solid geometry and trigonometry assuming Postulate (H) instead of Euclid's fifth postulate. They never encountered in this work a contradiction and believed therefore firmly that they had constructed a system different from Euclid's geometry which was logically perfect. They were therefore convinced that the geometry of Euclid was not the unique, logically possible geometry. Their conviction was based on the fact that a very long deductive chain did not lead to a contradiction; this, however, did not yet exclude with absolute certainty the possible occurrence of a contradiction at a later point and did therefore not guarantee the consistency of hyperbolic geometry.

The Consistency of Hyperbolic Geometry

To show the consistency of hyperbolic geometry, one has to build a model either in the Euclidean space or in the Euclidean plane. All axioms[7] of Euclidean geometry must hold in such a model with the sole exception of the parallel postulate. Instead of the parallel postulate the axiom (H) must be satisfied. We will discuss a model in the plane. It is constructed by renaming certain objects and certain relations in the Euclidean plane in such a way that the axioms of the hyperbolic geometry are satisfied. Then all theorems of hyperbolic geometry are valid in the model. Since the model is constructed in the Euclidean plane, it is possible to reformulate all these theorems in the terminology of Euclidean geometry. If hyperbolic geometry would contain a contradiction, then this contradiction would necessarily also appear in Euclidean geometry. Thus, the possibility of constructing such a model proves that hyperbolic geometry is as consistent, and therefore logically as admissible, as Euclidean geometry. Such a model shows at the same time that the parallel postulate is independent of the other axioms; that is, that the attempts to prove it were doomed to failure.

We describe briefly a model that was constructed by the German mathematician Felix Klein (1849–1925) and published in 1871. Consider a circle (Figure 16.12) in the Euclidean plane and call the interior of the circle the hyperbolic plane. The points of the interior of the circle are considered to be the points of hyperbolic geometry, the points on the circumference as well as the points outside the circle are excluded. The chords of the circle are called straight lines, and the end points of the chords are excluded since they are points on the circumference.

The available space does not permit us to verify here the statement that all Euclidean axioms, with the exception of the parallel postulate, hold in our model. To do this, one has to define when line segments and angles are considered to be congruent in the model. This requires some background in a branch of geometry, called projective geometry, and can therefore not be carried out here. However, we can see from Figure 16.12 that the hyperbolic parallel axiom (H) is satisfied. Let ℓ be a chord connecting the points U and V of the circle and let P be a point in the interior of the circle. Then the lines PU and PV are the two hyperbolic parallels to ℓ through P. All lines through P located in the angle UPV intersect ℓ while all the lines located in the angle $U'PV$ are ultraparallel to ℓ.

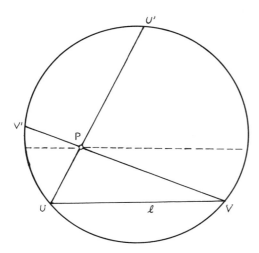

Figure 16.12. F. Klein's model for hyperbolic geometry.

There exist also other models in the plane as well as in space which can be used to show the consistency of hyperbolic geometry.

Subsequent Developments

In Euclidean as well as in hyperbolic geometry straight lines have infinite length. If the axioms that imply this are modified, one can also construct a geometry in which all straight lines intersect, that is, in which there are no parallels. A model for such a geometry can be constructed on the Euclidean sphere provided that diametrically opposite points are identified. In this model the great circles of the sphere play the role of straight lines. This geometry, called elliptic geometry, was first considered by B. Riemann in his inaugural address "Über die Hypothesen welche der Geometrie zugrunde liegen" (On the hypotheses that lie at the foundation of geometry), which he delivered in Göttingen in 1854. In this lecture Riemann went way beyond the construction of a new non-Euclidean geometry. He considered space as a multiply extended magnitude with a metric that could be different in different points. This Riemannian geometry became later one of the basic tools of the theory of relativity.

The emergence of non-Euclidean geometries led to a revision of the traditional views concerning the nature of the axioms. The fact that the Euclidean, the hyperbolic, and the elliptic parallel postulate were logically equally admissible made it impossible to maintain any longer that the axioms represented immutable truths. It was gradually realized that a mathematical theory deals with relations among concepts. Certain undefined notions (the fundamental concepts) are introduced and their properties are described by the axioms. The mathematical theory is developed by drawing conclusions from them. Every statement, except the axioms, is established in a strictly deductive manner. The axioms contain all the properties of the fundamental concepts that can be used in constructing the mathematical theory. In this sense the ensemble of axioms defines the set of fundamental concepts. This kind of definition is called implicit definition (or definition by axioms); it is in form but not in substance different from the explicit definitions used in the later parts of the theory. The understanding of the role of axioms had a profound influence on the development of mathematics and led to the use of the axiomatic method in numerous branches of the exact sciences.

We have not yet raised an important question that is as old as hyperbolic geometry, namely, the problem which geometry is valid in the physical world. One could attempt to decide this by carrying out experiments. Gauss surveyed a triangle formed by three mountaintops, the Brocken in the Harz mountains, the Inselsberg in the Thuringian forest, and the Hohenhagen near Göttingen. Lobachevski suggested that astronomical measurements could decide whether the sum of the angles of a triangle equals two right angles. In all such experiments the deviation from two right angles did not exceed the expected error of measurements. But could such an experiment decide which geometry is valid? In carrying out measurements one makes necessarily certain physical assumptions, for instance, one assumes the rectilinear propagation of light. A physical experiment can therefore only verify the complex of certain physical and geometric theories but can never decide which geometry is valid. One can modify either the physical or the geometrical assumptions. Hence, it is possible to use either the Euclidean or a non-Euclidean geometry in describing the physical world, if the physical theory is adjusted accordingly. This is exactly what happened in the theory of relativity which uses the general, non-Euclidean geometry of Riemann but modifies the physical theories. B. Riemann was probably the first to realize the need to distinguish between mathematical space and physical space. The mathematical space deals with deductions from arbitrarily assumed axioms while physical space is a scheme that is used in describing reality. This distinction was very clearly expressed by A. Einstein[8] who wrote: "As far as the propositions of mathematics refer to reality they are not certain, and in so far as they are certain they do not refer to reality."

In this paper we wanted to show the enormous importance of the discoveries of Lobachevski and Bolyai. Their first impact on mathematics was the realization that not only one geometry was logically conceivable. As a consequence the nature of the axioms as implicit definitions was understood, and the axiomatic method gave rise to important mathematical studies. These provided somewhat later important and powerful new tools for physics. But the effect of the discovery of non-Euclidean geometries was felt way beyond mathematics or physics. It dealt a devastating blow to Kantian philosophy and compelled us to revise our philosophical and epistemological views.

Notes

1. See T. L. Heath, *A History of Greek Mathematics*. Vol. 1, Oxford University Press, Oxford, 1921.

2. Some editions of Euclid refer to it as the eleventh axiom.

3. Quoted from R. Bonola, *Non Euclidean Geometry*. English translation by H. S. Carslaw, Dover Publications, New York, 1955, p. 98.

4. Quoted from ibid., p. 100.

5. This was preceded by a lecture given at the University of Kazan in 1826; the manuscript of this lecture is lost.

6. For readers who are familiar with some trigonometry and calculus, we can make a more specific statement. It can be shown that $\tan (\gamma/2) = e^{-h/k}$, where k is a constant that depends on the unit of length while e is the basis of natural logarithms.

7. From now on we shall use the terms axioms and postulates synonymously.

8. A. Einstein, *Geometrie und Erfahrung*. J. Springer, Berlin, 1921. Quoted from H. Weyl's *Philosophy of Mathematics and Natural Science*. Princeton University Press, Princeton, N.J., 1949.

S. M. Ulam

17 Infinities

Introduction

Revolutions in mathematics, changes in outlook on a Copernican scale, an upheaval in the attitudes or in the approach to mathematical ideas! How can this appear in a science that has grown since antiquity continuously and steadily by accretion, and where the notion of truth and falsehood has remained the same since its origins? Is it not true, as laymen think, that mathematics is all there, essentially fixed and finished except for elaborations and refinements? Certainly one does not expect changes in the foundations and the great lines of development of this science.

On the contrary. Very deep "conceptual" changes occur in mathematics from time to time, altering its very basis, its philosophical impact, its future impact on other sciences and the psychology at the root of creative intuitions.

In the following sections two somewhat connected revolutions in mathematics are described, one generated by Georg Cantor (1848–1918) and the other by Kurt Gödel, a German mathematician, now living, currently in the United States. The last section is concerned with the ideas of infinity in mathematics and in physical sciences.

Georg Cantor was a German mathematician whose principal works were published in the 1890s. They initiated a then new mathematical discipline now known as the theory of sets or, simply, as set theory. In turn, development of set theory generated a vigorous drive toward rebuilding, actually toward building, the conceptual foundations of mathematics. With several disciplines on its fringes (set theory, topology, mathematical logic, algebra) the work on foundations flourishes now and exercises a profound influence on the thinking of most now living mathematicians.

The Cantor quasi-Copernican revolution, to be described presently, consisted in changing drastically the views of infinity. Traditionally, more or less, infinity is just infinity. Contrary to this, due to Cantor, we now know that some infinities are "more infinite" than some others. Furthermore, Cantor found that, however infinite an infinity might be, there is always another infinity still more infinite. While somewhat delicate, the proofs of the two propositions of Cantor do not require any "previous knowledge" of mathematics and both are given in the next section.

The other quasi-Copernican revolution, due to Gödel, is more compli-

Department of Mathematics, University of Colorado, Boulder, Colorado 80302.

cated technically than the Cantor revolution and its description, in the third section, does not include any proofs. Briefly and roughly, the Gödel revolution consists in the abandonment of a very firmly established preconception that every meaningful mathematical proposition must be either right or wrong and that its correctness or falsehood can be proved. Contrary to this, Gödel proved that in every sufficiently large mathematical system meaningful propositions must exist which are undecidable.

The conceptual foundations of mathematics is a very active field of studies now, and outstanding contributions come from many countries.

Cantor's Revolution: Hierarchy of Infinities

In order to introduce the reader to the theory of sets, we must begin with terminology and basic concepts. The word "set" means simply a collection of any kind of things, whether books in a library, students attending a course of lectures, numbers of a specified category, points on a line or on its specified segment, points on the circumference of a circle, and so on and so on. Such words as "collection," "class," or "category" are frequently used as synonyms of "set." If an object a belongs to a set A, we say that a is a "member" or an "element" of A. If every element of A belongs also to another set B, we say that A is a "subset" of B. If A is a subset of B, but B contains elements that do not belong to A, then A is described as a "proper subset" of B.

The most commonly discussed sets of numbers are as follows:

(i) Set of "natural" numbers or the set of "naturals"; these are the positive integer numbers $1, 2, 3, \ldots, n, \ldots$ (without end).

(ii) Set of "rational" numbers, or of "rationals"; these are fractions a/b, where a and b are arbitrary integers, for example $\frac{1}{2}$, $\frac{3}{5}$, $\frac{5}{3}$, and so on.

(iii) Set of "real numbers" or of "reals." In order to get the idea of a "real" number, draw a line, say L (see Figure 17.1), mark on it the zero point (or the "origin") and a scale of distances. Next mark on the line L an arbitrary point x. Then the distance of that point from the origin, also denoted by the same letter x, is what is called a "real number." Depending on whether the selected point is on the right of the origin or on the left, the corresponding real number x is positive or negative.

It is easy to see that the set of naturals is a proper subset of the set of rationals. In fact any natural number n can be represented as a fraction, say $2n/2$. However, a fraction like $\frac{5}{3}$ is not an integer. In other words, the fraction $5/3$ is an element of the set of rationals but does not belong to the set of naturals. Similarly, the set of rational numbers is a proper subset of the set of reals. For any rational number, say r, one can find on the line L a point whose distance from the origin is equal to r. On the other hand, for example, $\sqrt{2}$ cannot be represented as a rational fraction a/b.

One of the frequent questions studied in mathematics is whether some entity A is, in some sense, equal to or whether it is larger or smaller than another entity B. Naturally, the same question is an object of study in the theory of sets, that is when the entities A and B are some sets. If both sets A and B are finite (set of books in a library, set of students, and so on), one way of solving the problem is to count the elements of A and those of B and to compare the two numbers so obtained. Unfortunately, one cannot count all the elements of an infinite set, and the comparison of "sizes" of two infinite sets must be based on a different procedure. Such procedure is suggested by a method of comparing finite sets which does not involve counting. For example, consider the question whether the "set of students" attending a given course is equal to or is smaller or larger than the "set of seats" in the classroom. Can we answer this question without counting both the students and the seats? Yes, we can. Simply, we can invite the students to occupy the available seats. Then, *if all the students are seated* and if *there are no empty chairs,* the two sets are equal. Otherwise, in an obvious way, one or the other set is the larger. It is this method, which involves no counting, that is the basis for the comparison of infinite sets. To understand this, it is important to be clear about the essence of what is happening when all the students select chairs to sit on and no chairs are left empty.

The essence of this happening is what is technically called the establishment of a "one-to-one correspondence" between the set of students, on the one hand, and the set of chairs, on the other. When a student (call him s_1) sits on a chair (call it c_1), then this student "corresponds" to this particular unique chair and vice versa. Obviously, while counting the elements of two infinite sets presents an embarrassment, to say the least, the establishment of a one-to-one correspondence between their elements might be a possibility.

One of the fruitful concepts introduced by Cantor is that of the "equivalence" of two sets. Two sets, say A or B are called "equivalent" if it is possible to establish a one-to-one correspondence between their elements. (The

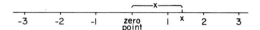

Figure 17.1. Concept of real numbers.

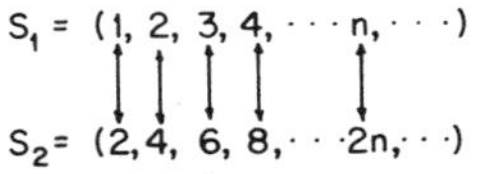

$$S_1 = (1, 2, 3, 4, \cdots n, \cdots)$$
$$S_2 = (2, 4, 6, 8, \cdots 2n, \cdots)$$

Figure 17.2. One-to-one correspondence between all positive integers and all even integers.

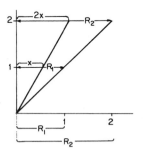

Figure 17.3. One-to-one correspondence between R_1 and R_2.

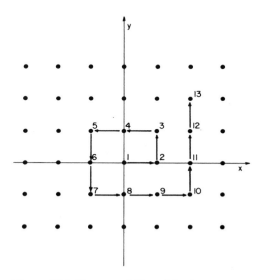

Figure 17.4. Denumerability of lattice.

terms alternative to "equivalent" are that A and B have equal "power" or equal "cardinality.") Otherwise, when it is impossible to establish a one-to-one correspondence between elements of A and those of B, then the "power" or "cardinality" of one of them is larger (or "higher") than that of the other. Here then, with reference to infinite sets the power (or cardinality) plays a role similar to that played by numbers for finite sets.

The application of the concept of power to particular cases brings out some surprises. For example, as measured by power, an infinite set may well be "equal" to its proper subset. The simplest example is the comparison in Figure 17.2 of the set, say S_1, of all natural numbers and of the set, say S_2, of all even natural numbers. Obviously S_2 is a proper subset of S_1. Yet the following two lines indicate a one-to-one correspondence between S_1 and S_2.

To every single natural number n there corresponds one specific even number $2n$ (and vice versa). Thus, even though the set S_2 might be considered as "one-half" of set S_1, their powers are equal, they are equivalent.

The same thing applies to two sets, say R_1 and R_2, of "real numbers," where R_1 stands for all reals x that are between 0 and 1 and R_2 for the set of those between 0 and 2. Again, in a sense, one of these sets, namely R_1, is one-half of the other. Yet, as indicated in Figure 17.3, a one-to-one correspondence between the elements of these two sets is easy to establish. Here, to every element of R_1, that is to every real number x between limits $0 \leqq x \leqq 1$, there corresponds one and one only element of R_2, namely $2x$. Thus R_1 and R_2 are of equal power.

The two examples just discussed may be described as "one-dimensional": the elements of the sets compared may be marked on a single line. The next example is "two-dimensional." Consider a plane and draw on it two perpendicular axes of coordinates $(0, x)$ and $(0, y)$ (Figure 17.4). Denote by L the two-dimensional "lattice" of points with integer-valued coordinates, x and y, each running from $-\infty$ to $+\infty$:

$$x = (0, \pm 1, \pm 2, \ldots, \pm n, \ldots),$$
$$y = (0, \pm 1, \pm 2, \ldots, \pm n, \ldots).$$

The question is how does the power of the lattice L compare to that of the set of natural numbers? In other words, the question is whether it is possible to establish a one-to-one correspondence between the elements of

L, on the one hand, and the naturals 1, 2, 3, . . . , n . . . , on the other. The answer is in the affirmative, and one of the methods of establishing such correspondence is indicated in Figure 17.4. Just follow the "rectangular spiral" indicated by the arrows and assign consecutive integers to the encountered lattice points. (The reader may enjoy looking for alternative methods of achieving the same goal.)

At this point it is appropriate to introduce a special term to designate those sets that are equivalent to the set of naturals. This term is "denumerable." As we have just seen, the elements of the lattice L can all be "numbered." Thus, L is denumerable. So is the set of rationals described earlier. Here again the reader may enjoy the experience of devising the one-to-one correspondence needed.

The multiplicity of cases where infinite sets are denumerable leads naturally to the question as to whether there are any infinite sets at all that are not denumerable. This question was asked and answered in the affirmative by Cantor. Prior to that the general routine of thought prevailed that infinity is just infinity. Here, then, is a case of a quasi-Copernican revolution in mathematics: an unfounded preconception was abandoned in favor of something "more pleasing to the mind," which was followed by very impressive developments.

The particular question asked by Cantor was whether the set of all real numbers between zero and unity, say X, is denumerable. The proof, customarily labeled the "diagonal proof," is by contradiction. We begin by supposing that X is denumerable and that, therefore, all the real numbers greater than zero and less than unity can be arranged in an infinite column, each number corresponding to a positive integer n. Then we show that a real number x, also between zero and unity, must exist that differs from all the reals in the column. In other words, in whichever way one tries to number all the reals between zero and unity, there will be always a real number of this class unnumbered.

Obviously, every element of X can be written down as a decimal fraction with an infinite sequence of digits. For example, the real number one-half ordinarily written as 0.5 can also be written as 0.500,000, . . . , and so on without end. The complication is that this same number can also be written as 0.499,999, . . . , and so on without end. In order to avoid duplication of

Table 17.1. Hypothetical Correspondence between All Real Numbers and Integers.

Positive Integers	Corresponding Reals					
1	$x_1 = 0.\boxed{\alpha_{1,1}}$	$\alpha_{1,2}$	$\alpha_{1,3}$	$\cdots$	$\alpha_{1,n}$	$\cdots$
2	$x_2 = 0.\alpha_{2,1}$	$\boxed{\alpha_{2,2}}$	$\alpha_{2,3}$	$\cdots$	$\alpha_{2,n}$	$\cdots$
3	$x_3 = 0.\alpha_{3,1}$	$\alpha_{3,2}$	$\boxed{\alpha_{3,3}}$	$\cdots$	$\alpha_{3,n}$	$\cdots$
.						
.						
n	$x_n = 0.\alpha_{n,1}$	$\alpha_{n,2}$	$\alpha_{n,3}$	$\cdots$	$\boxed{\alpha_{n,n}}$	$\cdots$
.						
.						
	$x = 0.\beta_1$	β_2	β_3	$\cdots$	β_n	$\cdots$

this kind we shall agree that, whenever the same real x can be written in two different ways, one with an infinite uninterrupted sequence of zeros and the other with a similar sequence of nines, we shall use the representation with zeros.

Now we proceed with aligning the reals x with integers to which they, hypothetically, correspond, as shown in Table 17.1.

Here each of the symbols α represents a digit, a numeral, either 0 or 1, or 2, . . . , or 9. Of the two subscripts attached to each α, the first indicates the integer to which the given number x corresponds. The second subscript indicates the order of the decimal. For example, the first number x_1 may be

$$x_1 = 0.357212;$$

then $\alpha_{1,1} = 3$, $\alpha_{1,2} = 5$, $\alpha_{1,3} = 7$, and so on.

Now we proceed to define a real number x between zero and unity which differs from all the numbers in the column. This number x can also be written as an infinite decimal fraction, say,

$$x = 0.\beta_1\beta_2 \ldots \beta_n \ldots, \text{ and so on, without end.}$$

To define x completely, it is sufficient to indicate the values of the consecutive digits β. This may be done in a variety of ways. We shall use only two different values for the β's, either 1 or 3. The rule is as follows. If $\alpha_{1,1} = 3$, then we write $\beta_1 = 1$. If the value of $\alpha_{1,1}$ is not 3, then we set $\beta_1 = 3$. In order to determine β_2, we inspect the expansion of x_2. If $\alpha_{2,2} = 3$, then we set $\beta_2 = 1$. Otherwise $\beta_2 = 3$. We proceed similarly to determine β_3, β_4, . . . , and so on. The general rule is this: if the nth digit in the expansion of x_n is $\alpha_{n,n} = 3$, then $\beta_n = 1$. Otherwise $\beta_n = 3$. The reader will notice that the definition of x depends only on those digits of the particular x_n which, in Table 17.1 are on the diagonal, namely on $\alpha_{1,1}$, $\alpha_{2,2}$, $\alpha_{3,3}$, . . . , $\alpha_{n,n}$, ... This explains the description of the proof, the "diagonal proof."

Obviously x so defined is a real number between zero and unity and, therefore, belongs to X. Also, x is different from all and every number x_n, the sequence of which is supposed to exhaust all the elements of X. Specifically, whatever x_n may be, the defined x differs from it at least in its nth digit $\alpha_{n,n} \neq \beta_n$.

The proof shows that the infinite set X of reals between zero and unity is not denumerable. Its power (or its cardinality) is greater than that of the set of natural numbers. Colloquially, one might say that the infinity of X is of higher order than that of the natural numbers. Several questions arise at this point.

One is whether it would be convenient to establish some conventional symbols, one to designate the power of the set of all positive integers (essentially, of all the denumerable sets) and another symbol to designate the power of set X? Obviously, the answer is in the affirmative. The symbol for the denumerable sets is $\aleph_0$ read "aleph zero." (Aleph is the name of the first letter of the Hebrew alphabet.) The symbol for the power of X, the set of reals, is C, connoting "continuum."

The next question is whether sets exist whose power is greater than $\aleph_0$ but less than C? If not, then it might be appropriate to replace the designation C by $\aleph_1$ (aleph one). Actually, Cantor conjectured that no infinity greater than $\aleph_0$ and less than C can exist, and this conjecture, called the *hypothesis of continuum,* has an interesting history, see later.

The third natural question to ask is whether there exist sets with power higher than C? More generally, given an infinite set E with some power, say $\aleph(E)$, does there exist another set, say K, whose power is greater than $\aleph(E)$? This question Cantor answered in the affirmative, thus showing that there is no limit of orders of infinity of sets: however high the cardinality of

E may be, there are sets of cardinality still higher. What a difference between this result and the earlier tradition that "infinity is just infinity"!

Properties of high orders of infinity, going even beyond Cantor's original imagination, were studied by the author of the present essay in connection with the so-called measure problem.

The proof of Cantor's proposition is very brief and may be stated in two or three sentences. However, some little effort is required to master it. Still, on the other hand, the individuals who enjoy conceptual (as contrasted with manipulative) mathematics very usually experience a considerable intellectual pleasure from this proof, once it is properly mastered. With these particular readers in mind, here goes the proof. (Other readers may wish to skip it!)

Consider then an arbitrary infinite set E whose elements will be generically denoted by e. Now, let $K(E)$ denote the class (or the set) of all subsets of E. Each such subset of E will be denoted by S, so that S is an element of $K(E)$. We shall show that the power of $K(E)$ is greater than that of E. Somewhat like in the diagonal proof earlier, the proof is by contradiction. We begin by assuming that by some method a one-to-one correspondence was established between the subsets forming $K(E)$ and the elements of E so that the two sets are equivalent. Thus, according to this assumption, to every e, element of E, there corresponds one element of $K(E)$, which we shall denote by $S(e)$. Conversely, the assumption that E and $K(E)$ are equivalent implies that, whatever subset S of E we take, there is an element of E, say $e(S)$, that corresponds to S.

Now, as in the diagonal proof, we will define a subset of E, which we shall denote by S^*, for which there will be no corresponding element of E. The definition is: an element e of E belongs to S^* if and only if the subset $S(e)$ does not contain e.

In order to appreciate this definition, it is convenient to visualize a table (Table 17.2, somewhat like the one involved in the diagonal proof) in which the first column is composed of all the elements e of E while the second column specifies the corresponding subsets $S(e)$.

In order to specify the subset S^*, we begin with the element e' and try to decide whether it belongs to S^* or not. For this purpose we inspect the subset $S(e')$ that corresponds to e'. If it happens that $S(e')$ contains e', then e' does

Table 17.2. Hypothetical Correspondence between
All Subsets and Individual Elements

Elements of E	Corresponding Subsets $S(e)$
e'	$S(e') = (e'_1, e'_2, \ldots)$
e''	$S(e'') = (e''_1, e''_2, \ldots)$
.	. . . .
.	. . . .
.	. . . .
$e^{(n)}$	$S(e^{(n)}) = (e^{(n)}_1, e^{(n)}_2, \ldots)$
.	. . . .
.	. . . .

not belong to S^*; on the other hand, if $S(e')$ does not contain e', then e' belongs to S^*. Similar procedure applied to other elements of E, that is to e'', e''', and so on, will determine which of them belong to S^* and which do not.

Now we consider more closely the assumption we made that there is a one-to-one correspondence between subsets of E and its elements e. This assumption means that to each and every subset S there corresponds a well-defined element of E. Since S^* is a subset of E, our assumption implies that there must exist an element, say e^*, which corresponds to S^* so that S^* must be located somewhere in the above column. In other words S^* must be identical with $S(e^*)$. But this is impossible. To see this, let us consider the question whether e^* is contained in S^* or not. If it is contained in S^*, then it must be contained in $S(e^*) = S^*$; however, by the very definition of S^*, this subset does not contain those elements of E that belong to subsets containing them. Thus, e^* cannot be an element of S^*. On the other hand, if we assume that e^* does not belong to S^*, it would follow that it does not belong to $S(e^*)$. Then, however, the definition of S^* would require that e^* belong to S^*. It follows that the assumption of S^* corresponding to any element of E leads to contradictions. The conclusion is that, in whichever way we try to establish a one-to-one correspondence between the elements of E and the subsets S of E, there will always be some subsets S left without the corresponding elements e. In other words, whatever the set E may be, the power of the set $K(E)$ of its subsets is greater than that of E itself.

Gödel's Revolution: Undecidability

And now let us look at Cantor's famous problem, the so-called *continuum hypothesis* which he could not solve and which David Hilbert ranked first in his celebrated talk on unsolved problems at the International Mathematical Congress of 1900.

Is the continuum the next order of infinity after $\aleph_0$ or are there perhaps sets of power greater than $\aleph_0$ and less than C, between the two? In other words do there exist sets of real numbers (parts of the continuum of all of them) that are not denumerable but of power less than the full continuum C? The statement that this is not so is known as Cantor's Continuum Hypothesis.

In this essay we shall now follow, to be sure in a simplified way, the history of this problem and its role in changing profoundly our conceptions of infinity in mathematics.

The constructions of mathematics, as already evidenced in ancient times by Euclid's formulation of geometry, became increasingly based on the *axiomatic* method in the course of the nineteenth century.

Aristotelian logic, perhaps surprisingly a priori, showed how starting with a few postulates and rules, a very small finite number indeed, one can, by repeated applications, obtain all of the elementary logical statements.

The axiomatic program had as its aim (most explicitly in the hands of the great German mathematician Hilbert) to express the entirety of mathematics starting with a finite number of symbols (or signs, on paper) and a finite number of rules for combining them, rules given in advance. Thus it would be possible to obtain all possible meaningful statements as consequences proved true or false by such finitistic procedure. In other words, mathematics could be regarded as one great game fundamental to all possible games. And, in theory at least, expressible by a finite automaton—a great computer— which would, given any statement that made sense according to original rules, prove its truth or its falsity. This indeed was Hilbert's hope. He expressed it in the statement: "There will be no *ignorabimus*[1] in mathematics."

Indeed, the progress of mathematics was and is marked by a succession of results, many of which constitute solutions of problems formulated earlier, sometimes long ago. New methods and new ingenuities manage either to prove the validity of a conjecture or disprove it by counterexamples. In set theory a comprehensive and generally used system of axioms for it was developed by the mathematicians Zermelo and Fraenkel.

Great was the mathematicians' surprise when Kurt Gödel showed in an epochal paper that there existed statements *in every sufficiently large mathematical system* which were *meaningful* and yet *undecidable*. This means that within any such system there are statements that cannot be proved or disproved using means of the system itself. This result destroyed all hopes for Hilbert's program.

For any such formal system Gödel constructs a statement meaningful in the system by means of an operation bearing a resemblance to the diagonal method. He does it by enumerating all statements of the formal system and constructing a statement within a system which can be interpreted as a statement *about* the system, as it were from the outside of it. This statement is then proved undecidable within the given system.

Of course, we cannot give here even an inkling of the very complicated construction that he employs. The reader could find such an account in a book by Nagel and Newman, for example.[2]

While it was true that such propositions, as originally constructed by Gödel, did not possess any immediate intuitive content, later developments have shown that some well-known mathematical problems are undecidable. Indeed, in a subsequent paper, Gödel showed that, within the currently accepted axioms of set theory, the Cantor continuum hypothesis cannot be disproved. What remained to be shown is that it cannot be demonstrated either. This was done in 1963 in a series of papers by the young American mathematician Paul Cohen.

Certain psychological elements behind Gödel's discovery are worth emphasizing. They have to do with what is sometimes called "routine of thought." If one grows up under the spell of some particular doctrine (the Hilbert doctrine in this case or the Ptolemaic astronomy in the case of Copernicus) one adheres to this doctrine routinely, and psychologically it is very difficult even to ask the question whether this doctrine is valid or not. Thus, for example, John von Neumann, himself the creator of an axiomatic system of set theory, tried to accomplish Hilbert's program by proving the decidability of mathematical theories. Had he questioned Hilbert's doctrine, his technical ingenuity and insight could have brought him to the discovery of undecidability. But he did not. It was Gödel who was first able to enlarge the horizon in daring to doubt the Hilbertian doctrine, much as had been the case earlier with Copernicus and the Ptolemaic system.

The principal idea of Gödel's proof is rather simple really. But it also involves a fantastically patient, detailed, and complicated travail of pushing through the labyrinth of all possibilities involved in the construction of a general axiomatic system.

Another subject of theory of sets covered by Gödel's work is the so-called axiom of choice, appearing both intuitive and innocuous, but in fact leading to consequences that appear paradoxical. Gödel's work implied the possibility of assuming this axiom without risk of contradiction. The converse of this result, that is, that the rejection of the axiom of choice, by itself, will not lead to contradictions either, was proved more recently.

What the results, relating to both the hypothesis of continuum and to the axiom of choice, amount to is that these statements are independent of the other axioms of set theory. Here we have a situation analogous epistemologically to the advent of non-Euclidean geometries when Bolyai and Lobachevski showed that the *fifth axiom of Euclid is independent* from the rest of them and constructed models where the postulate is not valid.

Infinities in Mathematics and in the Physical Universe
What could be the consequences of these new ideas for the problems of actual infinities in the physical universe? What is the importance of these discoveries in the realm of pure thought for our picture of the physical world? Speculations on the actual infinity of the cosmos are almost as old as astronomy itself. Is the universe of matter, of stars in particular, actually infinite? Is the space containing it infinite itself?

These questions have occupied and excited philosophers from early times on. Giordano Bruno, some three-quarters of a century after Copernicus speculated both about the infinities of the mathematical and of the physical world. His rejection of Aristotelian astronomy in favor of Copernicus and his adoption of the idea of the possibility of innumerable worlds became the chief causes of his downfall.

Now we know there are stages of hierarchy in the structure of matter: stars, collections of stars called galaxies, groups or clusters of galaxies, perhaps again collections of such clusters, which are called superclusters. Does this gradation continue into infinity? Is the universe bounded in space?

The theory of general relativity allows both the possibilities of the space being closed, in analogy to the three-dimensional surface of a four-dimensional sphere ("elliptic" metric), or an open manifold, infinite in extent, with a "hyperbolic" metric.

Is the infinite space asymptotically homogeneous, that is to say, are sufficiently large finite parts of it similar to one another, or is there a hierarchical structure as the one envisaged by the astronomer Charlier, for example?

The finiteness of the velocity of light prevents the possible interactions of arbitrarily far separated parts in finite time. But does time itself have an origin, or is it an infinite parameter in a physically meaningful way?

It is not easy to give precise mathematical formulation to these philosophical questions. Nor is it easy to describe physically relevant statements that would be undecidable on the basis of physical laws (playing the role of mathematical axioms).

The infinity of space in the large may have a counterpart in the actual infinity of ever smaller structures going into the atom, its constituent parts, the nuclei, the nucleons, and perhaps smaller parts of which they themselves are constructed, and so on ad infinitum.

It might be in such a case, that some properties of the "real world" on the basis of any finite collection of physical laws will be undecidable. Thus the all-embracing character of anthropogenic schemata would be in doubt, much as the belief in our occupation of the center of the universe was discarded by Copernicus.

But the meaning and value of the Copernican-like revolutions might really lie in the pragmatism of the usefulness and simplicity of new points of view.

In mathematics the introduction of what we imagine intuitively as infinity enormously simplifies many finite formulations and techniques. For instance, the infinitesimal calculus and the introduction of the limiting processes are so much more convenient and fertile than the operation with finite differences. The asymptotic statements in the theory of probabilities, in combinatorial analysis in general and in the theory of numbers itself, the statements about limiting densities of primes, and so forth, are both more elegantly stated and more suggestive. The Copernican description of the solar system is kinematically equivalent to the Ptolemaic description, but it is simpler and more elegant, and suggestive of more general and deeper knowledge brought later by Galileo, Kepler, and Newton.

In the same way new mathematical models including finite systems that mirror our intuitions of infinity will bring about greater understanding and mastery of the workings of our finite brain. In other words and to put it simply, the ideas of infinity and its study have had and will continue to have many practical functions.

Notes

1. Latin: "We shall never know."

2. Ernest Nagel and James Newman, *Gödel's Proof*. New York University Press, New York, 1958.

J. M. Hammersley

18   The Technology of Thought

Let me not to the marriage of true minds
Admit impediment.
> —Shakespeare, *Sonnet 116.*

Orientation
This article[1] has five separate themes: (1) the nature of technologies; (2) the role of mathematical symbolism in science; (3) Boole's inquiry into the laws of thought; (4) Babbage as the pioneer of the computer; (5) Copernican revolutions. The last part of the article glances at a few of the connections between these five themes, and concludes that symbolism is not really a Copernican revolution and computing only partly so. Denial of Copernican aspects may lend contrast and variety to this book: after all, a principal cause of indigestion is too many breakfasts with the same companion.

What Is a Technology?
A technology is a large-scale everyday means to a desired end. Fundamental research need not aim at the useful; but a technology aims from the outset at a preconceived goal, which is useful and desirable at least in the limited sense that the public wants (or has been persuaded to want) to use it. Men have always yearned to fly; and the technology of flight lies in discovering and perfecting techniques for making aviation a means of daily transport. A technology will often follow on the heels of research when matters of scale arise. No sooner have you created, say, a gram of liquid oxygen in the laboratory, than some guy demands a thousand tons of the stuff, and more. Put it on a production basis, he says. That calls for technology.

Suppose a mechanical shovel can scoop 100 tons of rock at each bite, in the time a man takes 20 spadefuls of 10 pounds each. The technological scale factor (speed multiplied by capacity) is 1000 in favor of the machine. If the Concorde flies 500 times faster than a man walks and carries 2000 times the load, its scale factor is 1,000,000. A computer can calculate a thousand million times faster than a man with pencil and paper, and store a million times as much data: its scale factor is 1,000,000,000,000,000.

Institute of Statistics, Oxford University, Oxford, England, OX13UL.

Mathematical Symbolism in Science

Some laymen say science would be plain if scientists used plain words instead of technical terms. Alas, science is not a plain subject: scientists themselves only manage to understand it with the aid of terminology built with patience and purpose to handle ambiguities and subtleties. To see the importance of notation, try multiplying MCMLXXIII by MLXVI without conversion to their Arabic forms 1973 and 1066. Mercifully, "to-day the Roman and his trouble are ashes under Uricon." [2]

The system of Arabic numerals is the most important single technical invention in the whole history of science; and it is no trivial invention, for it eluded all the great mathematicians of antiquity. Science is saturated with quantitative statements. We are so familiar with the arithmetic of decimals, negative numbers, fractions, powers and the like, that we scarcely pause to guess how science would wither if it could not refer to a power-supply of 25.6 kilowatts, or a temperature of $-164°C$, or a radioactive half-life of 10^6 years.

But arithmetic is not enough: science calls upon the whole of mathematics. An arithmetic statement about 25.6 kilowatts is precise or approximate in an obvious sense, while the algebraic statement about x kilowatts is imprecise in another sense: the former is specific, the latter general. Symbols, general in the latter sense, can formulate general laws of science, letting the mind shift from specific instances to patterns of behavior: or rather, the symbolism (as a technology or tool) was *invented for the desired purpose* of enabling this shift. To widen the scientific coverage, we deliberately evade being specific about what we mean. The odd result of this evasion is that symbols take on a life of their own and yield unexpected bounty.

Listen to J. W. L. Glaisher[3] on Napier's invention of logarithms:

Nothing in the history of mathematics is to me so surprising or impressive as the power it has gained by its notation or language. No one could have imagined that such "trumpery tricks of abbreviation" as writing $+$ and $-$ for "added to" and "diminished by" or x^2, x^3, . . . for xx, xxx, . . . , etc. could have led to the creation of a language so powerful that it has actually itself become an instrument of research which can point the way to future progress. Without suitable notation it would be impossible to express differential equations, or even to conceive of them if complicated, much less to deal with them; and even comparatively simple algebraic quantities could not be treated in combination.

Mathematics as it has advanced has constructed its own language to meet its need, and the ability of a mathematician in devising or extending a new calculus is displayed almost as much in finding the true means of representing his results as in the discovery of the results themselves. When mathematical notation had reached the point where the product of n x's was replaced by x^n, and the extension of the law $x^m x^n = x^{m+n}$ had suggested $x^{1/2} x^{1/2} = x$ so that $x^{1/2}$ could be taken to denote $\sqrt{x}$ then fractional exponents would follow as a matter of course and the tabulation of x in the equation $10^x = y$ for integral values of y might naturally suggest itself as a means for performing multiplication by addition. But in Napier's time, when there was practically no notation, his discovery or invention [of logarithms] was accomplished by mind alone without any aid from symbols.

Symbolism has grown apace from the early days of arithmetic with specific numbers, to algebra, and to vectors and matrices where a single letter stands for a whole collection of unspecified numbers; from + telling us to add two numbers, to Σ and $\int$ for adding strings of numbers or tiny bits of a whole. There are symbols for almost anything you care to think of; and even the magical ". . ." for what we do *not* care to think of, or for what we could never write down however much we wished.

Results interest a scientist more than the routines of getting them. A well-chosen algebraic notation can make algebra routine, just as Arabic numerals facilitate arithmetic. The point is well made by A. N. Whitehead in a famous passage:[4]

By relieving the brain of all unnecessary work, a good notation sets it free to concentrate on more advanced problems, and in effect increases the mental power of the race. . . . By the aid of symbolism, we can make transitions in reasoning almost mechanically by the eye, which otherwise would call into play the higher faculties of the brain. It is a profoundly erroneous truism, repeated by all copy-books and by eminent people when they are making speeches, that we should cultivate the habit of thinking what we are doing. The precise opposite is the case. Civilization advances by extending the number of important operations which we can perform without thinking about them.

The fashionable program of modern mathematics in the high schools errs if it tries to make children conscious of what they are doing when they do arithmetic. It is better to drill arithmetic into the child as second nature. Of course education should teach us how to think; but education is lopsided unless it also teaches us when and *how* not to think. You cannot ride a bicycle until you have forgotten how you balance.

George Boole and the Laws of Thought

The first person to really knit mathematics and logic together was Boole[5] in his famous book *An Investigation of the Laws of Thought,* of which Bertrand Russell said: "Pure mathematics was discovered by George Boole in his work published in 1854." That puts it a bit high, or else delimits pure mathematics rather severely; but, nevertheless, this book was a most original and audacious work of genius. Witness the clarion call of its opening paragraph:

The design of the following treatise is to investigate the fundamental laws of those operations of the mind by which reasoning is performed; to give expression to them in the symbolical language of a Calculus, and upon this foundation to establish the science of Logic and construct its method; to make that method itself the basis of a general method for the application of the mathematical doctrine of Probabilities; and, finally, to collect from these inquiries some probable intimations concerning the nature and constitution of the human mind.

In this book, Boole lays the foundations of what we call today the theory of sets, Boolean algebra, and the propositional calculus. There is not space here to describe the technical devices of Boole's theory: anyone wishing to pursue those, may turn to the literature of mathematical logic, or better to Boole's book itself. Suffice it to say that Boole assigned numerical truth-values to propositions and statements, and showed how these truth-values may be algebraically manipulated in a manner that models various accepted procedures of reasoning. Although quite like ordinary algebra, Boolean algebra also contained some innovations—for instance, the rule $x^2 = x$, which says that mere repetition of a statement cannot alter its truth. Of course, *belief* in a statement is different: the advertising profession knows that enough repetition will stupefy the public into believing anything.

Boole lived when nineteenth-century determinism flourished, and he wrote with an assurance no longer acceptable today. He held that the logical processes of the mind must be subject to intrinsic causal laws of some kind, just as the physical world seemed to him subject to absolute laws of Nature. He admitted that human beings were sadly irrational, wherefore

the mathematical laws of reasoning are, properly speaking, the laws of *right* reasoning only. . . . There is nothing analogous to this in the government of the world by natural law. The realm of inorganic Nature admits neither of preference nor distinctions. . . . The constitution of things without may correspond to that of the mind within. But such correspondence, if it shall ever be proved to exist, will appear as the last induction from human knowledge, not as the first principle of scientific inquiry. . . . The empire of Truth is, in a certain sense, larger than that of Imagination.

The last sentence in the foregoing quotation is the kernel of a long passage, where Boole skates delicately around the question of how far the laws of thought may influence human thinking about the Infinite (he shrinks from the word God). What, indeed, could orthodox Victorian theologians have thought about him? Fortunately, very few of them would have been able to penetrate the mathematics to the end of the book.

Although Boole approaches the correspondence of "things without" and "the mind within," he never really grasps the tautological side of his argument: that mathematical laws describe the processes of thought because we *choose* to think mathematically, because we have *directed* invention over the years toward an efficient mathematical symbolism. The grasping of that had to await the work of Gödel and others some seventy years after Boole. Just as the principle of uncertainty in physics toppled scientific determinism, so Gödel dealt hammer blows to the foundations of mathematics by showing that logical thought is relative and self-limiting; that logical systems are incomplete; and propositions may be neither true nor false but undecidable. Boole had written:

An almost boundless diversity of theorems, which are known, and an infinite possibility of others, as yet unknown, rest together upon the foundations of a few simple axioms; and yet these are all *general* truths. It may be added that they are truths which to an intelligence sufficiently refined would shine forth in their own unborrowed light, without the need of those connecting links of thought, those steps of often wearisome and painful deduction, by which the knowledge of them is actually acquired.

In 1901 Hilbert propounded some famous problems. His tenth problem asked for a general recipe for deciding which algebraic equations with integer coefficients had integer solutions: for example, $7x^3 + 2y^2 = 25$ has a solution $x = 1$, $y = 3$; whereas $x^2 + y^2 + z^2 = 23$ has no integer solution. Hilbert's contemporaries supposed that "an intelligence sufficiently refined" could find a recipe; and they would have been amazed at the answer which finally stemmed from Gödel's work. In 1970 a young Russian, Yuri Matijasevič, proved that no "intelligence sufficiently refined" could give a recipe because no such recipe exists. Readers seeking the technicalities of the subject, may consult M. Davis, *Computability and Unsolvability* (1958),[6] which also deals with the Turing machines to be mentioned presently. There is an English translation of Matijasevič's paper in *Soviet Mathematics—Doklady* (1970).[7]

Charles Babbage and His Analytical Engine

Charles Babbage pioneered the computer in the middle of the nineteenth century. There had been calculating machines before him: Pascal built one in 1642; but Babbage was the first man who conceived of computing as a technology, to be carried out on a large-scale production basis.

Babbage was born in 1791 at Totnes in England. He inherited a private fortune from his father and spent much of it on his machines. He moved in high society, much in demand for his mordant wit and brilliant conversation. He was Lucasian Professor of Mathematics (Newton's chair) at Cambridge for eleven years, though he never gave a single lecture. His fertile mind was always producing an endless flow of new ideas: he was the first to advocate operational research in industry; he founded the Royal Statistical Society; he invented the occulting lighthouse, the railroad dynamometer car, the ophthalmoscope;[8] he wrote a ballet, stood for Parliament, proposed the system of life peerages. Above all, he was a man of singular tactlessness, he made enemies in profusion, his failures with one after another of his overambitious and too far-reaching schemes made him a laughingstock, and he was a deeply embittered man at his death in 1871.

Babbage tried to build two machines. He finished neither of them because he was up against all the difficulties of pioneering a new technology. Precision engineering was rudimentary then. He had to develop his own special machine tools to obtain mechanical tolerances far finer than the common practice of his day permitted, and he had to train workmen to use these tools. His machine had many moving and interlocking parts, and he had to foresee which parts would be engaging which other parts at each moment of time; and for this, he had to devise entirely from scratch a system of algebraic symbolism for the mechanical drawings. Here he had to pioneer, as an ancillary sideline, a whole new theory of draftsmanship. He was far ahead of his time; and it is no surprise that he failed to finish either machine: the wonder is he ever got as far as he did.

He called his first machine the Difference Engine and began work in 1823. He spent ten years and about £8000 on it, before abandoning it when his chief toolmaker quit after quarrels over remuneration. In law the tools belonged to the toolmaker, who refused Babbage's offer to buy them. At that stage some parts of the Engine were built and working: whether the complete Engine would have worked is uncertain. Eventually most of it was melted down as scrap, though a few portions survive in British and American museums. In any case, Babbage's original mind had moved on to a much more ambitious project, his Analytical Engine.

The Difference Engine was only a glorified adding machine. The Analytical Engine was a true computer in the modern sense, though it was mechanical rather than electronic. The thermionic valve and the transistor had not yet been invented in the 1830s. However, the Jacquard loom was available; and it suggested the idea of punched cards. The Analytical Engine was to consist of three main pieces: a *store,* to hold 1000 numbers each of 50 decimal digits; a *mill,* to do arithmetic on numbers fed to it; and a *printer* to print the answers. Numbers in the store would represent intermediate results in the calculation; and punched cards would control the machine automatically, telling it which numbers should go from the store to the mill, what arithmetic should be done on them there, and where they should return to in the store after arithmetical treatment.

The concept, thus far, is imaginative; but the real measure of Babbage's genius is that he realized the machine would lack the true power of a computer if it did not possess a *stored program*. Elaborate computations consist mainly of simple arithmetic operations, performed over and over again with minor changes; and these changes depend upon the intermediate results as the calculation proceeds. An instruction (say, an order to move a certain number from the store and add it to a number in the mill) can be coded as a number; and so instructions themselves can be stored in the store; and now, if we let the mill work upon these stored instructions, we can make the machine modify its own instructions during the calculation. The machine also requires a power of recognition and a power of decision. Today we call these *conditional jumps* in the instructions; the machine will perform one or another operation, conditional upon its current state. All this Babbage foresaw, and only missed two features of modern computers; the first was electronic construction, and the second was binary arithmetic—the Analytical Engine was a decimal machine.

Much of our evidence on Babbage's ideas is indirect and unsure. He never wrote a complete account. But his principal collaborator, the Countess of Lovelace, published a long article; and correspondence between her and Babbage survives. Augusta Ada, Countess of Lovelace, was the daughter of the poet Lord Byron. She was a talented imperious woman, and a mathematician of international distinction. She was the world's first computer programmer. We have her original program for calculating the Bernoulli numbers on the Analytical Engine—she wrote it in pencil and the Earl of Lovelace, poor man, was told to ink it in.[9] In compiling this program, she discovered for herself the importance of the technical tricks now called *nested subroutines* and *address modification*. She was pleased with her notes and wrote to Babbage:

To say the truth, I *am* rather *amazed* at them and cannot help being struck quite *malgré moi* with the really masterly nature of the style. . . . I have made Lord L. laugh very much by the dryness with which I remarked "Well, I am very much satisfied with this first child of mine. He is an uncommonly fine baby, and will grow to be a *man* of the first magnitude and power."

Babbage must have already known much of what she discovered, but he was still impressed. Arguing a technical point with her, he wrote:

. . . a variable card never can be directed to order more than *one* variable to be given off at once because the mill could not receive it and the mechanism would not permit it. All this it was impossible for you to know by intuition and the more I read your notes the more surprised I am at them and regret not having earlier explored so rich a vein of the noblest metal.

To which she tartly replied:

. . . I cannot imagine what you mean about the variable cards; since I never either supposed in my own mind that one variable card *could give* off more than one variable at a time, nor have I (as far as I can make out) expressed such an idea in any passage whatever.

All that was in 1843; and just a hundred years later men of the caliber of Turing and von Neumann were delighted at their own rediscoveries of things like these.[10]

   She died in 1852, of cancer at the age of 36. Her last years were troublesome: she and Babbage had embarked on an "infallible" mathematical system for betting on horse races, which was disastrously unsuccessful. Twice she had to pawn the family jewels; and at the time of her death she was being blackmailed, for horse racing and gambling were not thought fit social activities for a Victorian peeress. Her daughter, Lady Anne Blunt, inherited her mother's gift for mathematics and her interest in horse racing. She became a leading Arabic scholar, traveling widely in Arabia and Mesopotamia where no European woman had ever been before; and she imported into England Arab horses to found the Crabbett stud, which is now the most famous in the world.[11] And Lady Wentworth, daughter of Lady Blunt and granddaughter of Lady Lovelace, has studied the mathematical genetics of horse breeding, using binary arithmetic—that computing trick which her grandmother and Babbage overlooked. Human genetics, like Byron's, are fascinating too.

This diversion on family genetics has a bearing on the thorny problem of an elite, which we shall presently meet in the next section on Copernican revolutions. Byron and his descendants provide a typical example of one of those family trees with a many-faceted sparkle of intelligent excellence, the result partly of culture and aristocratic leisure, and partly of natural selection with no impediment to the marriage of true minds. Society accepts the breeding of racehorses for speed and of fatstock for beef but frowns upon the breeding of people for those qualities of mind and character that down the ages have furthered and liberated most the human lot. Disincentives range from comprehensive schools and inheritance tax to the guillotine and the concentration camp.[12] Equality is the enemy of liberty and fraternity.

The Analytical Engine was never finished, being beyond mechanical feasibility in the nineteenth century; but its visionary concepts were entirely sound, and today Lady Lovelace's uncommonly fine baby is indeed a man of the first magnitude and power.[13]

But do computers have unlimited power? The answer is no; and was given by a young Englishman,[14] Alan Turing, in 1936 before computers became operational. Turing's work is philosophical and like Gödel's. Just as Gödel studied undecidability, so Turing dealt with *uncomputability*. He produced a theoretical design for a computer, now called a *universal* Turing machine, which would calculate anything that any other possible machine could calculate. On the other hand, he also proved that there were problems which this universal machine could not handle. Hence, such uncomputable problems will remain beyond the reach of all future machines. This does not mean that *actual* computers cannot be improved. His machine had a

store as large as you please, and it operated as fast as you please. Actual computers have limited storage, limited speed, and fall far short of the power
of a universal Turing machine.

What Is a Copernican Revolution?

William Golding wrote a book *The Hot Gates*,[15] whose title comes from the
location of the battle (Thermopylae) which Leonidas fought for the birth
of Greek civilization, and hence for the birth of our own civilization. In this
book Golding said: "The intuition of Copernicus was the intuition common to all great poets and all great scientists; the need to simplify and
deepen, until what seems diverse is seen to lie in the hollow of one hand."
That is true, but it does not explain why Copernicus so shook the world and
set it moving. For his fame rests on the effect rather than the content of his
work: he heralded the scientific age and the technological society.

A Copernican revolution is a major advance in human understanding,
made by one man against a dogmatic consensus of contrary preconceptions.
In historical retrospect it may be easy to discount these preconceptions and
discredit Copernicus's clerical opponents as ignorant fuddy-duddies. But we
may also look at today's preconceptions. For example, take the one in the
Declaration of Independence: "We hold these truths to be self-evident, that
all Men are created equal, that they are endowed by their Creator with certain unalienable Rights, that among these are Life, Liberty and the Pursuit
of Happiness. That to secure these Rights, Governments are instituted
among Men, . . ." We have only to consider their infinite variety to see that,
in any coldly scientific or factual [16] sense, men are not equal: and it is just
as well for the procreation of the species that men and women are unequal
in *some* respects. *Vive la différence!* [17] But the equality is deemed in a social
sense, equality before the law, parity of esteem, job opportunity, and so on.
The Declaration contains deliberate untruths, concocted for the very admi-
rable purpose of upholding human dignity and preserving the social structure of a nation. However, the maintenance of preconceptions and ideals in
the face of contrary facts leads to special arguments, and even to absurd
sophistry. For instance, there is the familiar but erroneous argument that,
if Jones is not equal to Smith, then Jones is either better or worse than
Smith. You might as readily deduce that chalk is better than cheese; but it
depends on whether you want to eat it or write with it on a blackboard. If

you are a schoolteacher writing a report on a child you may, when the situation warrants it, say the child is idle, dirty, dissolute, and intransigent; but there is one thing that you may *not* say. You may not say the child is unintelligent. That is a direct insult to the parents and unfair to the child who has little control over his home environment and none over his genetic makeup. Please notice that I am not talking about race and intelligence. Any difference (which may or may not exist) between the *average* intelligence of different races, for whatever reasons environmental or genetic, is microscopic in comparison with the differences between individuals. It is this large (and more dangerous) difference between individuals that concerns us, whatever our origins; and it is especially worrying in any advanced society that rests on the technology of thought in the broadest sense of the phrase. The intelligent man or woman is liable to be at the receiving end for remuneration, for envy, and for hostility—three things which, alas, go together nowadays. Universities, which are hothouses of intelligence, and whose business it is to follow the plain unvarnished truth wherever truth leads, are vulnerable to the forces of egalitarianism; for it is easier to level down than up. And yet the prosperity of the community at large depends upon cultivating talents, and upon the services of those who possess them. It is just because individuals are so very various and have such conflicting interests that "Governments are instituted among Men" to keep the peace. Education and the deployment of intelligence are already major concerns of any government, and will become more so as civilization bites deeper into the fruits (or else the poisons) of science and technology.

I mention this example of a contemporary preconception lest we should too readily throw stones at the sixteenth-century Establishment which saw Copernicus and his ideas as a disruptive danger to the tenets of society. After his revolution, the simplicity of Copernicus's thesis struck home with all the force of rebound from these unnecessarily complicated preconceptions, just abandoned. The preconceptions were not originally complicated but had become so because they were both wrong and firmly held; and twists and turns of special argument had been conjured forth to confute contrary evidence and bolster what must be firmly held. The preconceptions were so firmly held and so narrow because they rested upon the strongest motives known to man: vested self-interest and man's fostered belief in his own importance—in short, hierocratic politics. Man's inveterate habit of giving himself airs above his station in Nature leads him into accretions and contortions of erroneous thought. (Listen to some snatches from his own stored program at work on itself— ". . . What a piece of work is a man! How noble in reason! how infinite in faculty! . . . in apprehension how like a god! . . . . And yet . . . what is this quintessence of dust? Man delights me not . . ." [18]) Subsequent humility and release from these special pleadings, these former prejudices and dogmata, enhance his eventual delight in the simple unifying concept that underpinned the revolution. For example, with earth (man's home) no longer deemed the center of the universe, the planets acquire simple orbits under the sun's gravitation. Man, having accepted his descent from monkey and taking his due place in the evolving variety of living species, can enjoy the clear logic of natural selection. When the little piece drops sweetly home into the greater jigsaw, then we sense the edge of Occam's razor and admire the shaft of lucidity. *A Copernican revolution overthrows human egotism:* that is the touchstone from which all else follows.[19] As T. S. Eliot wrote in "East Coker":[20]

The only wisdom we can hope to acquire
Is the wisdom of humility.

Conclusions
Let us look at a few connections between our five main themes.

Mathematical symbolism and the computer are both *tools,* which scientists use to help shape and fabricate thoughts; both have been invented and developed for that desired end; both are used on a large scale; and accordingly both are technologies. Both are nothing more than tools for thought: the inanimate tool is under the control of the workman; and symbolism and computers are merely aids to thinking, and neither the author nor the governor of a scientist's thoughts. Both are *power tools:* they do more than relieve the mind of drudgery; they can handle arguments and calculations that the unaided mind could not by itself manage. They differ in that symbolism is an abstract tool, while the computer is a physical tool. Symbolism is more flexible and far-reaching and more sophisticated; whereas the computer deals, basically, with enormous quantities of humdrum arithmetic. You can admittedly do sophisticated mathematics with a computer; but the sophistication is in the program, not in the computer. The software of a computer is subject to the same principles as symbolism itself: a good compiler,[21] like a good algebraic notation, will free the mind to think at a higher level.

Boole set out to systematize logic in terms of *symbolism,* Babbage to systematize calculations on a mass-production basis by a machine. Gödel and Turing, respectively, found that there are very similar limitations to such systematics. The discovery of these limitations in the technology of thought is comparable to the discovery of the principles of uncertainty in physics, which finished off nineteenth-century determinism.

Boole might have offended theologians, but his mathematics deterred most of them from reading him. There is something to be said for being not understood by those who, if they try to understand, will misunderstand. Copernicus feared to offend the theological Establishment; but at least he wrote in Latin, which was incomprehensible to the populace. Today we write in the vernacular, and the mass media are everywhere: scholars may be glad that technical terms and symbolism deter laymen.

Copernicus would never have guessed what a profound effect he would have upon history; and this holds too for the originators of subsequent Copernican revolutions. The originator is just the man who makes the first crack in the wall of the dam, and thereafter it is the gathering outrush of the pent-up preconceptions which swirls through society. Accordingly, a Copernican revolution is less likely to spring from a technology than from pure science: for a technology aims at a preconceived goal, and hence the ensuing flood does not wash away preconceptions but fulfills them. Moreover, if the majority of society already desires the goal, the technology will not infringe a dogmatic consensus. And a successful technology, fulfilling a human want, will boost human morale, not humble it. On the other hand, the intrinsic scale factors in a technology may work the opposite way. If the scale factor becomes too large and the technology runs away with itself, so that the preconceived goal is overfulfilled, then undesired effects may come and humanity may feel itself dwarfed by the monster it has unwittingly spawned. That may have happened with the computer, which has acquired one of the largest scale factors in the history of technology. Babbage embarked on an Engine just for calculating mathematical tables and never dreamt of a computerized society.

Is the technology of thought a Copernican revolution? Mathematical symbolism is certainly a major advance in human understanding. But it was not one man's achievement. Symbolism is the work of many hands. According to Cajori,[22] no mathematician (with the exception of Leibniz and Euler) has invented more than two ideographs universally adopted in mathematics today; and even the relatively few symbols met in elementary algebra originate from over a dozen inventors. Nor was this advance made in the teeth of dogmatic opposition. Symbolism grew up naturally to fulfill a desired need for precise and succinct expression and to free the mind for wider issues. It was welcomed for its usefulness, for its technological success. To some extent it can provide a harmonious simple picture of the workings of Nature; but this feature is a little illusory. Symbolism can be an elegant language; but it is the message itself, not just its language, which must be deep and simple.[23] Finally, symbolism does not overthrow human egotism, quite the reverse. Mathematicians are not the humblest of men. Witness the tenor of Boole's thesis: that the laws of thought and "the empire of Truth" have their seat in the constitution of the human intellect. And even if it be that Gödel's work has overthrown that thesis, nevertheless the overthrow has not yet penetrated far into the public consciousness. No, mathematical symbolism is not a Copernican revolution: it is a long technological evolution.

Computing is a different case. It is a major advance in human understanding, this time initiated by one man and one woman. Babbage met opposition, but only because of his own tactlessness. He did not succeed in building his Engine: he was merely a ridiculous crank,[24] not a source of resentment or defensive rage or a danger to the Establishment. Nevertheless, he sowed the dragon's teeth of a Copernican revolution. The phrase "electronic brain" epitomizes the alarm some feel today. They ask these questions: "Can the computer think? Can it think better than I can? Will it make me redundant?" Yes and no.

The best-known general introduction to the comptuer is Lord Bowden's book *Faster than Thought,* first published in 1953 and today one of the classics of the literature.[25] It is a symposium of essays written by a couple of dozen people, prominent in the early days of computers; and Turing wrote the first half of Chapter 25. On the question of computers thinking, he says:

If one can explain quite unambiguously in English, with the aid of mathematical symbols if required, how a calculation is to be done, then it is always possible to programme any digital computer to do that calculation, provided the storage capacity is adequate.

Few experts would quarrel with that principle, save perhaps some philosophers who might want a gloss to cover uncomputable problems. But Turing.

of course, would have been more than familiar with the point at issue.[26]
Turing also asks:

Could one make a machine to play chess, and to improve its play, game by game, profiting from its experience?

and

Could one make a machine which would answer questions put to it, in such a way that it would not be possible to distinguish its answers from those of a man?

To these two questions he replies:

I believe so. I know of no really convincing argument to support this belief, and certainly of none to disprove it.

Since Turing wrote this, machines have been programmed to play a tolerable game of chess (tolerable at least by the standards of the ordinary man in the street): and no less a chess player than Botvinnik, a former world champion, has contributed.[27] Finally Turing asks, "Could one make a machine which will have feelings as you and I have?" and replies, "I shall never know, any more than I shall ever be quite certain that *you* feel as I do."

These quotations from Turing suffice as replies to the layman's question "Can the computer think?" The physiology of the human brain differs from the computer's. Virtually all modern computers are *serial* machines: that is to say (simplifying the matter somewhat without essentially falsifying it), the computer does only one thing at a time, and it carries out a long string of such actions one after the other at very high speed. A few machines, on the other hand, are *parallel* machines: they perform several actions simultaneously. In present or projected parallel machines, however, these simultaneous actions are all rather similar: there is not much *diversity* in the parallelism. The speeds of the chemical reactions in the nerve cells of the brain are much slower than the computer's electrical signals; but the parallelism of the brain far transcends that of the computer, especially in diversity. So far we know very little about how the human brain processes its thoughts. However, with the work of Gaze, Hubel, Wiesel, and others, we are beginning to unravel the fascinating and fantastically complex manner in which nerve fibers connect themselves up and transmit and organize visual and auditory stimuli.[28] If the richness of the interconnections of the brain has anything to do with richness of thought, it will be a long time before we design a computer that "thinks better" than a man.

That is not to discount specialized and limited activities at which a computer is very good. It is very good, much better than us at carrying out large amounts of arithmetic very quickly. It was designed for that, among other things. The mechanical shovel shifts more earth than the man with a spade; but the man is a better gardener.

There is also a subject called artificial intelligence. This is worthwhile in so far as it encourages us to think *accurately* about how the human intellect sets about solving problems, but we have far to go before we make a machine that (as Turing puts it) will yield answers indistinguishable from human answers. The possibility is there but the prospect remote; and even if the possibility becomes reality, the reality may not match our dreams. The medieval alchemist dreamt of transmuting gold; but when the transmutation of elements became a reality in the nuclear age, it was directed to very different ends than reproducing the contents of Fort Knox. Some people talk of superintelligent machines and superintelligent beings. Space probes do not encourage belief in clever green men in the solar system. But the universe as a whole provides a much more sporting chance—some people would say a virtual certainty. We know that other stars besides the sun have planets. The work of men such as Calvin and Oparin shows how organic and biochemical molecules may be spontaneously synthesized from the raw material of stars and interstellar gases. The universe is so vast and various that it would be remarkable if conditions for life, or some form of higher intellectual organization, did not exist elsewhere. Distances are enormous, and the chances of direct communication with or detection of extraterrestrial intelligence are slender. But even to entertain the suggestion of higher forms of life than man, *somewhere* in the universe, is a quelling of human egotism and a true ingredient of a Copernican revolution.

Will the computer make men redundant? If this is understood as asking if the human race will be superseded by a race of superintelligent machines, the answer is Delphic—not in *our* lifetimes. But if it asks whether some men will lose their jobs, the answer is yes—it happens already. Clerks and stocktakers, to look no further, can be affected. The opportunity to maintain computer records and to collate data can arouse irrational fears of a Big Brother to whom all our activities and possessions are just digits on an integrated chip. Technologies introduce social problems. Technologies, we have said, are the embodiments of mass production; and individual men and women fret when a technology treats them not as individuals or personalities but collectively as The Masses, subjected to the mass media, to the mass psychology of the advertiser, to the lemming hustle of the consumer society, to the impersonal hand of the bureaucrat—men and women like powerless flotsam borne willy-nilly by the moonstruck tides of democracy. These frustrations spring from a myriad different causes; and really the computer is little to blame. It is, after all, only one of many technologies. But it is a whipping boy, attracting an unfair share of the anti-intellectualism recently seen. Computers and computer records have been a noticeable target of university vandalism in various parts of the Western world. Computers do not strike most people as "pleasing to the mind." This opposition, which the computer has excited, differs in kind from the usual opposition to a Copernican revolution: it is not the reactionary opposition of a priesthood or an oligarchy straining for the status quo: it is the murmuring of the partly educated against a loss of individuality or against specialized skills felt to be beyond their reach. We have, perhaps, taken universal education too far, or rather made its higher levels too specialized, given too little attention to the broader aspects of history and the balances of civilization. This concerns all men of learning who teach. For civilization is precious but fragile, exposed to what Jung has called the Wind out of the East. The east wind howls over the steppes into the cloister and brings to Rome the Ostrogoth and Vandal.

Nevertheless there is a larger history, wherein the ebb and flow of civilization, the rise and fall of empires, is just a pulsebeat, a witness that the patient still survives. For suppose we ask this question: Are the transuranium elements artificial? Yes, you may say, man's artifice made them. Yet that but plays with words. If you accept the forging of inorganic atoms from raw hydrogen, the synthesis of organic molecules from interstellar inorganic gases, the whole long caravanserai through prebiotic soup, to DNA, to trilobites, to dinosaurs, to anthropithecus and Homo sapiens, to farmers and pharaohs and pharmacists, what is the driving force you see? Chance and necessity.[29] This welter of variety is the outcome over the aeons of random jostlings and joinings of this or that chance configuration subject to the harsh necessity of living under limited resources, natural selection through the survival of the fittest. In this little corner of the universe which we inhabit at this instant of time, this continuing process has yielded an organism capable of intellectual thought, able to build a cyclotron, able to produce the transuranium elements. These so-called artifacts are, like man himself, only the natural product of Nature's present and local whim

in this long evolutionary climb. In *Man and Superman*,[30] George Bernard
Shaw represents what he calls the Life Force, struggling and straining up-
wards toward intelligence, toward thought, toward the genius of a Mozart
who can summon up the chords of hell for Don Giovanni, up and onwards
to the grim austerity of the Ancients[31] in *Back to Methusaleh* (1919). All
that is Nature's version of her own technology of thought, waddling along
to whatever next like the unstoppable plate tectonics of *The Restless
Earth*.[32] On such a scale of time, the history of civilization is only the twitch
of an eyelid, and the dissolution of empires merely moments for dissemi-
nating this groping force of life. Unleash the Nazi persecution or sack Con-
stantinople, and the thistledown of knowledge floats afresh in the meadows,
and the seeds of thought germinate anew. Nature knows not pity, has no
compassion; and those who would search her face must look upon Medusa.

However, we are human, half not a part of Nature because men are
mortal. The glories of our blood and state are shadows, not substantial
things. On Nature's huge calendar, the life of a man is brief. But if he has
the humility to adjust himself to his proper station in the universe, to his
own human scale of time and his own human scale of values, then he may
enjoy liberty and the pursuit of happiness. In this human framework, pity
does flourish and compassion dwells. We have sympathy with others, both
like and unlike ourselves. We are each of us individuals; and in our individuality, in the splendid variety of civilization lies the zest of *our* lives. We
do not have to scale the ultimate truths, or even to burden ourselves with
too much sincerity or seriousness: the half-truths of faith, the finesse of po-
liteness, and the safety valve of laughter will temper our lives better. Let us
make the most and the best of the disparities of this world while we can,
pluck the flower while it blooms; for we shall, soon enough, "in the dust be
equal made with the poor crooked scythe and spade." [33] The only remem-
brances we can have are the *thoughts* we leave behind. The stone in Westminster Abbey records: here lies the mortal Newton.

Copernican revolutions are strong meat, and "human kind cannot bear
very much reality." [34] If it be true that Copernican revolutions quell human
egotism, and the last and only wisdom is the "wisdom of humility"; if it
be true that each man is only an insignificant member of an ephemeral
species on an obscure planet in the trackless wastes of space and time; if
Nature be, as Housman calls her, "heartless, witless Nature," prodigal with

life, careless of the individual, not even caring for the species, but holding only to her own System; if (as Boole said) "Nature admits neither of preference nor distinctions" and "the empire of Truth is larger than that of Imagination"; if these things appall, and the standards seem to be set too high, take heart from the puny but defiant cry of *individualism* in the last line of the following extract from Huxley's "Fifth Philosopher's Song":[35]

A million million spermatozoa,
    All of them alive:
Out of their cataclysm but one poor Noah
    Dare hope to survive.

And among that billion minus one
    Might have chanced to be
Shakespeare, another Newton, a new Donne—
    But the One was Me.

Notes

1. The article does not pretend to give complete historical coverage, but instead it highlights a few key incidents in the origins of the subjects discussed: in particular, there is no reference to the dominant role of the U.S. Government and American industry in developing computers in the period since 1950.

2. From the poem that begins "On Wenlock edge . . ." in A. E. Housman's *A Shropshire Lad.*

3. *Napier Tercentenary Volume.* Edited by C. G. Knott and published for the Royal Society of Edinburgh by Longmans, Green & Co., London, 1915, p. 75.

4. *An Introduction to Mathematics,* 1911, p. 59. Reprinted with permission from The Clarendon Press, Oxford.

5. George Boole was born in 1815 at Lincoln in England. He learned his mathematics from his father, a tradesman of studious bent. He became an assistant schoolmaster at the age of 16 and later, at the age of 34, professor of mathematics at Queen's College in the University of Cork in Ireland. Some of Boole's thoughts on teaching mathematics in schools appear in a delightful article by A. P. Rollett, *Mathematical Gazette,* Vol. *52,* 1968, pp. 219–241.

6. M. Davis, *Computability and Unsolvability.* McGraw-Hill Book Company, New York, 1958.

7. "Enumerable Sets Are Diophantine." *Soviet Mathematics—Doklady,* Vol. *11,* 1970, pp. 354–358. Originally published in *Doklady Akademii Nauk SSSR,* Vol. *191,* 1970.

8. Independently of Helmholtz.

9. "But, Oh ye Lords of ladies intellectual
   Inform us truly, have they not hen-pecked you all?"
   —Byron, *Don Juan*

10. Technical remark: The Analytical Engine could not automatically repunch cards while calculating; and hence it may be objected that it was not really a stored program machine. However, its current state could cause one card to replace another; and Lady Lovelace achieves address modification in this way (". . . when a certain number of cards have acted in succession, the prism over which they revolve must *rotate backwards* [her italics], so as to bring those cards into their former position; and the limits 1 to n, 1 to p, etc. regulate how often this backward rotation is to be repeated"). She is making the cards move backwards and forwards like the tape in a Turing machine. At one point, where she needs modifications stronger than this and more like repunching, she merely says "they are easily provided for." That (admittedly vague) remark must surely have made Babbage think about the mechanical implications, if he had not already appreciated them. Every programmer, particularly those from the early days of computing, will recall how his first program failed to work because he had not properly anticipated the machine's reactions to it; and we must remember that Lady Lovelace was writing a program for a machine which was not yet operational. Indeed, it was not until May 1949 that the world's first stored program computer (the EDSAC at Cambridge University under Wilkes's direction) actually performed its first fully automatic calculation.

11. For example, 90 percent of entries in the American Arab stud book are wholly or partly descended from Blunt-Wentworth stock.

12. ". . . the history of Russia is full of martyr intellectuals who hand on the torch, not to the many, but to others like themselves. For the trouble with political prisoners is that they cannot but constitute a classless elite, a *separate* body, with different standards and different ideas, and the masses are never much concerned with the fate of an elite—why should they be?" John Bayley, *The Listener*, Vol. *91*, 1974, pp. 194–196.

13. For more information on the lives of Charles Babbage and Ada Lovelace, turn to M. Moseley, *Irascible Genius*. Hutchinson, London, 1964.

14. He was 23 years old. For his family background, see the biography by his mother: Sara Turing, *Alan Turing*. Heffer, Cambridge, 1959.

15. *The Hot Gates and other occasional pieces*. Reprinted with permission from Faber and Faber, London, 1965, p. 40.

16. "15 February 1766: So far is it from being true that men are naturally equal, that no two people can be half an hour together, but one shall acquire an evident superiority over the other." James Boswell, *Life of Johnson*. Vol. 2, p. 13. Exercise for a student: produce a 30-minute dialogue for an imaginary conversation between Jefferson and Johnson.

17. Or, at a more metaphysical level,
    Just such disparity
    As is 'twixt Air and Angels' purity
    'Twixt women's love, and men's, will ever be.
—John Donne, *Songs and Sonnets*. Edited by Theodore Redpath, Methuen, London, 1967, p. 30.

18. William Shakespeare, *Hamlet*, Act II, Scene ii.

19. The reader will see that my definition of a Copernican revolution goes beyond the usual one adopted in this book. While agreeing that the abandonment of preconceptions is essential, I also maintain that these preconceptions be due specifically to human vanity.

20. From *The Complete Poems and Plays of T. S. Eliot*. Reprinted with permission from Faber and Faber, London, 1969, p. 179.

21. A *compiler* is a system (for example, Algol, FORTRAN, and so on) by which a program is fed into a computer and converted into the machine's own internal instructions.

22. F. Cajori, *A History of Mathematical Notations*, Vol. 2. Open Court Publishing Co., La Salle, Ill., 1929, p. 337.

23. "I am not yet so lost in lexicography as to forget that words are the daughters of earth, and that things are the sons of heaven." Samuel Johnson, Preface to his *Dictionary of the English Language*. London, 1755.

24. However, Dr. Alington, a former headmaster of Eton, said "It's the cranks that make the wheels go round."

25. B. V. Bowden, *Faster than Thought*. Pitman & Sons, Ltd., London, 1953.

26. According to a familiar definition, a philosopher is a man who *stamps* around in a quagmire of his own contriving.

27. M. M. Botvinnik, *Computers, Chess and Long-range Planning*. Springer Verlag, Berlin, 1970. See also D. Michie, "Programmer's Gambit." *New Scientist*, Vol. 55, 1972, pp. 329–332.

28. As a selection of introductory articles suitable for the layman, try David Hubel, "The Visual Cortex of the Brain," *Scientific American*, Vol. 209, No. 5, 1963, pp. 54–62; Alan E. Fisher, "Chemical Stimulation of the Brain." *Scientific American*, Vol. 210, No. 6, 1964. pp. 60–68; A. R. Luria, "The Functional Organization of the Brain." *Scientific American*, Vol. 222, No. 3, 1970, pp. 66–78; Lennart Heimer, "Pathways in the Brain." *Scientific American*, Vol. 225, No. 1, 1971, pp. 48–60.

29. Jacques Monod, *Chance and Necessity, An Essay on the Natural Philosophy of Modern Biology*. Collins, London, 1972.

30. *Man and Superman, A Comedy and a Philosophy*. Constable & Co., London, 1903.

31. "THE SHE-ANCIENT. The day will come when there will be no people, only thought." George Bernard Shaw, *Back to Methuselah, A Metabiological Pentateuch. Part V. As Far As Thought Can Reach: 31,920*. Constable & Co., London, 1921, p. 245.

32. Nigel Calder, *The Restless Earth: A Report on the New Geology*. British Broadcasting Corporation, London, 1972.

33. James Shirley, *The Contention of Ajax and Ulysses for the Armour of Achilles*, 1, iii, 1658.

34. T. S. Eliot, "Burnt Norton," in *The Complete Poems and Plays of T. S. Eliot*. Faber & Faber, London, 1969, p. 172.

35. Aldous Huxley, *Collected Poetry*, 1971, p. 106. Reprinted with permission from Mrs. Laura Huxley and Chatto & Windus, London, and Harper & Row, New York.

Part V

Study of Chance Mechanisms—A Quasi-Copernican Revolution in Science and in Mathematics

Summary

Primitive chance mechanisms are those involved in gambling: coin tossing, throws of one or more dice, roulette, and so forth. To each chance mechanism there corresponds a set of possible outcomes ("heads" or "tails" in coin tossing, six sides on which a die can fall, and so on). The outcome of any particular operation of a chance mechanism (of a particular "trial") is unpredictable. On the other hand, the frequencies of particular possible outcomes in a long series of trials characterize the chance mechanism used and are predictable. The first mathematical studies of chance mechanisms, in the early eighteenth century, have their roots in the problems of gambling: How frequently one can win by betting this way or that way? Reflecting the confusion between "frequent" and "probable," the mathematical discipline that arose is called probability theory.

Beginning with the nineteenth century, and increasing in the twentieth, science brought about "pluralistic" subjects of study, categories of entities satisfying certain definitions but varying in their individual properties. Technically such categories are called "populations." A "population" studied may be a population of humans (say, sons of fathers six feet tall), a population of galaxies (say of ellipticals, some bright and some faint), or a population of molecules of a quantity of gas in a container, all moving in different directions and with different velocities. With reference to this latter example, the subject of a pluralistic study would not be the happenings to any individual molecule (this would be an "individualistic" question) but rather the properties of the quantity of gas, as a whole, perhaps its temperature or pressure, which depend on how frequently the molecules move rapidly or slowly.

In a particular case the son of a tall father may be short or may be tall, but observations show that *frequently* he will be taller than the average. This circumstance, and similar frequency relations in other domains, indicated that the appropriate mathematical approach to phenomena of this kind is through the assumption that individual properties of population members are the outcomes of a chance mechanism. In turn the assumption of an underlying chance mechanism brought about a novel problem. This was the reverse problem of the theory of probability: given the frequencies characterizing a "population," to determine (or to estimate) the chance mechanism that could have produced them. The discipline that arose around this problem (and around some other similar ones) is called mathematical statistics.

The first essay in this part gives a general conceptual overview of probability theory and of mathematical statistics, with some illustrations. The second essay concentrates on very delicate probabilistic and statistical problems of physics, and has contacts with Part III. The last essay deals with the problem of experimentation with variable material, such as in agriculture, in biology, in medicine, or in meteorology. The typical problem is to find how a given experimental treatment affects the population of the experimental units, whether (in a sense) positively or negatively, or not at all. In order to find out, some of the available experimental units must be subjected to the treatments studied and some others left as controls. Paradoxically, quite often the experimenters act as the worst enemies of their own experiments: they have emotional attachments to particular treatments. Subconsciously they assign their favorites to experimental units they feel are "better" than the others. The first result is self-deception; the next result is deception of others. The remedy is to have the experimental units selected for the treatments through "randomization," that is, not by the experimenters themselves but by a reliable chance mechanism. A number of such mechanisms have been developed and are in use in many domains of experimentation. In other domains there is opposition, occasionally a fierce opposition.

Herbert Robbins

19   The Statistical Mode of Thought

Three hundred years passed from the publication of *De Revolutionibus Orbium Caelestium* to the discovery of the planet Neptune in the place predicted mathematically from the observed deviations of Uranus from its calculated orbit. Copernicus, Galileo, Kepler, and Newton had showed that the universe was designed on simple and beautiful mathematical principles, no less divine than the Holy Bible. The mathematical study of natural science could therefore be regarded as the highest form of religious inquiry, however subversive it might be to dogmatic beliefs.

The universe of science as revealed to its great discoverers from the Renaissance to the Victorian Age was a complex but strictly deterministic mechanism. Even Einstein's theory of relativity, which abandoned the basic Newtonian postulates of absolute velocity and time, remained within the framework of determinism. The objective of the intellectual revolution of which we shall speak was to understand the working of chance in the universe, by combining the mathematical theory of probability with the statistical analysis of mass phenomena.

Conceptual revolutions occur in mathematics as well as in the experimental sciences. Perhaps the most conspicuous example is the long process by which the ideal of mathematics as a body of geometric theorems deduced from Euclid's postulates was replaced by the supremacy of algebra and analysis. The mathematics of the eighteenth and early nineteenth centuries was dominated by the discovery and exploitation of the calculus, an instrument of surpassing power in the hands of Newton for explaining and predicting the motion of the planets. In the excitement of this triumph of mechanistic determinism and of its accompanying technological applications, the study of indeterminism and random variation was largely neglected. When the exact position of a hitherto unknown planet could be predicted by Adams and Leverrier from Newton's laws of motion, what matter that the height of a child could not be predicted from the heights of his parents? The creation of a mathematics of *indeterminism* and its application to the description and prediction of mass phenomena—physical, biological, and social—represent a revolutionary change in human thought.

When an experiment performed under the same conditions can have more than one outcome, and when instead of vainly trying to predict the *outcome of a single trial* we seek instead to find a mathematical model that will

Department of Mathematical Statistics, Columbia University, New York, New York 10027.

predict the approximate *frequencies of the various possible outcomes in a long series of trials,* we are in the domain of probabilistic mathematics, which deals with *populations* (conceptual or actual) rather than individuals. These populations may be, for example, molecules of a gas, stars or galaxies of a particular category, or groups of living organisms. Probability theory itself is a branch of deductive mathematics; given the probabilities of certain simple events (for example, of heads in a single toss of a coin), to find the probability of some complex event (for example, of obtaining at least 1 run of 10 successive heads in 100 tosses). "Statistical inference" is concerned with the formulation of rules for deciding on the basis of a sample of observed data what is the nature of the probabilistic mechanism that produced these data. For example, if of 100 diseased patients 28 are cured by a placebo and of another hundred 39 are cured by a drug, is this enough evidence to warrant deciding that the drug is more effective than the placebo, or should further trials be made? Any decision rule will sometimes lead to wrong decisions. Can one establish rules for making decisions under uncertainty which will reduce to tolerable levels the frequency of "wrong," and especially of "very wrong" decisions, when these decisions must be based on finite samples?

Probability theory had its origin in the attempt to explain the regularities observed in long run gambling experience. For example, in repeatedly tossing three dice, a sum of 10 occurs more often than a sum of 12. Why should this be so? Fragmentary writings on chances in gambling appeared in Europe only around A.D. 1500. It is not at all clear why the application of simple mathematical reasoning to games of chance came so late. Perhaps there was involved some notion that chance is a denial of God's all-embracing design, and hence an irreligious concept per se. At any rate, the theory of probability achieved mathematical respectability in 1654 when Pascal corresponded with Fermat concerning some questions about gambling which had been raised by the Chevalier de Méré, a man of the world with a taste for mathematics. Although further development of the theory was delayed by the commotion attending the birth of the calculus, the end of the eighteenth century saw the "probability calculus" established by the work of Jacques Bernoulli, De Moivre, Gauss, and Laplace as a legitimate branch of mathematics.

Bernoulli's celebrated "law of large numbers" showed mathematically why in a sufficiently long series of independent trials it is almost certain that the *proportion* of times in which a given event occurs will be very close to its *probability* p of occurrence in a single trial. Bernoulli was particularly interested in the converse problem: given that in n trials a given event has occurred x times, what can one say about the unknown probability p of this event? This question arises naturally whenever the considerations of symmetry and equal likelihood which suffice for games of chance no longer apply. Here, as noted earlier, we leave the domain of deductive logic and enter that of statistical inference, which is concerned with giving rules for behavior in the presence of uncertainty about the true state of nature, and with the profit or loss incurred in the long run by following such rules. Bernoulli gave no satisfactory answer to his problem, and none was forthcoming until 1763, in a posthumous paper by Thomas Bayes. His solution depended on assuming the existence of a "prior" probability distribution of p itself, before the value of x is observed. Since this prior distribution was necessarily subjective, Bayes's solution of the inference problem did not win universal acceptance, although it dominated thought on the subject for over a century and still has many adherents. Around 1930 Jerzy Neyman formulated the problem of inference from statistical data in behavioristic and operational terms, making no use of prior probabilities. Neyman's theory, developed in collaboration with E. S. Pearson, and formalized by Abraham Wald into a general theory of statistical decision making, coexists today with that of the Bayesian school.

The second source of the mathematics of indeterminism—the collection of empirical data concerning the population, wealth, and industry of a state—began to be carried out in a systematic manner during the sixteenth century. The registration of deaths in England was begun under Henry VIII in 1532. During outbreaks of the plague in London, weekly bills of mortality were prepared to help the wealthy decide when it was expedient to flee to a safer refuge in the countryside. John Graunt in 1661 and the astronomer Edmund Halley in 1693 studied birth and death statistics with a view to determining the purchase price of annuities. The first American census was taken in 1790. However, these purely factual collections of data were not at first seen to be subjects to which probability theory could be applied. Perhaps the first important link between probability theory and statistical data came from astronomy and geodesy. Suppose, for example, that an observer is concerned with measuring the angle from his position to two distant objects. Repeated measurements will give slightly different values, owing to uncontrollable random fluctuations in the conditions of the atmosphere, and so on, and it is necessary to reconcile these observations into a single "most reliable" estimate of the true value of the angle. To do this rationally requires some assumption about the nature of the random fluctuations, and in varying ways Gauss, Laplace, and others arrived at the "normal" distribution, the familiar bell-shaped curve of statistics textbooks (first derived by De Moivre as a useful analytic approximation to Bernoulli's algebraic formula for probabilities in coin tossing), as the most reasonable model for errors of observation. Legendre in 1805 and Gauss in 1809 devised the celebrated method of "Least Squares" for reconciling discordant data, and in 1812 Laplace's *Théorie Analytique des Probabilités* appeared; for some fifty years thereafter a succession of writers attempted to make the ideas and results of this difficult work available to the general scientific public.

It may be in order at this point to contrast the *deterministic* with the *probabilistic* approach to one simple physical process, that of radioactive decay. Suppose N atoms of a radioactive element are present at time $t = 0$. Since the atoms are unstable, the number $N(t)$ still present at any time $t > 0$ will decrease by unity at random instants and ultimately become 0. Let $F(t) = N(t)/N$ be the *fraction still present at time t.* If we write the differential equation expressing the fact that the rate of disintegration is proportional to the amount present, and solve this with the condition $F(0) = 1$, we obtain by calculus an explicit equation for $F(t)$ as a function of the time t, $F(t) = (1/2)^{T/t}$. The constant T is the "half-life" period of the element in question; $F(T) = (1/2)^{T/T} = 1/2$.

However, this can be only an approximation to the truth, since $F(t)$ is a discontinuous function, decreasing by $1/N$ as each atom disintegrates, and remaining constant in the time intervals between successive disintegrations. Moreover, the deterministic equation for $F(t)$ does not express the fact that disintegration is a *random process:* it is quite possible that at time T, for example, the value of $F(T)$ may be $> 3/4$ or $< 1/3$. To probe more deeply, therefore, we must ask for the *probability distribution* of $F(t)$, rather than for its *value*.

To solve this problem, we start from an initially unknown function $p(t)$, the probability that a given atom, present at time 0, will still be present at time t. On the assumption of no "aging" in intact atoms, it can be shown that $p(t)$ must have the form $p(t) = c^t$ for some constant $0 < c < 1$. To understand the physical meaning of c, we observe that $F(t)$, the proportion of atoms present at time t, can be likened to the proportion of heads obtained in N tosses of a coin, where the probability of heads in each single toss is $p(t)$. By a familiar result of probability theory, the random variable $F(t)$ has an expected or *mean* value $\mu(t) = p(t) = c^t$. Thus, if we define T to be the time at which $\mu(T) = \frac{1}{2}$, then $c^T = \frac{1}{2}$, and hence $\mu(t) = c^t = c^{T \cdot t/F} = (\frac{1}{2})^{t/T}$. Comparing this with the deterministic equation $F(t) = (\frac{1}{2})^{t/T}$, we see that, whereas the deterministic approach states that the fraction $F(t)$ present at time t is precisely $(\frac{1}{2})^{t/T}$, the probabilistic approach says only that this is the average value of $F(t)$. The relation between these statements is analogous to that between the (false) statement that in 100 tosses of a fair coin exactly half will be the heads and the (true) statement that the fraction of heads is a random variable f with a mean value of .50 in repeated trials of the experiment in question. The practical difference between these two statements depends on the magnitude of the random fluctuation of the random variable in question about its mean value. In the coin-tossing example the expected root-mean-square deviation of f from .50 can be shown to be .05, and from the normal approximation of De Moivre it follows that about 95 times in a hundred the observed value of f will be between $.50 - 2(.05) = .40$ and $.50 + 2(.05) = .60$.

In the case of atomic disintegration a similar analysis shows that the root-mean-square deviation of $F(T)$ from its mean value $\frac{1}{2}$ is such that there is about a 95 percent chance that at the half-life period T the value $F(T)$ will differ from $\frac{1}{2}$ by no more than $1/\sqrt{N}$, and about a 99.9 percent chance that it will differ from $\frac{1}{2}$ by no more than $1.64/\sqrt{N}$. Thus, for very large N the difference between the deterministic and probabilistic approaches is of little practical importance, though for small N it is considerable. In fact, a remarkable theorem of A. Kolmogorov implies that for $N \geq 100$ the chance that $F(t)$ will differ from its mean value $(\frac{1}{2})^{T/t}$ by no more than $1.36/\sqrt{N}$ *during its entire history* for $0 \leq t < \infty$ is about 95 percent. It took about a century of mathematical effort to get from the probabilistic law governing the behavior of $F(t)$ for a fixed t to the statement just given, which deals with the temporal evolution of the disintegration process for the totality of values $0 \leq t < \infty$, and is an example of the concept of "stochastic process" which dominates modern probability theory.

Although it may seem natural to us now to apply the methods of probability and statistics to mass phenomena in both the animate and inanimate world, the passage from games of chance with cards and dice to the laws of the universe of God and man is by no means an easy one. Even Einstein, a principal founder of quantum mechanics, refused to accept as final any physical theory in which chance is regarded as an ultimate and irreducible element. "Quantum mechanics is certainly imposing. But an inner voice tells me that it is not yet the real thing. The theory says a lot, but does not really bring us any closer to the secret of the Old One. I, at any rate, am convinced that He does not throw dice." [1] Certainly the idea that probability theory, despite its profane origin in gambling calculations, is in fact a basic and irreducible concept for understanding the physical world, represents a "Copernican" overthrow of the dogma of determinism.

Copernicus put the sun instead of the earth at the center of the solar system, Kepler replaced circles by ellipses, and Newton derived the entire scheme of astronomical and terrestrial motion from three simple laws. With each step toward a clearer understanding of the Old One's scheme of things, our power to control nature has increased, whatever the inconvenience of having to change our scientific and theological views. Even His seeming predilection for quantum mechanics need not disturb us, since large-scale phenomena are still more or less determinate, thanks to Bernoulli's law of large numbers.

The transitional figure in the application of statistical reasoning from the physical to the biological and social sciences was Adolphe Quételet. A mathematician and astronomer by training, director of the royal observatory at Brussels, he studied under Fourier and Laplace, and was on friendly terms with most of the leading scientists of Europe. Convinced of the importance of probability theory in all branches of science, it was at his initiative that the first International Statistical Congress met in Brussels in 1853. It is easy to see why probability theory and statistical analysis were needed in the biological and social sciences. Given the initial position and velocity of a planet or a cannonball, its future trajectory is determined by the Newtonian differential equations of motion. But given the birth of a baby, will it be a prelate or a criminal?

In 1877 H. P. Bowditch published a bivariate frequency distribution of height and age, and graphs of average height and weight for each year of age, for a large group of Boston schoolboys. He was concerned with finding a way to measure the strength of the relationship among these three quantities; for example, given the age alone, what was the nature of the distribution of weights, and how did this change if both age and height were given? Such questions would now be regarded as ones of partial and multiple correlation, and to them he found no satisfactory answers. While making his study Bowditch was in correspondence with Francis Galton, a cousin of Charles Darwin. Galton's main interest was to elucidate in quantitative form the laws that govern human heredity. One great question for Galton was, if offspring tend on the average to have the same characteristics, physical or mental, as their parents, but with appreciable random variations, how can a population remain stable during many generations despite the cumulative effect of these successive random variations?

Galton collected data on the bivariate frequency distribution of the height of parents and their children. Taking the "midparent's" height as x and the adult son's (or daughter's adjusted for sex) height as y, he found that it was not true that the mean value of y for a given x was x itself; instead, if μ denotes the mean height of the entire population, which for simplicity we shall assume constant from one generation to the next, and if $\mu(x)$ denotes the mean height of sons whose midparent's height is x, then $\mu(x) - \mu$ will on the average be equal to $b(x - \mu)$, where b is about $\frac{1}{2}$. Thus there exists a "regression" or "tendency toward mediocrity" in human inheritance: children of exceptional parents tend on the average to be only half as far from the general population mean as were their parents. The amount of variation about $\mu(x)$ in the distribution of heights of sons with midparent's height x is nearly the same for all values of x, and Galton found that this variation is about $\frac{3}{4}$ of the total variation of all height about the general population mean μ. Thus only $\frac{1}{4}$ of the variation in the heights of sons is "explained" by their midparents' heights, the rest being due to other causes.

That the "regression coefficient" b is about $\frac{1}{2}$ was of no mere academic interest to Galton, for it meant to him that unless some form of selective breeding were instituted for the human race there must be an inexorable drift toward mediocrity, not only of height but of intelligence, energy, and morality, which would present a fatal obstacle to all plans for progress toward a higher form of society.

Galton's empirical discovery of regression led him to seek some mathematical form for the bivariate frequency distribution of any two weakly or strongly related quantities x and y which would exhibit the general characteristics that he had observed in his studies of inheritance of height.

I asked him (J.D.H. Dickson, Tutor of St. Paul's, Cambridge) kindly to investigate for me the Surface of Frequency of Error that would result . . . from the various shapes and other particulars of its sections that were made by horizontal planes, inasmuch as they ought to form the ellipses [cf. Kepler!] of which I spoke . . . I certainly never felt such a glow of loyalty and respect toward the sovereignty and wide sway of mathematical analysis as when his answer arrived, confirming by purely mathematical reasoning my various and laborious statistical conclusions with far more minuteness than I had dared to hope, because the data ran somewhat roughly and I had to smooth them with tender caution.[2]

Thus entered into natural science the ideas of regression, correlation, and the bivariate normal distribution. By 1893 Karl Pearson had worked out the general theory for an arbitrary number of interdependent quantities, and thenceforth it was possible to deal mathematically with the question to what extent the values of a given set of quantities $x_1, \ldots, x_r$ *influence* (not determine) the values $y_1, \ldots, y_s$ of another set of quantities. Current studies of social, economic, educational, and environmental phenomena use essentially the techniques originated by Galton and Pearson.

For example, to what extent do genetic factors, home environment, schooling, and so on influence a child's success in life, however that be defined? Since schooling is perhaps most easily manipulated, it was disappointing to find in 1966 from the Coleman Report that school characteristics seem to have little effect. A recent study by C. Jencks et al.[3] suggests that the same is true of genetic and other factors, leaving almost the same variability when all identifiable factors are held constant as in the general population; thus "success" in life may be largely a matter of chance, to be explained by neither heredity nor environment. However, the study of complex social and biological problems by statistical analysis of data not obtained from designed and controlled experiments is a risky business, and the last word has not been spoken.

The idea of using probabilistic models in human affairs is a comparatively recent development of the "Copernican" replacement of the deterministic-individualistic by the probabilistic-statistical method. The creation and application of probabilistic models to social, economic, and environmental problems appears now, as it did to Galton, to be a necessary condition for the continued existence of human society. Unfortunately, there is no over-supply of genius in this field, and no great understanding as yet of its essential role in future progress. I quote at this point from a letter written to Galton in 1891.

Dear Sir,

Sir Douglas Galton has given me your most kind message; saying that if I will explain in writing to you what I think needs doing, you will be so good as to give it the experienced attention without which it would be worthless. . . . I am not thinking so much of Hygiene and Sanitary work, because these and their statistics have been more closely studied in England than probably any other branch of statistics, though much remains to be desired: as e.g. the result of the food and cooking of the poor as seen in the children of the Infant Schools and those of somewhat higher ages. But I would—subject always to your criticism and only for the sake of illustration—mention a few of the other branches in which we appear hardly to know anything, e.g.

A. The results of Forster's Act, now 20 years old. We sweep annually into our Elementary Schools hundred of thousands of children, spending millions of money. Do we know:

(i) What proportion of children forget their whole education after leaving school; whether all they have been taught is *waste?* The almost accidental statistics of Guards' recruits would point to a large proportion.

(ii) What are the results upon the lives and conduct of children in after life who don't forget all they have been taught?

(iii) What are the methods and what are the results, for example in Night Schools and Secondary Schools, in preventing primary education from being a *waste?*

If we know not what are the effects upon our national life of Forster's Act is not this a strange gap in reasonable England's knowledge?

B (1) The results of legal punishments—i.e. the deterrent or encouraging effects upon crime of being in gaol. Some excellent and hardworking reformers tell us: Whatever you do keep a boy out of gaol—work the First Offender's Act—once in gaol, always in gaol—gaol is the cradle of crime. Other equally zealous and active reformers say—a boy must be in gaol once at least to learn its hardships before he can be rescued. Is it again not strange in practical England that we know no more about this?

B (2) Is the career of a criminal from his first committal—and for what action—to his last, whether (a) to the gallows, or (b) to rehabilitation, recorded? It is stated by trustworthy persons that no such statistics exist, and that we can only learn the criminal's career from himself in friendly confidence. . . .

In how many cases must all our legislation be experiment, not experience! Any experience must be thrown away.

B (3) What effect has education on crime?

(a) Some people answer unhesitatingly: As education increases crime decreases. (b) Others unhesitatingly: Education only teaches to escape conviction, or to steal better when released. (c) Others again: Education has nothing to do with it either way.

C. We spend millions in rates in putting people in Workhouses, and millions in charity in taking them out. What is the proportion of names which from generation to generation appear the same in Workhouse records? What is the proportion of children de-pauperised or pauperised by the Workhouse? Does the large Union School, or the small, or 'boarding out' return more pauper children to honest independent life? . . . Upon all such subjects how should the use of statistics be taught? . . . .

I have no time to make my letter any shorter, although these are but a very few instances. What is wanted is that so high an authority as Mr. Francis Galton should jot down other great branches upon which he would wish for statistics, and for *some teaching how to use these statistics in order to legislate for and administer our national life* with more precision and experience.

One authority was consulted and he answered: "That we have statistics and that Government must do it." Surely the answering question is: The Government does not use the statistics which it has in administering and legislating—except to "deal damnation" across the floor of the H. of C. at the Opposition and *vice versa*. Why? Because though the great majority of Cabinet Ministers, of the Army, of the Executive, of both Houses of Parliament have received a university education, what has that university education taught them of the practical application of statistics? Many of the Government Offices have splendid statistics. What use do they make of them? . . . .

You remember what Quételet wrote—and Sir J. Herschel enforced the advice—"Put down what you expect from such and such legislation; after—years, see where it has given you what you expected, and where it has failed. But you change your laws and your administering of them so fast, and without inquiry after results past or present, that it is all experiment, see-saw, doctrinaire, a shuttlecock between two battledores."

Might I ask from your kindness—if not deterred by this long scrawl—for your answer in writing as to heads of subjects for the scheme? Then to give me some little time, and that you would then make an appointment some afternoon, as you kindly proposed, to talk it over, to teach, and to advise me? Pray believe me, Yours most faithfully,
Florence Nightingale[4]

As Karl Pearson comments,

Florence Nightingale believed—and in all the actions of her life acted upon that belief—that the administrator could only be successful if he were guided by statistical knowledge. The legislator—to say nothing of the politician—too often failed for want of this knowledge. Nay, she went further: she held that the universe—including human communities —was evolving in accordance with a divine plan; that it was man's business to endeavour to understand this plan and guide his actions in sympathy with it. But to understand God's thoughts, she held we must study statistics, for these are the measure of his purpose. . . . Thus it came about that for Galton, and for Florence Nightingale, the end and the means were the same: men must study the obscure purpose of an unknown power, —the tendency behind the universe; and the manner of our study must be statistical. Therein, according to Francis Galton, lay the way to that unsolved riddle of 'the infinite ocean of being'; therein, according to Florence Nightingale, lay the cipher by which we may read 'the thoughts of God.'[5]

The issues raised by Florence Nightingale are, of course, still with us today. To cite some current examples:

Crime

A four-column story in the *New York Times,* August 5, 1973, is headlined "Black Murder Victims in the City Outnumber White Victims 8 to 1." In the article one finds that in fact the death *rate* from homicide for black New Yorkers was 48 per hundred thousand, while for whites it was 6 per hundred thousand. The assertion in the headline would therefore be true only if there were as many black New Yorkers as white, an implausible demographic assumption. The homicide rate estimates quoted were obtained from a sample of 100 of the 1466 police homicide reports in New York City for 1971. The information in this sample was transferred to punched cards that could be read by the *Times* computer (an adaptation of the 1903 Mergenthaler linotype machine?) which found that intraracial homicides accounted for 82 percent of the (unspecified) number of cases in the sample of 100 for which such information was available. A spokesman for the Police Department explained that, since crime is an individual act, divulging such data would prey on unwarranted racial fears. A picture of the pavement on which the chalk outline of a murder victim had been drawn was appended, along with the information that statisticians who were consulted had said that the sample of 100 was large enough to permit broad conclusions about who kills whom in the city, although small percentages would be less exact than large ones. The entire article is well worth reading for those interested in the advances since 1891 in the statistical analysis of crime.

Pollution and Energy Sources

A North Central Power Project recently proposed by the Bureau of Reclamation and utility companies would involve the strip mining of coal in northeast Wyoming to generate 50,000 megawatts of electric power, more than is now produced in Japan or Great Britain. The coal to be burned would emit more nitrogen oxides, sulfur dioxide, and particulate matter than all sources in New York City and the Los Angeles air basin combined. An evaluation of the environmental impact of this proposal has yet to be made. If the energy is deemed to be needed, how would the NCPP compare with nuclear power plants that produce no smoke, a "small" amount of radioactivity, and have an unknown potential danger of malfunction and catastrophe? We are here faced with a problem of relative risk that can be analyzed only by indirect statistical methods. For an illuminating example of the pitfalls inherent in such problems, see the Epilogue of the Health-Pollution Conference in the *Proceedings of the 6th Berkeley Symposium on Mathematical Statistics and Probability*.[6]

Testing Drugs

Since the thalidomide tragedy it has been difficult to obtain government permission to market new drugs of potential benefit but unproved safety. What rational standards should prevail here for the public benefit, and how should clinical trials be conducted and analyzed? Current articles in medical journals tend to be of the "Headaches Found More Common Among Users of Aspirin" variety. The present author has recently been concerned with the problem of designing clinical trials for comparing two treatments with a view to minimizing the number of subjects given the (initially unknown) inferior treatment, and with interpreting retrospective studies designed to evaluate the relative risk of stroke, and so on among smokers or pill takers as compared to abstainers. He can testify that present statistical knowledge of how these things should be done is in a primitive state. Spending time and money on collecting data, while laudable in itself, does more harm than good when, as is usually the case, it is followed by a pennyworth of routine "statistical analysis" and the publication of authoritatively misleading conclusions.

References

1. Quoted in *Einstein* by R. W. Clark, World Publishing Company, New York, 1971, p. 340.

2. Francis Galton, *Natural Inheritance*. London, 1889.

3. C. Jencks et al., *Inequality*. Basic Books, Inc., Publishers, New York, 1971.

4. Karl Pearson, *The Life, Letters and Labours of Francis Galton*. Cambridge University Press, Cambridge, England, 1924, pp. 416–418.

5. Ibid.

6. *Proceedings of the 6th Berkeley Symposium on Mathematical Statistics and Probability*, Volume 6. University of California Press, Berkeley, California, 1972. Reprinted in *Bulletin of Atomic Scientists*, September 1973, pp. 25–34.

M. Kac

20 The Emergence of Statistical Thought in Exact Sciences

Introduction

One does not think of the nineteenth century as a century of scientific revolutions but as a period of growth and consolidation, and in the main this view is right.

It is during the nineteenth century that the mechanics of Galileo and Newton achieves ultimate perfection in the work of Hamilton and Jacobi. Astronomy liberated by the Copernican revolution a little more than three centuries back goes from triumph to triumph, culminating in the feat of predicting by calculation the position of a hitherto unobserved eighth planet of our system. The theory of heat, still primitive in the early days of the century, flowers and comes to maturity through an inspired succession of remarkable ideas of Carnot, Clapeyron, Robert Mayer, Joule, Clausius, Kelvin, and our own Josiah Willard Gibbs. Last but not least, Maxwell creates his magnificent theory of electromagnetic phenomena which brings new light to light itself.

But with the possible exception of Maxwell's theory, which in some ways was revolutionary and which carried in it seeds of yet greater revolutionary changes, there were no clearly visible breaks in continuity of progress and no radical assaults on tradition and orthodoxy.

And yet there was a revolution in the making—a revolution that ultimately changed science in a way as profound as any except perhaps the Copernican one.

Mechanics Dominates the Nineteenth Century

The nineteenth century is dominated by mechanics. Even the electromagnetic theory of Maxwell is couched in mechanistic terms. The downfall of the caloric theory of heat is brought about by proofs that mechanical energy and heat are equivalent, thus hinting that perhaps thermal phenomena too can be brought within the reach of mechanics.

Mechanics achieves almost the stature of geometry—a model of rigor, precision, and intellectual purity.

Mechanics is also complete; if we only knew the positions and velocities of all bodies today, we could by solving the equations of motion predict the future course of the world.

What a magnificent design for the universe!

The Rockefeller University, New York, New York 10021.

True, there are little difficulties here and there. For example, to know the positions and velocities of bodies, one must measure them, and measurements are subject to error. But this is set aside as only a technical matter. We must merely keep perfecting the instruments, something that in principle should be entirely feasible. And so at the end of the century Albert A. Michelson sees the future of physics as being devoted to making more and more refined measurements and not much else.

Enter Thermal Science
By the late sixties, thermal science (or thermodynamics as it is now called) is in essence also complete. It rests safely and firmly on two laws, one embodying the aforementioned equivalence of heat and mechanical work and the other though of earlier origin becomes known as the second law.

It is a much subtler principle, and as formulated by Clausius, it reads: "Heat can never pass from a colder to a warmer body without some other change, connected therewith, occurring at the same time."

What is meant is that passage of heat from a colder to a warmer body cannot happen spontaneously but requires external work to be performed on the system. The law as stated allows no exceptions. The "never" means absolutely never!

The two laws yield a harvest of results and consequences so rich and varied as to be almost beyond belief. Hardly a corner of science (and engineering!) remains untouched or uninfluenced, and when some years later physics begins its struggle with quanta, the second law is the one beacon of light on the dark and uncharted seas.

Only one fundamental problem remains, and that is to fit the second law into the mechanistic frame. It is in facing up to this problem that one runs into difficulties whose resolution precipitates a crisis.

Are There Atoms?
The laws of thermodynamics describe thermal properties of bodies without regard to the *structure* of matter.[1] To strict thermodynamicists like Ostwald and Mach, thermodynamics was a closed subject albeit incomplete to the extent that, for example, equations of state, which describe how pressures and densities of substances are related at constant temperatures, had to be taken from outside the theory.

Thus the familiar Boyle-Charles law

$$pV = RT,$$

relating the pressure p and volume V of an ideal gas at constant temperature T is an *additional* empirical law as far as thermodynamics is concerned.[2]

It seems clear that to *derive* equations of state one must have some idea of how matter is put together, and it is here that the atomistic view enters the picture.

The idea that matter is composed of tiny invisible atoms has its origins in antiquity, and it surfaces throughout history to be repeatedly discarded and nearly forgotten. It is a strange hypothesis with no evidence to support it and nothing but philosophical fancies speaking in its favor.

But with John Dalton (1766–1844) there comes a change. For Dalton makes a fundamental discovery that chemicals combine to make other chemicals only in precisely defined ratios. Thus 16 grams of oxygen will combine with 2 grams of hydrogen to form water, but in *any other ratio* either some oxygen or some hydrogen will be left over.

It is a difficult law to reconcile with the picture of matter being continuous, and Dalton proposes a simple and compelling explanation on the basis of an atomistic view.

It suffices only to assume that oxygen is composed of little units whose weight is sixteen times as great as the weight of similar little units which make up hydrogen, and that two units of hydrogen combine with one oxygen unit in order to "explain" completely the strictness of the 16:2 ratio.

The revival of the atomistic hypothesis by Dalton marks the beginning of modern chemistry, and from now on the atomistic view is never without adherents and partisans. There are, of course, detractors and opponents, and they are, in fact, in the majority during the nineteenth century, but the atomists are not easily silenced even if the opposition does include figures of the magnitude of Wilhelm Ostwald and Ernst Mach.

There is, by the way, an analogy in the ways in which the atomistic and the Copernican hypotheses entered science. The arguments for both were based on the simplicity of description they provided. Neither could at first be decided upon on the basis of irrefutable experimental or observational evidence. Both contradicted the "evidence" of the senses. Both were strongly, almost violently, rejected by the "establishment." Both when finally accepted opened great new vistas for science.

Can One Derive Laws of Thermodynamics from Those of Mechanics?
If one accepts the view that matter is composed of particles (atoms or molecules), then its behavior and properties should follow from the laws of mechanics. In particular, the second law should be derivable from these laws, provided one knows (or assumes) something about the forces between the constituent particles.

To derive the second law from the laws of mechanics is an ambitious and difficult task that Ludwig Boltzmann set himself as a goal a little over a hundred years ago.

But already in 1867 Maxwell saw a hitch. In a letter to his friend Peter Guthrie Tait, Maxwell pointed out that if the atomistic (or molecular) point of view is taken seriously, the second law *cannot* be a consequence of mechanics.

Here is how Professor Martin J. Klein[3] the foremost authority on the history of thermodynamics and kinetic theory puts it:

Maxwell considered a gas in a vessel divided into two sections, *A* and *B*, by a fixed diaphragm. The gas in *A* was assumed to be hotter than the gas in *B*, and Maxwell looked at the implications of this assumption from the molecular point of view. A higher temperature meant a higher average value of the kinetic energy of the gas molecules in *A* compared to those in *B*. But, as Maxwell had shown some years earlier, each sample of gas would necessarily contain molecules having velocities of all possible magnitudes. "Now," Maxwell wrote, "conceive a finite being who knows the paths and velocities of all the molecules by simple inspection but who can do no work except open and close a hole in the diaphragm by means of a slide without mass." This being would be assigned to open the hole for an approaching molecule in *A* only when that molecule had a velocity less than the root mean square velocity of the molecules in *B*. He would allow a molecule from *B* to pass through the hole into *A* only when its velocity exceeded the root mean square velocity of the molecules in *A*. These two procedures were to be carried out alternately, so that the numbers of molecules in *A* and *B* would not change. As a result of this process, however, "the energy in *A* is increased and that in *B* diminished; that is, the hot system has got hotter and the cold colder and yet no work has been done, only the intelligence of a very observant and neatfingered being has been employed." If one could only deal with the molecules directly and individually in the manner of this supposed being, one could violate the second law.

A few years later in another letter (to John William Strutt later Lord Rayleigh), Maxwell states the nonmechanistic nature of the second law in even clearer terms:

> The 2nd law of thermodynamics has the same degree of truth as the statement that
> if you throw a tumblerful of water into the sea, you cannot get the same tumblerful
> of water out again.

Maxwell's profound and prophetic insight into the nature of the second law came as a result of his discovery that velocities in a gas are distributed according to a statistical law. Already in 1859 Maxwell wrote (in a letter to Stokes):

> Of course my particles have not the same velocity, but the velocities are distributed according to the same formula as the errors are distributed in the theory of least squares.

In this quiet unobtrusive way, the revolution against the reign of determinism begins.

To Maxwell the statistical approach appeared quite natural and not at all revolutionary.[4] A system composed of an enormous number of particles should be capable of exhibiting spontaneous deviations from the average (fluctuations) and in this way violate the second law in its dogmatic form. But this was not at all clear to Maxwell's contemporaries, who continued the search for a mechanistic derivation of the second law.

In 1872 Boltzmann published a paper in which among others he used an assembly of hard spheres (small billiard balls) colliding elastically in a vessel as a model of a monoatomic gas and showed that a state of equilibrium will be reached in a way entirely consistent with the second law. Since he seemingly used only the laws of mechanics, he could claim that at last he had derived the second law from mechanics.

Boltzmann's memoir, one of the landmarks in the history of exact sciences, precipitated a long, sometimes bitter, controversy. Only a few years after its appearance, Loschmidt pointed out that the derivation could not be entirely right, for the laws of mechanics are time reversible, while the approach to equilibrium is unidirectional and shows what is called the "arrow of time." In defense of his thesis Boltzmann was forced to appeal to a statistical interpretation of his results, but his explanations were not fully understood, and other difficulties of a logical nature were yet to come.

Of these, a difficulty that became known as the recurrence paradox seemed decisive. It was raised by Zermelo (a student of Planck's who inherited his suspicions of the validity of Boltzmann's approach from his teacher), who invoked a theorem of Poincaré which implied that a mechanical system of the kind Boltzmann considered will show a quasi-periodic behavior, returning repeatedly to states arbitrarily close to the initial one unless the initial state was of a most exceptional character.[5] This was in such an obvious contradiction with the approach to equilibrium that Zermelo was quite ready to dismiss the whole mechanistic approach to thermodynamics.

Paradoxes Resolved by Introducing the Statistical Point of View
Boltzmann again invoked his statistical interpretation but again failed to convince the opposition. And yet that a statistical treatment can resolve the difficulties and the paradoxes becomes clear by considering an idealized (and artificial) model of temperature equalization which was proposed in 1907 by Paul and Tatiana Ehrenfest.[6]

Here is a description of the model which is taken from an article that I wrote for the *Scientific American* in 1964:[7]

Consider two containers, A and B, with a large number of numbered balls in A and none in B. From a container filled with numbered slips of paper pick a numeral at random (say 6) and then transfer the ball marked with that number from container A to container B. Put the slip of paper back and go on playing the game this way, each time drawing at random a number between 1 and N (the total number of balls originally in container A) and moving the ball of that number from the container where it happens to be to the other container.

It is intuitively clear that as long as there are many more balls in A than there are in B the probability of drawing a number that corresponds to a ball in A will be considerably higher than vice versa. Thus the flow of balls at first will certainly be strongly from A to B. As the drawings continue, the probability of finding the drawn number in A will change in a way that depends on the past drawings. This form of dependence of probability on past events is called a Markov chain, and in the game we are considering, all pertinent facts can be explicitly and rigorously deduced. It turns out that, on an averaging basis, the number of balls in container A will indeed decrease at an exponential rate, as the thermodynamic theory predicts, until about half of the balls are in container B. But the calculation also shows that if the game is played long enough, then, with probability equal to 1, all the balls will eventually wind up back in container A, as Poincaré's theorem says!

How long, on the average, would it take to return to the initial state? The answer is 2^N drawings, which is a staggeringly large number even if the total number of balls (N) is as small as 100. This explains why behavior in nature, as we observe it, moves only in one direction instead of oscillating back and forth. The entire history of man is pitifully short compared with the time it would take for nature to reverse itself.

To test the theoretical calculations experimentally, the Ehrenfest game was played on a high-speed computer. It began with 16,384 "balls" in container A, and each run consisted of 200,000 drawings (which took less than two minutes on the computer). A curve was drawn showing the number of balls in A on the basis of the number recorded after every 1,000 drawings. As was to be expected, the curve of decline in the number of balls in A was almost perfectly exponential. After the number nearly reached the equilibrium level (that is, 8,192, or half the original number) the curve became wiggly, moving randomly up and down around that number. The wiggles were somewhat exaggerated by the vagaries of the machine itself, but they represented actual fluctuations that were bound to occur in the number of balls in A.

Behavior predicted by thermodynamics thus appears to be only the *average* behavior, but as long as the fluctuations are small and times of observation short compared with the "Poincaré cycles" (that is, times it takes for the system to come back near its initial state), it can be trusted with a high degree of confidence. Significant deviations from the average are so unlikely that they are hardly ever observed. And yet fluctuations are absolutely essential if the kinetic (molecular) and the thermodynamic views are to be reconciled.

What Price Atoms? Difficulties and Paradoxes
The price for the atomistic view was thus rejection of determinism and an acceptance of a statistical approach, and to the scientific establishment

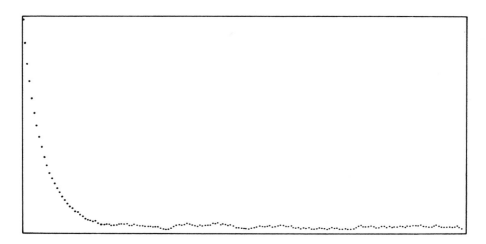

Figure 20.1. The Ehrenfest game played on a high-speed computer.

of the nineteenth century it was too high a price to pay. It was too high especially since no one has ever "seen" atoms or molecules, and there were no known violations of the second law. Why exchange classical simplicity and elegance of thermodynamics for a discipline full of difficulties, uncertainties, and paradoxes? ("Elegance," said Boltzmann bitterly, "should be left to shoemakers and tailors.")

But as the battle for the atomistic and therefore also nondeterministic view seemed on the verge of being lost, help came from a nearly forgotten discovery of an Irish botanist Robert Brown.

In 1827 Brown observed that small (but visible under a microscope) particles suspended in a liquid perform a peculiarly erratic motion. It was not however until 1905 when, independently of each other, A. Einstein[8] and M. Smoluchowski explained this phenomenon on the basis of kinetic theory. Brownian motion, they have shown, is a result of the particle being "kicked around" by the molecules of the surrounding liquid, and while the liquid appears to be in thermodynamic equilibrium on the macroscopic scale, it is on the microscopic scale in a state of disorganized thermal agitation. It is this state that contradicts strict thermodynamic behavior and calls for a probabilistic description.

The theory of Einstein and Smoluchowski had not only explained Brownian motion but it made a number of predictions (for example, that the mean square displacement of a Brownian particle during a time t is proportional to t, the coefficient of proportionality known as the diffusion coefficient being simply related to the temperature and viscosity of the liquid)[9] which have been confirmed, within a few years, in a series of beautiful experiments by Jean Perrin.

Brownian motion has thus made thermal fluctuations visible and in this way the hitherto hypothetical atoms became real.

Fluctuations and Why the Sky Is Blue

Although the fluctuations are small, some of the effects they can produce are quite striking. Of these the most striking is the blueness of the sky.

On a sunny day when it is impossible to look directly at the sun, the color of the sky is determined by the *scattered light*. When a scatterer is small compared to the wavelength of incident, electromagnetic radiation is scattered with the intensity *inversely proportional to the fourth power of the wavelength* (the Rayleigh law), and hence in the case of visible light, it scatters much more light from the violet (short waves) end of the spectrum than from the red end (long waves). But what are the small scatterers that do the job?

At first it was thought that tiny dust particles are the culprits, but this explanation had to be abandoned when it became clear that after a storm the air is much clearer of dust and yet the sky appears, if anything, even bluer.

The correct answer was finally provided by Smoluchowski and Einstein (about 1910), who pointed out that it was the individual atoms of gases of which air is composed that are the scatterers. But this is not enough for if the scatterers were arranged in a regular way (as in a crystal), there would be hardly any scattering at all. It is only because the scatterers are distributed in a kind of random fashion that preferential scattering results in the blue color of the sky! [10]

Another way of saying it is that density fluctuations provide the mechanism of light scattering.

If in a gas of average (number) density of v molecules/cm^2 we consider a cube of side a cm, it will contain *on the average* va^3 molecules. However, the actual number of molecules is not known precisely and is what one calls a *random variable*. In a nearly ideal gas the molecules are almost independent, and the deviation from the average can be shown to be of the order of $\sqrt{va^3}$.[11]

The relative deviation is thus of the order

$$\frac{1}{\sqrt{va^3}},$$

which for v of the order of 2.5×10^{19} (corresponding to a gas at $0°$ under 1 atmosphere of pressure) and a of the order 5×10^{-5} cm (corresponding to the wavelength of visible light) is only a little less than one-tenth of 1 percent (0.1 percent).

Visible light will therefore "see" air not as a homogeneous medium but as one having an irregularly granular structure. It is these deviations from regularity that are responsible for the scattering of light resulting in the blue color of the sky.

The End of the Revolution

In the preface to the second part of his monumental work *Lectures on Gas Theory*[12] Boltzmann wrote (in August 1898): "In my opinion it would be a great tragedy for science if the theory of gases were temporarily thrown into oblivion because of a momentary hostile attitude toward it. . . ." And in the next paragraph he continued: "I am conscious of being only an individual struggling weakly against the stream of time." Only sixteen years later Marian Smoluchowski, the greatest Polish theoretician since Copernicus, delivers an invited address in Göttingen under the title "Gültigkeitsgrenzen des Zweiten Haupsatz der Wärmetheorie" (Limits of Validity of the Second Law of Thermodynamics) that only a few years earlier would have been considered subversive. In the address impressive experimental and theoretical evidence against the dogmatic view of the second law is carefully reviewed, and the *"never"* is replaced by "well, hardly ever."

The revolution had come to an end.

The status of the second law that for so long dominated the science of heat phenomena as it emerged from the struggle is best described in the words of Martin J. Klein:[13]

Smoluchowski's brilliant analysis of the fluctuation phenomena showed that one could observe violations of almost all the usual statements of the second law by dealing with sufficiently small systems. The trend toward equilibrium, the increase of entropy, and so on, could not be taken as certainties. The one statement that could be upheld, even in the presence of fluctuations, was the impossibility of a perpetual motion of the second kind. No device could ever be made that would use the existing fluctuations to convert heat completely into work on a macroscopic scale. For any such device would have to be constituted of molecules and would therefore itself be subject to the same chance fluctuations.

Or to use an analogy with games of chance which Smoluchowski himself liked to use: one can win occasionally, one can even amass a fortune, but one cannot design a *system* that would guarantee it.

Notes

1. It is this seeming limitation that proves to be a source of strength in the early days of the quantum revolution. For while quantum theory radically changed our views on the structure of matter, the changes could never go far enough to violate either the first or the second law.

2. On the other hand, it *follows* from the laws of thermodynamics that the internal energy of a gas obeying the Boyle-Charles law depends only on temperature (that is, is independent of the volume), an important fact found experimentally by Joule.

3. Martin J. Klein, "Maxwell, His Demon, and the Second Law of Thermodynamics." *American Scientist,* Vol. *58,* No. 1, 1970, pp. 84–97.

4. Perhaps it was because Maxwell was more interested in deriving *experimentally verifiable* properties of dilute gases rather than in reconciling the approach to thermal equilibrium with the laws of mechanics—the fundamental problem that Boltzmann pursued with passion throughout his whole scientific life.

5. In technical terms, these exceptional states form a set of measure zero in the set of all allowable states, the measure in question being determined by the laws of mechanics.

6. Paul and Tatiana Ehrenfest coauthored in 1911 a fundamental article for the *Encyclopedie der Mathematischen Wissenschaften* in which they explained with great clarity and precision Boltzmann's views. The article is available in English translation (by M. J. Moravcsik) published by the Cornell University Press in 1959.

7. "Probability." *Scientific American,* Vol. *211,* No. 3, September 1964, pp. 92–108 (in particular p. 106).

8. Einstein's first paper on the subject (published in *Annalen der Physik* in 1905) was entitled "On the Movement of Small Particles Suspended in a Stationary Liquid Demanded by the Molecular-Kinetic Theory of Heat" (I am quoting from an English translation of Einstein's papers on Brownian motion by A. D. Cowper which first appeared in 1926 and has been republished by Dover Publications Inc. in 1956). When Einstein wrote this paper, he was unaware that the movement he so brilliantly analyzed had actually been observed.

His second paper "On the Theory of Brownian Movement," published in 1906, begins with the words:

"Soon after the appearance of my paper on the movements of particles suspended in liquids demanded by the molecular theory of heat, Siedentopf (of Jena) informed me that he and other physicists—in the first instance, Prof. Gouy (of Lyons)—had been convinced by direct observation that the so-called Brownian motion is caused by the irregular thermal movements of the molecules of the liquid."

9. Even more importantly, the coefficient contains the so-called Boltzmann constant $k = R/N$, where R is the universal gas constant and N the "Avogadro number," that is, the universal number (6.02×10^{23}) of molecules on a mole of *any* substance. It was thus possible to determine N from Brownian motion experiments and compare the value thus obtained with determinations based on different principles. The agreement was excellent.

10. At sunset when we can look at the sun directly, the sky appears reddish because now we do not see the scattered (blue) light and hence see the complementary color.

11. This is a special case of what is technically known as the "weak law of large numbers." A more familiar consequence of this law is that in a series of n independent tosses of a coin the excess of heads over tails (or vice versa) is, for large n, of the order $\sqrt{n}$.

12. I am quoting from the English translation by Stephen G. Brush, University of California Press, 1964.

13. See Note 3.

W. G. Cochran

21 The Vital Role of Randomization in Experiments and Surveys

An Early Agricultural Experiment

To begin with experimentation, much of this is comparative. The investigator applies a number of different agents or procedures, often called the treatments, to his experimental material—human beings, animals, clouds, chemical reactions, for example—and takes measurements in order to compare the effects of the different treatments.

In planning an experiment so as to obtain accurate comparisons, the natural attitude of the laboratory experimenter is to keep everything constant except the difference in treatments, leaving nothing to chance. Any difference found in the result for treatment A and that for treatment B must then be due to the difference in the effects of the two treatments. But keeping everything constant is rarely feasible even in the laboratory. It is impossible when we are comparing human beings (identical twins differ in many respects), animals, or a seeded with an unseeded cloud system.

For coping with the variability present in the great bulk of experiments, the revolutionary concept introduced by R. A. Fisher was the idea that something *must* be left to chance.[1] In putting forward the device of randomization, he insisted that the final choice of the plan for the experiment be determined by the outcome of a series of coin tossings or by some similar gambling device.

At first sight, the deliberate use of gambling as an essential step in an experiment sounds most unscientific, to say the least. In order to see what randomization does for us, we need to consider the difficulties encountered in drawing conclusions from experiments on variable material, and the techniques developed historically for handling these difficulties.

The difficulties can be illustrated from the results of seven trials conducted in 1764 by the great English experimental agronomist Arthur Young, at the age of 23.[2] He wished to compare the relative profitability of sowing wheat by drilling the seed in rows ("the new husbandry") as against broadcasting the seed ("the old husbandry") on seven fields that he owned. The total area in any single trial was either an acre or a half-acre. Each area was divided into two equal parts—one drilled, one broadcast. Young states: "the soil exactly the same; the time of culture, and, in a word, every circumstance equal in both." (Here, Young is trying to keep everything constant except the differ-

Department of Statistics, Science Center, Harvard University, Cambridge, Massachusetts 02138.

ence in methods of sowing.) All expenses for ploughing, seed, sowing, weeding, and harvesting were recorded minutely on each half in pounds, shillings, pence, hapennies, and farthings (the monetary system that the British have only recently abandoned). At harvest (on the same day) a sample from each half was sent to market to determine the selling price.

Table 21.1 shows the differences in profit per acre (drilling minus broadcasting) in the seven trials.

Table 21.1. Difference $(D - B)$ in Profit per Acre (pounds sterling).

Trial	1	2	3	4	5	6	7	Average
$D - B$	−0.3	−0.3	−2.3	−0.7	+1.7	−2.2	−3.0	−1.01

The range of results, from £1.7 in favor of drilling to £3.0 in favor of broadcasting, is typical of experimentation with variable material. Despite his statement: "the soil exactly the same," Young knew well that two half-acres do not give the same yield even under the same treatment and husbandry. In fact, in his introduction to the long series of experiments that he conducted, he writes "he (the reader) will no where find a connected train of experiments invariably successful enough to create suspicions."

In drawing conclusions from a limited number of trials with discordant results, Young had no standard methods available and had to confine himself to giving his descriptive impressions of the relative performances of the two treatments. He also made two general observations that are as relevant today as in 1764.

The first simply amounts to saying that experimenters are human rather than coldly objective. In reviewing previous experiments, he noted how often the investigator had a "favorite," as he called it, among the treatments. Writing of Jethro Tull, the famous inventor of the seed drill, Young states "Mr. Tull . . . lets nothing escape his pen to destroy his measure (i.e., favorite treatment)."

Young realized further that any conclusions drawn from a series of trials apply only to the type of situation under which the trials were conducted. He warned that his results could not be trusted to hold on a different type of soil, or under different farm management, or in different years with their changing weather. This caution is well illustrated by the seven trials that Young conducted in the next year 1765. This time, drilling won in six trials out of seven, with an average superiority of £0.57 per acre. Young's warning might be expressed as: "Experiments must be capable of being considered a representative sample of the population to which the conclusions are to be applied." Clearly, conducting experiments so that conclusions are widely applicable is not easy.

Developments in Astronomy and Probability

Attempts to draw conclusions from the results of series of trials of this type suggest the need for two objective techniques: (1) a technique to assist in appraising the strength of a claim that broadcasting was definitely superior to drilling under the 1764 conditions, (2) a measure of the accuracy or reliability of the average gain of £1.01 per acre for broadcasting in these seven trials.

One fascinating aspect of science is that advances needed in one branch may come from quite different branches. In the present case, techniques helpful for summarizing agricultural experiments were developed from the study of errors of measurement in astronomy and from the mathematical theory of probability, mainly during the nineteenth century.

Astronomical measurements are often difficult to make accurately; when the astronomer repeats the measurement, he gets a different result. This variation is ascribed to errors of measurement, since the astronomer is measuring the same quantity on each occasion. The practice arose of repeating the measurement several times, taking the average of the measurements as presumably more accurate than any single measurement. This naturally led to the question: How accurate is the average of 3, 5, or 10 measurements? It is clear that we cannot hope to give a simple definite answer to this question, such as "the average is too large by 0.39," for this would require a knowledge of the quantity being measured. Some other way is needed.

Errors of measurement may be studied by repeating a measurement many times until the average can be regarded as accurate. The difference between an individual measurement and the average is then the size of the individual *error* of measurement. When these errors are classified by their sizes in a frequency distribution, the frequencies can often be well approximated by a bell-shaped mathematical curve, symmetrical about zero error. In this curve, the relative frequency of errors with sizes between x and $(x + dx)$ is

$$\frac{1}{\sigma\sqrt{2\pi}} e^{-\frac{1}{2}\frac{x^2}{\sigma^2}} dx. \tag{21.1}$$

The symbol σ, the *standard deviation* of the distribution, determines how widely spread the errors are. Strictly, σ^2 is the average of the squares of the sizes of the errors and is called the *variance* of the distribution.

A property of this curve is that the absolute error (ignoring sign) is equally likely to exceed or be less than 0.675σ. This quantity, the *probable error*, was often used to describe the accuracy of a measuring process. Further, the absolute error has only a 5 percent chance of exceeding 1.96σ. The curve came to be called the Normal Law of Error. In standard tables, the quantity tabulated is x/σ, which has mean 0, standard deviation 1. The tables give the probability that x/σ exceeds any given value.

If y is a measurement of a quantity μ, the error of measurement is of course $x = y - \mu$. Hence, for measurements following the Normal Law of Error, the frequency distribution of the measurements themselves is, from Equation 21.1,

$$\frac{1}{\sigma\sqrt{2\pi}}\, e^{-\frac{1}{2}\frac{(y-\mu)^2}{\sigma^2}}\, dy. \tag{21.2}$$

By the symmetry of this curve, μ is the mean of the distribution.

Thus far, we have noted that repeated measurements of a quantity μ are often approximately normally distributed about μ. Combined with this observation, a theorem in mathematical probability provided objective statements about the accuracy of the average of a number of estimates. The theorem goes as follows.

Suppose we have made n measurements $y_1, y_2, \ldots, y_n$ of μ and can regard them as independent and following the normal law, with mean μ. (The term "independent" means that the error $(y - \mu)$ of a measurement is not influenced in any way by the other measurements made.) The theorem states that their average $\bar{y}$ is also normally distributed with mean μ, and standard deviation $\sigma/\sqrt{n}$. Hence, the quantity $(\bar{y} - \mu)/\sigma/\sqrt{n} = \sqrt{n}(\bar{y} - \mu)/\sigma$ follows the standard normal tables.

From this result, objective statements in probability terms can be made about the accuracy or reliability of the mean $\bar{y}$ of several independent, normally distributed measurements. For example, there is a 50 percent chance that the error in $\bar{y}$ lies between $-0.675\sigma/\sqrt{n}$ and $+0.675\sigma/\sqrt{n}$, a 95 percent chance that the error in $\bar{y}$ lies between $-1.96\sigma/\sqrt{n}$ and $+1.96\sigma/\sqrt{n}$, and so on for any chosen level of probability. It is the variability of the measurements that forces us to use probability statements in describing the accuracy of their averages.

These statements require a knowledge of σ, which we seldom possess. From n independent measurements, it was known, however, that an estimate of σ^2 is given by

$$s^2 = \sum_{i=1}^{n} (y_i - \bar{y})^2/(n-1).$$

Thus in practice, statements about the accuracy of the sample mean $\bar{y}$ have to be based on the quantity $t = \sqrt{n}(\bar{y} - \mu)/s$ rather than on the quantity $\sqrt{n}(\bar{y} - \mu)/\sigma$ that follows the standard normal tables. In an important step forward in 1908, W. S. Gosset, a chemist in the Guinness brewery in Dublin, obtained and tabulated the distribution of t for a series of n independent and normally distributed measurements. By replacing the standard normal tables by the t-table, probability statements can be made that take account of the number of measurements on which the estimate of σ is actually based. Since Gosset wrote under the pen name of "Student," his table became known as Student's t-table.

The possibility of using these ideas in drawing conclusions from the results of experiments in agriculture and biology began to be realized toward the end of the nineteenth century. We can see how we might try to make statements about the accuracy of Young's average difference of £1.01 per acre in favor of broadcasting. He has seven measurements $y_1, \ldots, y_7$ of this difference μ, giving $s = 1.607$ and $t = \sqrt{7}(1.01 - \mu)/(1.607)$. From the t-table for seven measurements, there is a 50 percent chance that t lies between ± 0.718, or that μ lies between $1.01 \pm (1.607)(0.718)/\sqrt{7} = (0.57, 1.45)$. Similarly, limits for μ corresponding to any desired degree of certainty can be found. We have had to assume, however, something that we do not know, namely that the measurements y_i can be regarded as normally and independently distributed about μ.

Uniformity Trials in Agriculture
Although experiments in agriculture had been conducted regularly at least since the early 1700s, it is remarkable that serious studies of methods for

conducting field experiments did not begin until around 1910. The studies used an ingenious device called *uniformity trials*. (Mercer and Hall,[3] Wood and Stratton,[4] Montgomery.[5]) In these, an area suitable for experiments was divided into a large number of small plots, harvested and weighed separately. The treatment and the farm husbandry were the same on *all* plots. Thus, the variation in yield from plot to plot was a direct measure of the experimental error due to the heterogeneity of the soil.

The goal was to find the best size and shape of plot, number of plots (replications) per treatment, and plan for the layout of the experiment on the field. For any proposed plan, "dummy" treatments $\overline{A}, \overline{B}, \overline{C}, \ldots$ were assigned to the plots to see which plans gave the most accurate comparisons among the dummy treatment means A, B, C, and so on.

Some indications from uniformity trials were as follows.

1. The fertility pattern over a field is complex and varies from field to field.
2. Not surprisingly, plots near one another tend to be more alike in yield than plots far apart.
3. A rising or falling trend in yields along either side of the field is often seen.
4. Since the amount of variation from plot to plot changes with the field and with the crop, each experiment must provide its own estimate $s/\sqrt{n}$ of the standard deviation of the treatment averages $\overline{A}, \overline{B}, \overline{C}, \ldots$

From these indications, a good plan for accurate experimentation, according to Student,[6,7] and Beavan[8] places different treatments *A, B, C*, and so on, on plots *near* one another, so that the comparisons among them are more accurate. For the same reason, replications of the same treatment should be scattered as widely as possible across both the length and breadth of the field, so that the averages for all treatments would be affected in the same way by rising or falling trends parallel to the sides. Various checkerboard plans that seemed to meet these objectives were recommended.

For comparing two varieties of a cereal, a simple and convenient plan was Beavan's Half-Drill strip, which used plots extending the complete length of the field. The order across the breadth was *ABBAABBA . . . ABBA*. The advantage of this order, as is easily verified, is that any straight-line trend across the breadth is completely eliminated from the error of the comparison $(\overline{A} - \overline{B})$.

Enter Randomization

When Fisher began work on field experiments at the Rothamsted Experimental Station in the early 1920s, he found systematic plans of the preceding type advocated and widely employed. Their purpose was the obviously important one of providing accurate comparisons among the treatment means $\overline{A}, \overline{B}, \overline{C}, \ldots$. He realized also that objective conclusions from the results of an experiment could be stated only in probability terms. These probability statements required that the experiment provide a valid estimate of the variance and hence of the standard deviation of comparisons between the treatment means. A further assumption required was that experimental errors were independent and followed a normal distribution.

Fisher did not at first consider trying to satisfy all these diverse requirements when planning an experiment. He concentrated on the provision of a valid estimate of error. The problem, he noted, was that the *actual* errors of the comparisons between $\overline{A}, \overline{B}, \overline{C}, \ldots$ arose from the differences in fertility between plots treated *differently,* while the estimate of error must come from differences in fertility between plots treated *alike*. How can the experimenter ensure that his estimate of error measures the actual errors? As Fisher saw it,[9]

An estimate of error so derived will only be valid for its purpose if we make sure that, in the plot arrangement, pairs of plots treated alike are not nearer toegther, or further apart than, or in any relevant way distinguishable from pairs of plots treated differently. Systematic plans deliberately placed plots treated differently as near together as possible, while plots treated alike were scattered as far apart as possibile, thus violating these conditions.

Suppose that k treatments, each with n replications, are to be compared on $N = nk$ plots. One way of obtaining a valid estimate of error, noted Fisher, is to arrange the plots *completely at random*. The plot numbers 1 to N can be written on cards and mixed thoroughly. The n plots to receive treatment A are drawn, and so on. Tables of 1 million of the digits 0 to 9 in random order have been published to facilitate this process.[10]

This method clearly satisfies the desired condition. Whether a pair of plots receives the same or different treatments depends only on the result of the gamble, and in no way on the properties or positions of the plots. Algebraically, the method has the following property. Let the plot yields due to soil fertility be any amounts $y_1, y_2, \ldots, y_N$. For any treatment A, it can be proved that, if we average over all possible random arrangements, the actual error variance of $\bar{A}$ equals the usual estimate of the error variance of $\bar{A}$, both averages being equal to

$$\sum_{i=1}^{N} (y_i - \bar{y})^2/n(N - 1).$$

The importance of this result is that it holds for *any* y_i, that is, for any fertility pattern.

The experimenter may object that this method, complete randomization, makes no use of any knowledge or judgment that he has about the nature of the variation in his material, leaving the arrangement of the experiment entirely to chance. On the contrary, Fisher pointed out that the introduction of randomization did not prevent the experimenter from using such knowledge to increase the accuracy of the comparisons between the treatment means.[11] But he must do this in ways also guaranteed an unbiased estimate of his actual error variance. To compare k treatments, the experimenter could first form groups or blocks of k plots (or units) that he judged to be closely similar in fertility. The plan, a *randomized blocks* design, puts each treatment once in every block, its position in the block being determined by random numbers, independently in every block. For this plan the average actual error variance of a treatment mean over all possible randomizations again equals the average estimated variance. Both averages now equal

(Average variance within blocks)$/n$.

Thus, if the experimenter's judgment was correct that his blocks were highly uniform, he obtained more accurate comparisons. In any event he obtained unbiased estimates of error. Since each treatment is in every block, this plan automatically eliminates all differences in fertility between block means from the errors of the *actual* comparisons. These differences must be eliminated also in estimating the error variance. This is done by a simple calculation known as the analysis of variance, developed by Fisher.

More flexible plans using the device of blocking in one or in two directions at right angles were produced and are given in books on the design of experiments (for example, Fisher[12]). Each plan has its appropriate method of randomization and method of estimating the error variance.

Randomization and Probability

The act of randomization, as Fisher[13] showed later, has much more basic consequences. It can supply its own foundation for the probability statements made in the conclusions, without requiring any gratuitous assumption by the experimenter that his experimental errors are independent and normally distributed. Two examples will be given.

The first example, Fisher,[14] considers a simple experiment to test the claim of a lady that by tasting a cup of tea she can tell whether the milk or the tea infusion was added first. The plan consists of preparing four cups with milk added first, four with tea added first, and presenting them in random order to the lady for tasting and judging. The lady is told that there will be four cups of each kind.

Suppose we agree to admit the lady's claim if she classifies all eight cups correctly. How likely are we to be wrong? It is known that there are $8!/4!4! = 70$ distinct orders in which the eight cups can be presented. If she has no discriminating ability, she has only a 1 in 70 chance of naming the correct order. What if she is correct on three of the "milk first" cups, wrong on the fourth? We would not then admit her claim, since probability calculation shows that the random order gives her 17 chances out of 70, or nearly 1 in 4, of getting this or a better result entirely by blind guessing. This type of argument, a *test of significance,* is used in appraising a claim that there is a real difference between the effects of two treatments. As a working rule it is common to regard this claim as not established if the observed $(\overline{A} - \overline{B})$ difference or an even larger one could have occurred by chance with a probability greater than 1 in 20, or 5 percent, if the two treatments had identical effects.

Good experimental technique—ensuring that the tea is of the same strength in all cups—makes the "cups of tea" experiment more sensitive, while careless technique—making some cups of China and some of Indian tea—decreases the sensitivity by increasing the "error" variation in taste between cups. It is the randomization that justifies the probability argument used in the test of significance.

As a second example, consider four *ABBA* sandwiches (sixteen plots) in a half-drill strip experiment. This plan assumes that, apart from a possible straight-line trend in yields, the variation from sandwich to sandwich is normal and independent. In effect, systematic plans always assume that Nature does the randomization for us.

However, to take an extreme example, suppose that Nature does not cooperate at all, the trend in yields being quadratic with no other variation. On the "null" hypothesis that A and B have exactly the same effect, the plot yields can be represented by the squares of the numbers from 0 to 15, as in Table 21.2.

In every sandwich, the difference $(\bar{A} - \bar{B})$ is 2. Thus the overall actual

Table 21.2. Simulated Yields of 16 Plots.

	Yields	$(A - B)$ In Pairs	$(A - B)$ In Sandwiches
A	0		
		-1	
B	1		
			2
B	4		
		5	
A	9		
A	16		
		-9	
B	25		
			2
B	36		
		13	
A	49		
A	64		
		-17	
B	81		
			2
B	100		
		21	
A	121		
A	144		
		-25	
B	169		
			2
B	196		
		29	
A	225		

error of $(\overline{A} - \overline{B})$ is also 2. With a continuation of this quadratic trend, the actual error remains 2 no matter how many sandwiches there are (that is how large the experiment).

What about the estimation of error variance? Student suggested two different methods involving different assumptions.[15] His preferred method was to estimate the error variation from the variation among the $(\overline{A} - \overline{B})$ means for each sandwich. Since every sandwich mean is 2, Student's estimate of the error variance is 0 for any size of experiment, as against an actual error of 2 in $(\overline{A} - \overline{B})$ due to the variation in plot yields. An experimenter might object that this example is unrealistic in assuming a completely systematic pattern of variation. This is true, but the pattern can contain major systematic features of which the experimenter is unaware.

How can a randomized plan handle this extreme situation? With blocks of two neighboring plots, a randomized blocks design is not advisable if a rising or falling fertility trend is suspected. The randomization might by chance give A preceding B in every block, clearly undesirable. It is better to have A precede B in four of the eight blocks, chosen at random, with B preceding A in the other four. This plan, the Cross-Over, is much used in medical research in which drugs A and B are given to each patient in succession to obtain intrapatient comparisons, and an effect of the order in which the drugs are given is feared.

As with the cups of tea, there are 70 possible random choices of the four blocks in which A precedes B. If T is the $(A - B)$ total, algebra shows that Student's t always increases whenever T increases, being easily calculated from T. Thus, if the true difference in effect between A and B is zero (the null hypothesis), the randomization distribution of Student's t can be obtained by writing down the frequency distribution of the 70 values of T. On the null hypothesis, the value $T = \pm 56$ is nearest to that required for significance at the 5 percent level, the probability of this or a greater value (± 64) being $4/70 = 0.057$. Similarly, if this were an actual experiment in which we were estimating the true difference μ between the effects of A and B, we could find limits μ_L, μ_H such that the statement "μ lies between μ_L and μ_H" has a 5.7 percent chance of being wrong.

For practical work, Fisher did not recommend the randomization test as a routine replacement for the use of Student's t table. It is slower, and conclusions usually agree reasonably well by the two methods. In this example, with errors neither random nor normal, Student's table gives $P = 0.046$ corresponding to our $P = 0.057$ for $T = 56$. The randomization test, however, dispenses with the assumption that experimental errors are normal and independent and serves as a fundamental check.

As comparative experimentation spread to a multiplicity of areas other than agriculture, randomized plans were speedily adopted. This is not surprising. The patterns of error variation in such areas were diverse and often poorly known apart from a few obvious features. A randomized plan enabled the experimenter to use his judgment in attempting to increase the precision of his comparisons, and provided a foundation for valid conclusions whether his judgment was correct or not.

Randomization in an experiment by no means removes all the complexities in drawing conclusions from the results. The probability statements from the randomization set apply strictly to the data found in the experiment. They do not remove Young's warning about inferences to broader populations. The psychology students in a randomized experiment in University A are not necessarily representative of psychology students in general, even less so of young adults in general. This point can sometimes be handled by devising an experiment or a series of experiments that are done on a random sample of the members of the population to which we wish our conclusions to apply, though questions of feasibility and cost usually prevent this. Randomization at least sets us on the right road.

Consequences of Failures to Randomize

The sources of error variation in an experiment fall into three classes: (1) Those controlled by devices like blocking; they are removed from the actual errors of the comparisons and should be removed from the estimated errors. (2) Those randomized; they affect equally the actual and the estimated errors of the comparisons of treatment effects. (3) Those neither controlled nor randomized; these are the sources that can make the probability statements incorrect and misleading in the conclusions. The most frequent consequences are biased comparisons.

For example, in a large experiment in 1930, 10,000 school children in Lanarkshire, Scotland, received 3/4 pint of milk daily while another 10,000 children in the same schools received no milk, as controls in a study of the effects of milk on height and weight. The teachers selected the two groups of children, in certain cases by ballot (presumably random drawing) and in others on an alphabetical system. Student comments:[16] "So far so good, but after invoking the goddess of chance they unfortunately wavered in their adherence to her for we read: 'In any particular school where there was any group to which these methods had given an undue proportion of well-fed or ill-nourished children, others were substituted in order to obtain a more level selection.' "

Student continues: "This is just the sort of after-thought . . . which is apt to spoil the best laid plans." At the start of the experiment the selected controls were definitely superior in height and weight to the "feeders" by an amount equivalent to three months growth in weight and four months growth in height. In "improving" on the random selection, it looks as if some teachers unconsciously tended to assign the milk to those who seemed to need it most.

In this connection, a useful property of randomization is that it protects the experimenter against a tendency to bias of which he is unaware. If, instead of randomizing, he picks out first from the batch the ten rabbits to receive treatment A, he may have a tendency to select the larger rabbits first, so that treatment A is given to larger and perhaps sturdier rabbits than treatment B. Medical experimenters are so well alerted to the danger of personal bias that if feasible they supplement randomization by another proviso called double blindness—meaning that neither the patient nor the doctor who is measuring the effect of each drug shall know which drug is being measured. A medical researcher who says: "My experiment was randomized and double-blind" may sound to the layman to be groping in the dark, but there are sound reasons for this introduction of chaos.

The consequences of failure to randomize are seen most markedly in the important class of comparative studies in which randomization is not feasible. The illness and death rates of samples of cigarette smokers and non-smokers are compared in order to measure the effects of cigarette smoking (if any) on health. The two samples are usually found to differ in their age distributions. Smokers and nonsmokers may also differ in their eating and drinking habits, amounts of exercise, and in numerous other variables that can affect their health. About all that the investigator can do is: (1) Take supplementary measurements to discover whether such disturbing differences exist between the two samples. (2) If they do, try to remove their effects on $(\overline{A} - \overline{B})$ by statistical adjustments. Such adjustments again require a mathematical model with assumed random and independent residual variation. (3) Give his opinion about the effects of disturbing variables for which he has been unable to adjust.

Since, as Young observed, investigators often have "favorites" among the treatments, other investigators may give contrary opinions or point out sources of disturbing variation of which the investigator was apparently unaware. The conclusions become a matter of debate, rather than anything firmly established. (This is what happened in the Lanarkshire Milk Experiment.) For this and other reasons, the Report of the President's Commission on Federal Statistics contains a section, by Light, Mosteller, and Winokur, advocating greater efforts to introduce controlled, randomized studies in evaluations of the effects of public policies.[17]

Sample Surveys
In the second half of the nineteenth century, census bureaus in a number of countries began experimenting with sampling as a tool for estimating the population means and totals of important statistics. Sometimes a sample (perhaps 1 percent, 10 percent, 20 percent) of the latest census returns was being preserved for further analyses. Or certain specific information—demographic, agricultural, or social—was wanted for a geographic area or a large town, and collection from a sample promised to save time and labor.

If results from the sample were to be trustworthy, the sample must obviously be representative of the population from which it was drawn. Two approaches to what was called Representative Sampling were developed.

Random Selection of Units
The units might be houses, farms, or marriages or tax returns on lists. Random selection gave every unit an equal and independent chance of being included in the sample. This method depended on an even more remarkable property of randomization, the Central Limit Theorem. Averages per unit calculated from large random samples tend to become normally distributed about the population mean, whether the original data are normal or not. This result is proved in probability theory and can be verified in the classroom by experimental sampling. Thus, the normal distribution provided probability statements about the size of the error in any average estimated from the sample.

A further development was stratified random sampling. The units in the population were first arranged in subpopulations or strata, the objective being to have each stratum internally homogeneous. Thus, in an urban study of household incomes and expenditures, the strata might be suburbs, arranged from poorest to wealthiest. An independent random sample of households was taken from each stratum, the number m_i drawn in the ith stratum being made proportional to the total number M_i in the stratum. This device ensured that the sample was representative of the different economic levels present in the city.

Purposive Selection of Groups of Units

It was realized that labor could be saved by sampling complete groups of units, whole blocks in a town instead of individual households. However, a random sample of 200 blocks out of the 2000 in a town was recognized as less representative than a random sample of 1 household in 10. Purposive selection attempted to overcome this defect. The first step was to choose certain control variables related to the subject matter of the study. If census data were available, the 200 blocks were deliberately selected so that the distributions of the control variables in the sample were similar to those in the whole town. It was thought that this method would make the sample more representative for other variables as well.

As in the planning of experiments there were two rival candidates: random selection and purposive nonrandom selection. After considerable practical experience had been gained with both methods, the International Statistical Institute appointed a commission in the 1920s to report on methods of representative sampling. The summary of the report by Jensen described both methods, stating:[18]

The particular advantage of the random method is that one can always be sure of the degree of accuracy with which one is working, as any precision required can always be attained by including a suitable number of units in the sample. The weak point in this method lies in the difficulties of carrying out in practice the strict rules which are demanded by the application of the law of large numbers.

(The primary difficulties mentioned were the need for an exact definition of the unit of sampling, for example, a household or a farm, the need for strictly independent random sampling, and the fact that failure of certain types of people to answer the questionnaire could bias the results. The latter problem is always present in sample surveys, but cannot be blamed on the random selection.)

The summary continues:

The advantage of the purposive method is that it is capable of being applied to practically every field of research, even when the conditions of selection at random are lacking. Furthermore, a saving in time and labour will often be possible owing to the fact that the units which are included in the sample are selected by groups. Whether it may broadly be said that this method is inferior to the random selection as regards precision is debatable; but at any rate there is this difference, that in the purposive method the precision can generally not be so easily measured by mathematical means and in some cases some essential feature in such measurement cannot be ascertained.

(The last comment recognized the same problem as arose with a systematic experimental design. The precision of a purposive sample could be estimated only by making unverifiable assumptions about the nature of the variation in the population.)

On balance, this 1926 report seemed to favor purposive selection as more widely applicable and often labor saving. Purposive selection was, however, already on the way out. For example, in November 1926 the Italian statisticians Gini and Galvani had to select about a 15 percent sample of the 1921 Italian census, to be preserved for further analyses. The census returns were arranged by communes (8354 in number), grouped into 214 districts. They decided to select 29 districts purposively, in such a way that the sample means for seven control variables approximately equaled their means for the whole country. For a noncontrol variable y, they used the data from the 29 districts to construct a linear relation that predicted y from the control variables x_i. From this relation, they predicted y for all 214 districts and hence its population mean. They found, however, that their method gave poor estimates of the population means for important noncontrol variables. Their final judgment was that stratified random sampling, with the commune as a sampling unit, would have given much better results, although they noted that it would be difficult to decide on the best choice of strata.

The paper that was most effective in displaying the limitations of purposive selection was published by Neyman in 1934.[19] He showed that even if the Gini and Galvani sample consisted of all 214 districts, their prediction method would give the correct population mean for a noncontrol variable

y only if a particular linear relation between y and the control variables x held throughout the 214 districts. Further, the Gini and Galvani sample estimate was the most precise of its type only if a further relation about the nature of the variability of y in the population also held. Both relations might occasionally hold in practice, but, as he illustrated from survey data, they could not be assumed to hold with any generality. Thus, the user of purposive selection was gambling on knowledge about the population that he did not possess.

In the same paper Neyman also developed the theory needed for the practical use of stratified random sampling to best advantage. This method, with further developments such as unequal probabilities of selection of different units, and the use of ratio estimates to known control variables, became the basic sampling tool in surveys all over the world.

Without the element of random selection, even a very large sample can mislead. The *Literary Digest's* prediction of a Landon victory over Roosevelt in 1936 was based on a sample of more than 2 million questionnaires. The trouble was that the sample was drawn from lists of telephone and automobile owners, who tended to vote Republican in those days.

Conclusion

In both experiments and sample surveys, methods with randomization as an essential component challenged and replaced methods depending on deliberate choice by the investigator. Unlike the Copernican situation, systematic or purposive methods did not have behind them centuries of authoritative support, since the serious study of how to plan either experiments or sample surveys with variable material took place only within the last century. But the idea that in an experiment or survey the investigator must rely to some extent on chance was revolutionary. The natural tendency of investigators was to think that as experts they could surely do better than reliance on blind chance. But this may not be so even in simple problems. In studies of the growth in height of wheat plants, agronomists selected samples after careful inspection of the plants.[20] When the wheat was 2 feet high, they tended to select samples with a positive bias in height. When the wheat was 4 feet high, their selected samples had a negative bias in height.

Randomization is also "pleasing to the mind" in its insistence that data must be collected in a way that justified the probability statements to be used in drawing conclusions from them. As Fisher put it with regard to experiments,[21]

Owing to the fact, however, that the material conduct of an experiment had been regarded as a different business from its statistical interpretation, serious lacunae had been permitted between what had in fact been done, and what was to be assumed for mathematical purposes. . . . It was necessary to treat the question of the field procedure and that of the statistical analysis as but two aspects of a single problem, and an examination of the relationship between the two aspects showed that once the practical field procedure was fixed, only a single method of statistical analysis could be valid, and, what was of more practical importance, that its validity depended on the introduction of a random element in the arrangement of the plots.

References

1. R. A. Fisher, *Statistical Methods for Research Workers*. First ed., Oliver and Boyd, Edinburgh, 1925.

2. Arthur Young, *A Course of Experimental Agriculture*. Exshaw et al., Dublin, Vol. *1*, 1771, pp. 136–155.

3. W. B. Mercer and A. D. Hall, "The Experimental Error of Field Trials." *J. Agric. Sci.*, Vol. *4*, 1911, pp. 109–132.

4. T. B. Wood and F. J. M. Stratton, "The Interpretation of Experimental Results." *J. Agric. Sci.*, Vol. *3*, 1910, pp. 417–440.

5. E. G. Montgomery, "Variation in Yield and Methods of Arranging Plots to Secure Comparative Results." *Nebs. Agric. Exp. Sta., 25th Ann. Rpt.*, 1911.

6. Student, "Appendix to Mercer and Hall's paper on 'The Experimental Error of Field Trials.'" *J. Agric. Sci.*, Vol. *4*, 1911, pp. 128–132.

7. Student, "On Testing Varieties of Cereals." *Biometrika*, Vol. *15*, 1923, pp. 271–293.

8. E. S. Beavan, "Trials of New Varieties of Cereals." *J. Minist. Agric.*, Vol. *29*, 1922.

9. R. A. Fisher, "The Arrangement of Field Experiments." *J. Minist. Agric.*, Vol. *33*, 1926, pp. 503–513.

10. Rand Corporation, *A Million Random Digits*. Free Press, Glencoe, Ill., 1955.

11. Fisher, *Statistical Methods for Research Workers*, 1925.

12. R. A. Fisher, *The Design of Experiments*. First ed., Oliver and Boyd, Edinburgh, 1935.

13. Ibid.

14. Ibid.

15. Student, "On Testing Varieties of Cereals," 1923.

16. Student, "The Lanarkshire Milk Experiment." *Biometrika*, Vol., *23*, 1931, pp. 398–406.

17. R. J. Light, F. Mosteller, and H. S. Winokur, Jr., "Using Controlled Field Studies to Improve Public Policy." In *Federal Statistics*, Report of the President's Commission, Vol. *II*, 1971, pp. 367–402.

18. A. Jensen, "Report on the Representative Method in Statistics." *Bull. Int. Statist. Inst.*, Vol. *22*, 1926, pp. 359–377.

19. J. Neyman, "On the Two Different Aspects of the Representative Method: The Method of Stratified Sampling and the Method of Purposive Selection." *J. Roy. Statis. Soc.*, Vol. *97*, 1934, pp. 558–606.

20. F. Yates, "Some Examples of Biased Sampling." *Ann. Engrg.*, Vol. *6*, 1935, pp. 202–213.

21. R. A. Fisher, "The Contributions of Rothamsted to the Development of the Science of Statistics." *Rothamsted Exp. Sta. Report*, 1933, pp. 1–8.

Part VI

Technology

Summary

That developments of new technologies can revolutionize human life is not subject to doubt. Whether a sudden development of a novel significant technology must be preceded by the abandonment of a preconception, that is, by an intellectual quasi-Copernican revolution, is a different question. What appears necessary for a sudden outburst of a new technology is for someone to visualize the potentialities and to decide that the thing "can be done, that it must be done and that it shall be done." (Quotation from a speech of Franklin D. Roosevelt.) Occasionally, such a decision may involve a quasi-Copernican revolution (as the present writer feels must have been the case with Babbage—see essay on "The Technology of Thought," Chapter 18 in Part IV) but not necessarily. At the time of the momentous decision not only all the scientific elements of the new technology may be available but also its desirability and prospects may be clear. The new technology results from the decision and from the subsequent concentrated effort at implementation.

The first essay, Chapter 22, describes the origins and development of aviation, from the Wright brothers to jets to rockets. At a superficial glance it may seem that the feat of brothers Wright must have involved the abandonment of a preconception that prevented their many predecessors from achieving the goal of a powered air vehicle capable of flying in the directions chosen by the pilot. However, the true story appears different.

The second essay, Chapter 23, describes the current efforts toward space astronomy, astronomy based on observations using telescopes carried by satellites. One such space observatory, named after Copernicus, was launched in August 1972. The continuing effort is expected to be completed in the 1980s through the launching of the so-called LST (large space telescope). LST depends upon the development of several new technologies necessary to ensure that on commands from the earth the telescope be pointed at the desired direction and that the results of the observations be appropriately transmitted to the earth.

A most inspiring revolutionary development in technology, with a distinctly Copernican flavor, is described in the essay on lasers, Chapter 24. The background is the properties of atoms and of light, symbolized by the names of Albert Einstein, of Max Planck, and so on (see Part III). The substance of the developments consists of a variety of traditional ideas that the tools to perform certain activities must be made of rigid materials such as steel, and so forth. Now these preconceptions are being abandoned in favor of the idea that these same functions can be more satisfactorily performed by appropriately manipulated light. Thus, from times immemorial, the customary tool of a surgeon has been a knife, first a rather primitive knife and currently a rather specialized scalpel. The novel idea is that the surgical operations can be better performed using a laser beam of light. Again, in order to transmit electric power, the necessary implement has been a metal wire. The revolutionary technology now being developed aims at the transmission of electricity by laser beams.

The fourth essay, Chapter 25, which is also the last in the volume, sketches the history of modern civilization in its intimate connection with struggles for mastery of sources of energy. A quasi-Copernican revolution may occur due to a single individual. But it can also result from a slow process extending over generations. The essay describes a series of such revolutions, each resulting in a novel energy utilization technology. Then the novel technology determines changes in the structures of contemporary human societies. The range covered is broad: from societies depending upon food-gathering technology (if you call it "technology"), through domestication of animals and stable agriculture, through choices between two slaves and a donkey as sources of energy, through waterwheels and windmills, to coal and steam engines and, eventually, to the current efforts to master nuclear energy. When this mastery is achieved, will it turn our planet into a Garden of Eden? Or will this happy ending be prevented by a spectacular nuclear holocaust destroying our civilization?

H. Guyford Stever

22 Man Takes Wings

Introduction

On December 17, 1903, late in the afternoon, Orville Wright sent from Kitty Hawk, North Carolina, the following telegram to his father, the Reverend Bishop M. Wright, in Dayton, Ohio: "Success four flights thursday morning all against twenty one mile wind started from Level with engine power alone average speed through air thirty one miles longest 57 seconds inform Press home Christmas."

To Wilbur (1867–1912) and Orville (1871–1948) Wright this experiment was one step, one giant step, in their steady development of the practical airplane, something that had interested them in early life and occupied most of their time since 1899.

In the words of Gibbs-Smith:[1]

Wilbur and Orville Wright were the first men to make powered, sustained, and controlled flights in an airplane and land on ground as high as that from which they took off. They were also the first to make and fly a fully *practical* powered airplane, one that could take off and land without damage to itself or its occupant, and could fly straight, turn and circle with ease. Finally, they were the first to make and fly a practical passenger-carrying airplane. The first of these achievements was brought about on December 17, 1903; the second by the autumn of 1905; and the third in 1908. The Wrights were also first to build and fly properly controllable gliders, and the first to master glider flight. All these successes—and more—are fully documented and established, and are unequivocally accepted by all modern aeronautical historians.

The date December 17, 1903 marks the beginning of a new era in the life of humanity. At first glance it may appear as marking a Copernican revolution. However, on closer examination, the revolution that began on that day was not Copernican in character; no well-established preconception was abandoned. Rather, the success of brothers Wright was due to their careful use of a mass of results of other workers, accumulated over a period of more than a century, and to their collation of these results with those of their own brilliant research and development.

Since that giant step in 1903 by the Wright brothers in the manned-flight revolution, the airplane has taken its place in our everyday life. Many of the early dreamers of manned flight thought of it in terms of escape. "How I yearn to throw myself into endless space and float above the awful abyss,"

National Science Foundation, Washington, D.C. 20550.

were the words of Goethe, and express succinctly the early thoughts of men contemplating the flight of birds and thinking about ways man could emulate them. But other leaders of dreaming and constructive thinking about manned flight had different objectives. Sir George Cayley, one of the great pioneers of manned flight, in 1809 wrote,

I may be expediting the attainment of an object that will in time be found of great importance to mankind; so much so, that a new era in society will commence from the moment that aerial navigation is familiarly realized . . . I feel perfectly confident, however, that this noble art will be brought home to man's convenience, and that we shall be able to transport ourselves and families, and their goods and chattels, more securely by air than by water, and with a velocity of from 20 to 100 miles an hour.

Today there are aircraft for the escape into endless space. Many do use the airplane as a sport, an avocation affording relaxation and a detached view of earth. But mostly the airplane is used commercially in our daily lives. The manufacture of commercial aircraft for passenger and freight hauling is an ever-growing activity. Most long-range and medium-range passenger travel for business and for pleasure is accomplished by aircraft. As far as the transportation of people across the oceans is concerned, Cayley's prediction has already been achieved, and the great ocean liners of earlier days are used primarily for slow vacation trips rather than quick carrying of passengers to their business or pleasure destinations.

But many of the early thinkers and dreamers of manned aircraft also thought in terms of power—military power. Lord Tennyson in 1842 wrote:[2]

Saw the heavens fill with commerce, argosies of magic sails,
Pilots of the purple twilight, dropping down with costly bales;
Heard the heavens fill with shouting, and there rain'd a ghastly dew
From the nations' airy navies grappling in the central blue.

He went on to predict a world federation as a result of this kind of aerial warfare, something that is still part of man's hopes and dreams.

The nature of the manned-flight revolution and its analogue to the Copernican revolution is a matter of debate. There is no question that the manned-flight revolution has had great impact on the daily life and thinking of mankind. There is no question that the Wright brothers were giants in the manned-flight revolution and were, like Copernicus, at the turning point in the revolution. Perhaps the best way to judge the nature of the revolution is to look at manned flight, using the Wright brothers' accomplishment as a fulcrum, looking both back and forward from that point. In doing so, it should be possible to show how gigantic the Wright brothers' accomplishments were, and at the same time to give full credit at least to a few of the many men who contributed ideas before the Wrights, and then to show how explosively great was the technical revolution of manned flight following the Wright brothers' time.

The Early Era of Manned Flight
The first era of the manned-flight revolution begins probably before recorded history and certainly in the time when myth and reality were hard to distinguish. It ends about the close of the eighteenth century. Many of the good ideas that the Wrights and other pioneers used actually stem from this earlier era, but most of the scheming and experimenting and writing of the time was based on adventure and imitation of bird flight without good technical thinking. In those cases where there was good technical thought, there were always many gaps that prevented the achievement of any great milestones for others to build upon.

The literature is filled with myths and dreams from very early days. And the dreams mixed up three kinds of flight: flight like that of the birds, with wings flapping to gain speed or height or with wings extended for gliding and soaring; or flight by floating in the airy ocean like the fish floating motionless in the water of the sea; or propelled flight into the sky, like the shot of an arrow from a bow. There is the well-known tale of Daedalus and Icarus flying to escape from Crete, using wings made by Daedalus of wax and bird feathers. Was the myth based on real experiments? And was the idea that Icarus disobeyed Daedalus and flew too close to the sun with a melting of the wax and collapsing of the wings a mythical way of explaining an actual, tragic, misconceived attempt to fly like the birds?

So much is made of the poorly conceived attempts to fly, often tragic, that much is missed in the early history of flight. Many of the important ideas of modern flight were well accepted. The idea of the aerodynamic lift of wings, either shaped to imitate those of birds or in the form of the surfaces of kites of many shapes, was common throughout the early era and came from many different countries. The idea of a propeller for propulsion also stemmed from this early era. The helicopter toy in which a propellerlike blade is set into rapid rotation by pulling a string wound up around a spindle was originated at least in the Middle Ages and possibly before. The helicopter blade figured in many designs of aircraft, and in 1768, Paucton, a mathematician, apparently made the first suggestion of a propeller for horizontal propulsion.

The light weight for the aircraft also came from an earlier era when men used for construction their lightest materials: paper and strong cloth, slender wooden rods, together with wire and string. Men also devised engines for fitting into aircraft or tried bicyclelike pedals to harness man power.

But there were some missing links in the early era, although the helicopter type of flight was well thought out, and the work of Leonardo da Vinci is often cited along that line. It is true that his designs of helicopters were very interesting and probably triggered some others' thinking. However, the fact that his complete engineering notebooks really did not come to light until the early part of the twentieth century meant that most of his thinking did not influence as many pioneers as is often thought. Furthermore, Leonardo was strongly dedicated to the ornithopter flight of birds in which the power, the forward thrust, results from the flapping of the wings. This kind of flight has yet to be achieved mechanically.

Another missing link in the early period had to do with stability and control, a subject of great interest during the nineteenth century. It was this lack that caused more accidents than any other factor.

The Dawn of Flight Technology
The nineteenth century was a golden era for many burgeoning technologies, including that of flight. Progress was made at a reasonably rapid and steady pace. Though scientific and technological eras generally do not have sharp boundaries, certainly the contributions of Sir George Cayley (1773–1857) could be picked as a major turning point. Cayley's contribution to start off the nineteenth century's accomplishment in the flight revolution is described by Stever, Haggerty, and the editors of *Life*[3] as follows:

Cayley has been called by historians "the true inventor of the airplane." The title may be an exaggeration, but incontestably Cayley laid the groundwork for later aeronautical research. He was the first to assemble in theoretical form the many elements necessary for practical flight. He thought of the wing not only in terms of lift but also of drag—the resistance produced by a body moving through air. He investigated the amount of lifting surface which supported a given weight in birds, and he recognized that the lifting proper- ties of wings varied with the angle at which the wing moved through the air mass. He proposed a mechanical power system—an engine, which he called a "first mover." He realized that it should necessarily be light and he suggested that it might operate "by the sudden combustion of inflammable powders or fluids"—the internal combustion engine. Furthermore, he recognized the need for providing stability and control.

Those authors continue: "In a remarkably succinct statement, Cayley summed up the task confronting aerodynamicists. 'The whole problem is confined within these limits: to make a surface support a given weight by the application of power to the resistance of the air.' "

This remarkable understanding of flight came a full century before the Wright brothers' first great achievement. Clearly, the stage was being set. Cayley's substantial pioneering work in flight was accomplished over the first half of the nineteenth century. This chapter will concentrate heavily on the work of the Wright brothers, which started practical flight and de- serves great credit, though there were many other workers at the same time and following. Similarly, one can say the same of Cayley, since his work is symbolic of and leading in his time. It can give the reader a good idea of what the nineteenth century was like in flight technology. Cayley and all who were interested in science, technology, adventure, and the affairs of the world were excited because in the late eighteenth century man had actually ascended in lighter-than-air craft—that is, in balloons—filled with either hot air or hydrogen, which had been discovered in 1766 by Henry Cavendish to be lighter than air. There was a great incentive then for men to try to cap- italize on the earlier dreams of flight by using lighter-than-air craft. But Cayley decided following the work of this early ballooning flight, to con- centrate on the following: "To make use of the inclined plane propelled by a light first mover."

Cayley, like so many before, studied bird flight, and he, as so many before, realized that the birds' wings had lifting characteristics; but he also recog- nized and did research on the aerodynamic drag of objects moving in the air. His work included the building of model wings to study the distribution of pressure on them, both those pressures which result in lift and those pres- sures which result in drag. Figure 22.1 is a diagram of forces that must be

·

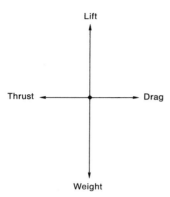

Figure 22.1. Forces that must be in balance for successful flight.

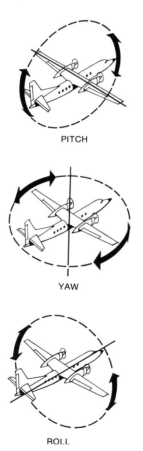

Figure 22.2. The three basic rotational movements of an airplane.

in balance if an aircraft is to maintain steady, uniform flight. The thrust of the engine must equal the aerodynamic drag of the entire aircraft, and the lift of the wings and other lifting surfaces must equal the total weight of the aircraft and its contents. This force diagram illustrates very well the state of Cayley's understanding. Incidentally, many people still do not realize that the magnitude of the thrust and the drag of aircraft in relatively low-speed flight are, in fact, considerably less than the magnitude of the lift and the weight. That is, the lifting force generated by aircraft motion through the air is much greater than the drag that motion creates. Thus, the forcing thrust of the engine and propeller (or jet engine) is much less than the weight of the aircraft. Such a wonderful leverage. For example, a highly efficient low-speed wing can have a lift-to-drag ratio as high as 30. Transport aircraft have a lift-to-drag ratio in the range of 15, and at very high speeds in the supersonic range, where the drag increases very rapidly, one might still achieve lift-to-drag ratios as high as 8.

Cayley also set a pattern for systematic research and experiment in the nineteenth century. He built a model glider as early as 1804, after considerable work, and was so successful with it that he then went to a full-sized, unmanned version. He tried to put a "first mover" into his glider, but the only engine available was the steam engine, and steam engines with sufficient power turned out to be much too heavy for this size of aircraft. He did think of other forms of propulsion. He continued work on aerodynamic streamlining; he made studies to reduce the resistance to the air; he made studies of movable tail surfaces for climb and dive control and also did some work on naturally stable designs. In addition, he tried to achieve manned-glider flight, though he did not fly them himself—probably a great mistake.

This description of Cayley's work on stability or controllability brings up the second major technical point, something that many of the nineteenth-century pioneers worked on. It was never completely mastered, however, until the Wright brothers achieved it. Figure 22.2 shows the three basic rotational movements of which an airplane is capable—called pitch, yaw, and roll—based on rotation around the three axes. In pitching, the nose rotates up or down about the lateral axis of the airplane through the wings. A yawing airplane rotates around the vertical axis with the nose moving to one side or the other. When rolling, the wings dip in rotation around the long axis of the fuselage. For a uniform, stable flight of an aircraft it is not only necessary to have a balance of forces, as shown in Figure 22.1, but it is also necessary to have a balance of torque about each of these axes. Furthermore, in order for the aircraft to be controllable in flight direction either up or down, resulting in climb or dive, or to turn right or left, the rotation about these axes must be controllable at will. It is also interesting that some degree of natural stability must be built into the aircraft so that if a shift of torque in an air gust or a shift in the load center of gravity, or some other change in the aircraft flight takes place, it is possible to set up a corresponding returning torque to get back on an even keel. As we will see later, the Wright brothers' greatest technical contribution to manned, powered, controlled flight was in the field of controllability.

The Wright Brothers' Debt to Their Predecessors in the Nineteenth Century

Perhaps a straightforward way to show the numerous accomplishments of the nineteenth century would be to describe how the Wright brothers got started in aviation and what they learned from their predecessors to help in their design of the first successful machine. It has already been pointed out that one of the characteristics of many of the experimenters of the nineteenth century, and even before, was the building of models. An aviation pioneer, Penaud (1850–1880), built, among many things, a naturally stable glider model powered by rubber bands and demonstrated it. It is said that in 1878, Bishop Wright bought one of Penaud's rubber-powered airplane models for his two sons, Wilbur and Orville. Also, the frequent mention of flying experiments in newspapers, magazines, and books stimulated the interest of the Wrights when they were young.

Wilbur Wright has written[4] that he started his serious work at the time of the death of Otto Lillienthal in the failure of a gliding experiment. Lillienthal (1848–1896) was clearly the greatest pioneer in experimental flight up to the time of the Wright brothers. He made one towering contribution to manned flight. In more than 2000 actual man-carrying flights that he flew in gliders built by himself and his associates, he developed more experience in piloting than anyone had done until the Wright brothers. The Wright brothers believed very strongly in the need for skillful piloting of both gliders and manned aircraft.

Wilbur Wright pointed out that his active interest in aeronautics started from the death of Lillienthal, which was well publicized, as were his many successful gliding flights. They were published in magazines and newspapers throughout the world. Wilbur apparently believed that Lillienthal's death came because a gust upset the flight of his glider in such a way that he was unable to recover in a roll. Wilbur had observed that the birds maintained their lateral balance in gusty wind by twisting the tips of their wings in opposite directions. One of the most important contributions of the Wrights thus came at the very beginning of their work, for they invented the twisting of wings which provided roll control in their early manned and unmanned gliders and in their powered machines as well. They kept this idea from the beginning until the end. This twisting, by the way, was interpreted as wing-warping, and that is the description of their invention which is often given today. From Lillienthal they also adopted the idea that manned experiment with gliders was important. Lillienthal's death led them to develop one of their most important contributions of roll control; Lillienthal's great work on aerodynamic design of the wing for lift and of the horizontal tail for pitch control was also of tremendous help to them.

By 1899, Wilbur Wright had interested his brother Orville in flight, and from that point on began a classic case of technology advancement—design, experiment with models and with components, testing in glider flight, and successful accomplishment of first man-controlled powered flight. Initially, they sought the ample literature of the experimenters of the nineteenth century. They wrote in 1899 to the Smithsonian Institution of Washington to get information about books and papers on flying. Soon after that they learned more of other experiments, first through correspondence, and later by actual exchange of visits with Octave Chanute, who had written extensively of his own and others' experiments in gliding and flying models and manned craft. Chanute's publication in 1894, *Progress in Flying Machines,* was a particularly good summary of the time. In their experimentation over the next four years, until 1903, they not only had access to much aerodynamic data of Lillienthal's and the many experimenters of the nineteenth century, but they also developed ways of calculating their own data and built a wind tunnel to discover the lift and drag characteristics of wings and control surfaces. They worked on questions of center of pressure and of optimizing the wing shapes.

The Achievement of the Wright Brothers

The Papers of Wilbur and Orville Wright[5] is a splendid collection of the work of the Wright brothers themselves. From it one can see the care with which they experimented, designed, tested, and then flew their gliders and powered machines.

One important design feature of the Wright brothers' gliders and powered, man-carrying airplanes was the biplane wing. The great advantage, of course, of the biplane and multiplane designs is that with struts and braces a truss can be designed to give much greater strength to the wing structure, and both planes of a biplane can contribute to the lift, adding effective wing area. Many others, including Cayley and Stringfellow, had experimented with multiplane design.

Another interesting part of the Wright brothers' experimentation concerned the propulsive thrust that was provided by a 12-horsepower gasoline piston engine, driving two counterrotating pusher propellers. Bicycle chains were used as drive mechanisms from the engine, which was located near the center of gravity of the airplane. Perhaps it was their confidence in the machine age that an engine would be available at the right time, for it was one of the last features to be developed and added to their designs. In fact, they tried to buy an automotive engine on the market but discovered that the necessary horsepower was found only in much too heavy engines. So, they designed their own. During the last year before successful flight they made the engine that produced 12 horsepower for 200 pounds weight. Though it was adequate, it was by no means an outstanding engine. Professor Samuel Pierpont Langley (1834–1906), who was experimenting before and during the Wright brothers' successful development period with powered, manned aircraft, had available—based on a contracted engine designed by Stanley M. Balzer and built by his principal engineering assistant Charles Manly—a much better engine that produced 50 horsepower for about the same engine weight. Both of these engines really owed their development to a series of engine developments for automobiles. This was started with the gasoline engines of Otto and Daimler in the 1860s that, in turn, had been preceded by the gas engine of Lenoir in 1860. Other experimenters in the nineteenth century had tried steam engines for powered flight tests with some degree of success; they included leading figures such as Cayley, William S. Henson (1812–1888), Sir Hiram Maxim (1840–1916), and Clement Ader (1841–1925). The use of the propeller to convert the power of the engine to thrusting force, as has already been pointed out, was an inheritance from earlier times.

A most important feature of the success of the Wright brothers came from their realization that many of the serious experimenters of the nineteenth century had tried to achieve too much natural stability in the fore and aft direction in pitch and had tried to build in too much natural stability in roll by putting dihedral in the wings; that is, having the wings on each side of the aircraft at an angle, which gives an oscillating, falling-leaf stability. Wilbur immediately recognized that Lillienthal's death came, in spite of some built-in stability, from lack of sufficient controllability in roll to counter a large wind gust. They, the Wright brothers, put all of their confidence in controllability of slightly unstable aircraft, together with skillful piloting, and they were correct in this trend.

To make this latter point more clear, the example of the flight of the arrow can be used. An arrow has stabilizing surfaces, usually feathers, at the tail end of the shaft, well to the rear of the center of gravity of the arrow. If an air gust hits the arrow, or if the arrow is oscillating because of a slightly imperfect launch from the bow, the restoring moment caused by the relatively small feather is quite large because of its long lever arm. If one puts the force generated by an aerodynamic surface well behind the center of mass in flight, stable, straight flight is assured. In the case of the airplane, the wing with the largest contribution to lift must be close to the center of mass in order to lift that mass. To balance these, there is a stabilizing torque from the horizontal tail placed well aft of the center of mass.

Even the Wrights had not solved all the problems of stability and control. One of their stability design features gave them continuing trouble, that of the horizontal stabilizer-controller being placed forward instead of aft of the center of gravity. This feature of their early models resulted in pitching divergences that could be overcome only by very skillful piloting. They and others later abandoned the forward placement of the horizontal control surface and placed it to the rear.

The Wright brothers recognized that controllability could best be obtained if the degree of stability was low, or even slightly unstable, but the degree of controllability high. Their designs were slightly unstable but controllable by skillful piloting. Possibly their bicycle-building and operating experience made them realize that imbalance overcome by steering was the way to go; it was in any case a crucial insight by the Wright brothers.

In controllability, the Wright brothers used their wing-twisting or wing-warping technique for roll control; they used the well-developed theory of plane tail surfaces controllable in angles of attack for pitch control; and they used a steerable rudder for yaw control. In their experiments in glider flights and their early powered flights, they experienced a great deal of difficulty in trying to make controlled turns until they finally discovered that both the roll and the yaw controls—that is, the elevator and the rudder—had to be synchronized to produce a smooth, effective turn. The dynamics of this can be worked out easily, and such synchronization can be built into the aircraft or, better, handled in flight by a skillful pilot.

For this description of the manned, controlled, powered flight revolution it is impossible to list all of the many contributions of earlier pioneers, or even to name but a fraction of the pioneers. Suffice it to say that a century of experimentation laid an essential groundwork. The genius of the Wright brothers was, first, to be inspired by the partial successes that went before them; then to respect the data in experiments that many had gathered in all aspects of flight, analyze the causes of failure, introduce several concepts of their own, and finally put together all of this into successful designs. This they did, and success was theirs.

In America, as in other countries, there were many who contributed greatly to flight, and there were some who claimed some form of first success. A controversy raged for years concerning the work of Professor S. P. Langley, a distinguished scientist who started his work on powered flight well before the Wright brothers and, in fact, attempted a manned, powered flight in the months just previous to the first successful manned, powered flight by the Wright brothers. His aircraft, *The Aerodome,* flown by his mechanic, Manly, crashed on take-off from a boat on the Potomac River, supposedly from the failure of the catapult mechanism or from a structural failure. The success of the Wright brothers, who had not been supported by anybody but themselves, combined with the fact that Langley, after very careful scientific work, had received a major contract from the War Department several years before—a contract for $50,000 to build a full-sized, man-carrying airplane—brought sufficient ridicule on Langley that he gave up his experiments and, according to story, died a deeply disappointed man a few years later. His proponents supported the reconstruction of his aircraft by Glenn Curtiss about a decade later, and it did fly successfully. There was a great argument as to whether Curtiss actually added the necessary control-in-roll to make it a controllable flight. The important thing, however, is that Manly did not himself train to become a skilled flyer as the Wright brothers did, and most people believe that Langley had not solved the roll-control problem. However, no one should denigrate his accomplishments. The Wright brothers themselves learned much from Langley, and many other features of his aircraft were great triumphs, which fully qualify him for an important place in aeronautical history.

After the Wrights' successful flight they went right on designing, first the practical airplane and next a practical man-carrying airplane. Then their successes, particularly as proved in 1908, began to be widely recognized. There are many claims by other inventors that they successfully flew first. A host of experimenters flew in one form or another, but none could lay undisputed claim to completely successful, powered, manned, controllable flight. The Wright brothers were not lucky, as is often intimated; they were skilled, brave, careful, brilliant workers. Following their successful flight, they did hold back a bit on pictures and publication of all of their techniques, simply because they wanted to be protected. The basic patent of the successful airplane—especially including their wing-twisting invention, which was applied for in 1902—was not granted until 1906.

Flights Since the Wright Brothers

All that has happened since the Wright brothers is familiar history. The component technologies that the Wrights put together developed at rapid rates. The engine, for example, went through a to-be-expected development to ever-increasing horsepower, ever improved power-to-weight ratios: the efficiency of the propeller lifted from the approximately 65 percent efficiency of the Wright brothers to 85 percent today. A different form of engine, the gas turbine, which had been invented in 1893 by Charles Parson, was developed into even more powerful thrusting engines than the several-thousand-horsepowered internal combustion engine. The practical turbojet era in military flight began during World War II, and the practical turbojet era for commercial flights started in the late 1950s.

Materials and structures have progressed unbelievably. No longer does one need the spruce spars and wire braces and cloth surfaces of the Wright brothers. The development of more effective trusses for fuselage and wings has continued. In the 1920s the monocoque fuselage—that is, the tubular fuselage—was introduced. Also introduced in the materials revolution were the light metals, aluminum and magnesium alloy, which soon took over from wood, cloth, and wire. The strength of the new metal was sufficient so that monoplane wings could replace biplanes. As the flight speeds went up, other materials, including titanium and steel, were found to be necessary to take the high frictional heats generated at supersonic speeds.

Aerodynamics started as a science early, and this science was used by experimenters of the nineteenth and early twentieth centuries. It became a major experimental triumph in the twentieth century, and the shape of aircraft today is largely determined by the combination of complex experimental and theoretical aerodynamic developments to minimize the drag and maximize the lift of the wings, fuselage, and control surfaces.

There is one major component of the technology of modern flight about which the Wright brothers knew practically nothing—possibly it should be described as two—that of automatic or aided control and that of guidance. Today's modern airliners have autopilots that successfully fly the aircraft in controlled flight; in fact, they can even be landed and taken off by such autopilots, though they are not quite trusted to do that in everyday commercial flying. In the air, instead of the early pilotage by looking at familiar landmarks, we now have radio, radar, and inertial guidance systems, even employing modern satellite communication systems that can guide aircraft anywhere and everywhere through good weather and bad. This technology was started by pioneers only a decade or so after the Wright brothers achieved their first flight. But the twin technologies of automatic control and guidance really reached their greatest forward movement during and after World War II. Flight control and guidance are products of the large instrumentation and electronics industry characteristic of the twentieth century.

The Wright brothers' successes capitalized on so much of the good work that went before them, and were the prerunners of so much of heavier-than-air flying which followed, that they can indeed be considered the giants of the development of flight, a revolution not only in technology but also in our everyday living.

Ballooning

Going back well before the Wright brothers, as was already pointed out, in the late eighteenth century man achieved flight to quite high altitudes and to some distance by ballooning. Although lighter-than-air craft never had an impact on society approaching that of heavier-than-air craft, clearly the age-old dream of flight became a reality when man learned to employ the buoyancy of either hot air or hydrogen gas in balloons. Lighter-than-air craft are today used for sport, and have a few other uses, but as far as the development of flying is concerned, their greatest effect may have been to encourage those pioneers who were working on heavier-than-air craft to work harder.

On November 21, 1783, on a well-announced occasion, the Montgolfier brothers, Etienne and Joseph, successfully launched a hot-air balloon with passengers—Jean François Pilatre de Rosier and the Marquis d'Arlandes. It is said that the first flight was accompanied by a number of tense moments when small fires from the basic energy source to heat the air also set small fires in the fabric of the balloon. However, a successful twenty-five minute ride for a distance of more than five miles over Paris was achieved. This manned balloon flight had been preceded on September 19, 1783, when three animals—a sheep, a duck, and a rooster—were sent aloft from Versailles on the occasion of a command performance by King Louis XVI, who had heard of the Montgolfier brothers' earlier experiments with large but unmanned balloons.

The start of the Montgolfier brothers' interest in hot-air balloons is a little uncertain, there being many tales as to why they became interested. The brothers were papermakers, and possibly an accidental happening with some of their paper products gave them the first idea. They were not quite sure why hot air lifted balloons, but they did conduct some experiments and made a public demonstration with a small hot-air balloon of about thirty-five feet in diameter on June 4, 1783.

A more scientific approach to ballooning succeeded shortly after the Montgolfiers' manned flight when a balloon filled with hydrogen, which had been designed and built by the scientist Charles, was launched before an even larger crowd than had witnessed the Montgolfiers' flight. Charles and a companion made a two-hour, twenty-seven mile flight.

It was well over a century in advance of manned heavier-than-air craft flight that ballooning began. In that time, there were many accomplishments in the lighter-than-air craft field—for one thing, balloons were used to develop all sorts of components that helped heavier-than-air craft later. Also, in 1797, the first manned descent in a parachute was made by Jacques Ganerian.

The lighter-than-air craft, of course, were very sensitive to the wind so that they were not really maneuverable. Nevertheless, tethered balloons were used very shortly after the development of the balloon for military observation. Naturally, there was a great deal of thought about putting power into these lighter-than-air ships, and on September 24, 1852, Giffard made a tentative flight with a steam engine to power a lighter-than-air craft. The powered, steerable balloon reached the peak of its development in the twentieth century when commercial airship service was inaugurated in Germany. Count Ferdinand von Zeppelin, who was the father of the rigid airship—that is, a balloon built around a stiff frame—developed the *Graf Zeppelin,* which was 775 feet long. In 1936, the *Hindenberg* was built, which was 804 feet long, a hydrogen-inflated, rigid airship. It began transoceanic service and made sixty-three flights, thirty-seven of them across the Atlantic. However, the important days of the dirigible ended in a disaster when the *Hindenberg* caught fire, either from a lightning discharge or a static electricity discharge, as it prepared to moor at Lakehurst, New Jersey. Before that time there had been many developments in the rigid airship, and it was brought to a fairly high state of the art. The development of engines for the dirigibles and the accompanying propellers to convert the power to thrust was, of course, an important development for heavier-than-air flight.

Manned Space Flight
The last of the three types of manned flight, that of being projected into space, was achieved on April 12, 1961, when cosmonaut Yuri Gagarin orbited the earth in his spacecraft, Vostok I. His spacecraft had been launched into earth orbit by a multistage liquid-propellant rocket, and it was successfully recovered by being parachuted back to earth. The 10,400-pound spacecraft ejected a spherical reentry capsule after one orbit of the earth and 1.8 hours of flight. The cosmonaut, Gagarin, remained within the capsule until landing 400 miles east of Moscow.

Manned spacecraft reached a second great peak on July 20, 1969, almost a decade later, when astronaut Neil A. Armstrong, the spacecraft mission commander, and the lunar module pilot, astronaut Edwin E. Aldrin, Jr., stepped out of their Apollo XI lunar module onto the surface of the moon. They had left astronaut Michael Collins piloting the command module while awaiting their return from the surface in order that the command module could return to earth, as it did—safely—being parachuted into the Pacific Ocean.

Before the occurrence of these two great manned spacecraft achievements, plus numerous other ones of similar nature, the spaceflight era had been introduced on October 4, 1957, with the launching of Sputnik I into earth orbit. Sputnik I was a simply-instrumented, 23-inch aluminum sphere weighing 184 pounds that radioed back to earth temperature, density, cosmic ray, and meteoroid data for twenty-one days. It stayed in orbit for about three years before its orbit decayed, and it destroyed itself with the heat from air friction as it reentered the earth's atmosphere.

Although many component technologies reached peak performances in these great space flight achievements, there is no question but that the key technology was rocket propulsion. Pyrotechnic rockets, like the ones used in fireworks displays, were known in antiquity, being a natural outgrowth of the experiments with combustible and exploding mixtures, such as black powder. The real origin of these rockets is obscure, though most historians agree that they were probably a Chinese invention. Certainly the first time their use was mentioned in any chronicle was in the year 1232, the occasion being a siege of the Chinese city, Kai-fung-fu, by the Mongols. By then, they had been developed into some kind of a simple rocket bomb. From that point on, the use of rockets as military devices grew in various armies in Europe and in Asia, and the powder rocket—solid propellant rocket—reached a new peak of development by Colonel William Congreve in the late eighteenth century. During the nineteenth century military rockets were used in waxing and waning degrees and are used to this day. The modern liquid propellant rocket, which has been vital in the manned spaceflight story, had its origin in the work by Robert H. Goddard. He was interested in exploring the atmosphere to high altitudes, and he proved by various calculations, not necessarily original, that a multistage rocket was the way to do this. After first trying solid propellant rockets, he switched to rockets fueled by liquid oxygen and gasoline, and among the patents he obtained in the 1920s were the fundamental ones on which today's liquid propellant rocketry is based.

Liquid propellant rockets were used in rocket weapons during World War II and developed to a reasonably high state. As military weapons after World War II they were developed even higher—to the point where they could generate enough thrust and vehicles could be built sufficiently light in weight so that intercontinental flight was possible. It was then that the several programs working toward first unmanned and then manned space-flight were started.

To date, the manned space flight achievement has not taken as big a place in everyday life as has manned heavier-than-air craft flight. Still, there are already routinized uses of unmanned spacecraft for detecting global weather patterns and for relaying communication signals all around the globe. Soon other uses will become more common: surveillance of earth's resources from space and navigation of ships and aircraft based on a satellite navigation craft. Certainly there will be scientific research based on satellites and space-craft penetrating more deeply to other parts of our solar system, to Mars and to Venus. In 1972 an orbiting astronomical observatory, *Copernicus,* was launched to make astronomical observations from orbit, unhindered by the blanket of earth's atmosphere.

Looking into the future, one can expect continued use of man's capability in manned spacecraft, perhaps in ways that will eventually prove to have the greatest impact of all three modes of flight. Certainly, it has already begun to change man's concept of the earth as the only place where man can live—another giant step in man's development—just as Copernicus contributed to a better understanding of the place of man's earth in the cosmos.

Conclusion

Now all three forms of flight, which have shaped the thoughts, writings, and experiments of many men throughout history, have been achieved. Experimentation will continue, of course, as long as man's inherent curiosity and search for knowledge manifest themselves.

This was not a revolution of the Copernican type, in the sense that the brilliant perception of one man precipitated an abrupt change in scientific thought, because many people, known and unknown, contributed to its success. Still, each of the three forms of flight had its giants who, like Copernicus, brought their field of knowledge to a turning point and helped man see both his world and his universe with greater clarity. As each form of flight continues to develop, all must pay due homage to these giants.

References

1. Charles H. Gibbs-Smith, *Aviation—An Historical Survey from its Origins to the End of World War II*. Her Majesty's Stationery Office, London, 1970.

2. Alfred, Lord Tennyson, "Locksley Hall," lines 119–123, *Poems*, 1842.

3. H. Guyford Stever, James J. Haggerty, and the Editors of *Life, Flight*. Life Sciences Library, Time, Inc., New York, 1965.

4. Marvin W. McFarland, ed., *The Papers of Wilbur and Orville Wright*. 2 vol., McGraw-Hill Book Company, New York, 1953.

5. Ibid., p. 11.

Donald W. Goldsmith

23 The Copernicus Satellite in the New Era of Space Astronomy

Introduction: Astronomy as an Earthbound Science

For thousands of years human beings have looked through the skies of our own planet to see the celestial objects around us. Our atmosphere, which contains less than one hundred-billionth of the total mass of the earth, provides an essential part of our requirements for life. From the atmosphere we obtain oxygen to breathe, rain to replenish the cycle of water, and an invisible barrier against deadly radiation from space. Astronomers respect these life-giving functions of our atmosphere and have always had a vested interest in maintaining its clarity, but their admiration has been tinged with an ironic regret: the atmosphere prevents us from seeing so much we would like to see but cannot from inside our planetary aquarium.

Astronomers use large telescopes to increase the light-gathering power of the unaided human eye, and large antennas to receive radio waves that human bodies cannot sense at all. But even the largest earthbound telescopes cannot overcome some barriers to a thorough analysis of the radiation from celestial objects. Our atmosphere poses two great obstacles to light from distant sources. First, it removes most of the radiation with wavelengths other than those that characterize visible light and radio waves. Second, the constant motion of the atmosphere tends to blur the images of faraway objects, thus reducing our ability to see fine details and to observe the faintest objects. Of these two effects, the first is the more important, because entire classes of objects that radiate little or no energy in the form of visible light or radio waves have turned out to be powerful emitters of electromagnetic energy with other wavelengths. In addition, many stars and galaxies that do shine in visible light may yield their secrets to us if we can observe their radiation at other wavelengths screened from us by our atmospheric shield. Today we have launched observational platforms that circle our planet above the atmosphere, providing us with information never before available, and in many cases never suspected, until the age of rocketry promoted such space telescopes from a dream of isolated humanity to the next step on the way to the exploration of the stars and planets around us. A key step forward in humanity's exploration of the heavens came with the launching of the Copernicus satellite in August 1972.

Department of Earth and Space Sciences, S.U.N.Y., Stony Brook, New York 11790.

The Effect of the Earth's Atmosphere on Astronomical Observations
The air that we breathe consists (roughly) of four-fifths nitrogen molecules
(N_2), one-fifth oxygen molecules (O_2), and trace amounts of water molecules
(H_2O), carbon dioxide molecules (CO_2), and less abundant constituents that
include various human-made pollutants like sulfur dioxide (SO_2). All the
electromagnetic radiation that reaches us consists of tiny wavelike particles
called *photons*. The success or failure of these photons in penetrating the
atmosphere depends on the properties both of the photons and of the
molecules the photons encounter. We should esteem this penetration process, for it allows us life on earth.

Each of the photons that form light waves (and all other forms of electromagnetic radiation) possesses a definite frequency, energy, and wavelength.
The *intensity* of a beam of radiation depends on the number of photons in
the beam, but each photon has a characteristic frequency of vibration, a
characteristic wavelength of vibration, and a characteristic energy, and almost all of the information about a photon can be given by naming any one
of these quantities. For any photon, the product of the frequency times the
wavelength always produces the speed of light, a universal constant nearly
equal to 300,000 kilometers per second. The photon's energy is always a
constant number times the frequency. Thus, larger energies and larger frequencies are associated with shorter wavelengths, and lower frequencies go
with longer wavelengths, for every sort of electromagnetic radiation.

Our sun produces most of its radiation as photons with wavelengths between 3×10^{-5} and 7×10^{-5} centimeter. Not by chance, our eyes are most
sensitive to photons with wavelengths within this interval, and we call such
photons "visible light." But in addition to the photons that we can see, the
sun also emits a fraction of its energy as *photons of ultraviolet light,* which
have wavelengths somewhat less than those of visible-light photons. (Ultraviolet light, by definition, has wavelengths from about 10^{-6} to 3×10^{-5}
centimeter.) Such ultraviolet photons each have an energy and a frequency
that exceeds those of visible-light photons.

When a photon collides with an atom or with atoms that are bound together as molecules, the photon may be able to "excite" the atom or molecule
into a new state of greater energy or even to break apart the system by
separating the constituent particles completely. The photon's ability to excite or destroy atomic systems depends on the energy it possesses. The con-

servation of energy inherent in collisions between photons and particles demands that the photon must furnish the atom or molecule with the difference in energy between the undisturbed state of affairs and the state that exists after the photon hits. The photon disappears in the collision process, yielding its energy to the atom or molecule. In order to produce the disruption of an atom or molecule, the photon must have an energy that at least equals the "binding energy" of the system. For example, the simplest atoms, hydrogen, each consist of an electron in orbit around a proton. If photons of visible light encounter hydrogen atoms when the atoms are in their lowest-energy state, the photons will pass among the atoms with no significant interaction, that is, no excitation or disruption of the atoms. This occurs because no photon of visible light has enough energy to disrupt a hydrogen atom or to excite the atom even into its next-highest energy state.

In a similar way, molecules made of several atoms can or cannot be broken apart ("dissociated"), by photons; again the question hinges on the photons' energies and wavelengths. Photons of visible light cannot destroy molecules of oxygen, nitrogen, water vapor, and carbon dioxide, and they pass through our skies unhindered. But high in the atmosphere, ultraviolet photons from the sun are disrupting such molecules, and the chief actor in this drama of dissociation is oxygen. An ultraviolet photon with a wavelength less than 1.5×10^{-6} centimeter has enough energy to break an oxygen molecule into its two component atoms when it collides with the molecule. Sometimes these atoms recombine with another stray oxygen atom to form an O_2 molecule that once again is destroyed by photon collision, but sometimes one of the atoms links up with an oxygen molecule to form ozone (O_3). Before too long, the ozone molecule also dissociates under photon impact, but at any moment some ozone molecules do exist at altitudes of 15 to 150 kilometers.

Ozone molecules are important because the three atoms that form them are more loosely bound together than the two atoms in an ordinary oxygen molecule. Photons with wavelengths greater than 1.5×10^{-5} centimeter cannot dissociate molecular oxygen or nitrogen, but many of them can dissociate ozone. As a consequence, the nitrogen and oxygen molecules in our atmosphere absorb those ultraviolet photons with wavelengths shorter than 1.5×10^{-5} centimeter, while ultraviolet photons with wavelengths between

1.5 × 10^{-5} and 3.0 × 10^{-5} centimeter disappear because of their interaction with ozone molecules, some of which are always being formed 15 to 150 kilometers above us, only to be destroyed in their turn a few seconds later.

The absorption of ultraviolet light by molecules of nitrogen, oxygen, and ozone removes all photons with wavelengths less than 3 × 10^{-5} centimeter from the sunlight we observe, and this is definitely a plus for us: such photons would dissociate the molecules in our bodies' outer layers, leaving us incapable of carrying on with civilization. Our bodies consist of complex molecular chains made of atomic elements such as carbon, oxygen, nitrogen, phosphorus, and sulfur. If photons with wavelengths less than 3 × 10^{-5} centimeter strike these molecules, they will break some of the linkages and prevent their unhindered re-formation. To prevent this eventuality, molecules high in our own atmosphere must die (anthropomorphically speaking) that we may live. These molecules thus provide a barrier to photons with dangerously short wavelengths. A much-reduced flux of solar photons with wavelengths between 2.5 × 10^{-5} and 3 × 10^{-5} centimeter does reach lower altitudes, and the increase in their numbers as we go from sea level to heights of even a few kilometers can be noticed by skiers and hikers at high altitudes. Astronauts must take special precautions to shield themselves against ultraviolet photons once they leave our earth's protection to travel on their far more vulnerable spacecraft. And even here on earth we should not consider our own shield against ultraviolet radiation to be immutable: someday we might awake to find that we had inadvertently added enough new material at high altitudes to alter the ozone layer that filters out the wavelength interval that contains most of the sun's harmful ultraviolet emission.

To complete the picture of our atmosphere's effect on photons, we should note that molecules of water vapor, carbon dioxide, and oxygen absorb most radiation with wavelengths between 8 × 10^{-5} and 2 × 10^{-1} centimeter. Such photons cannot destroy these molecules, but they can and do excite them to states of higher energy, sacrificing their own existence in the process. Some infrared and microwave photons can penetrate almost to sea level because their particular wavelengths do not correspond to the right energy needed to excite some particular transition to a state of higher energy in a molecule of water vapor, carbon dioxide, or oxygen. The absorption of photons with wavelengths between 8 × 10^{-5} and 2 × 10^{-1} centimeter

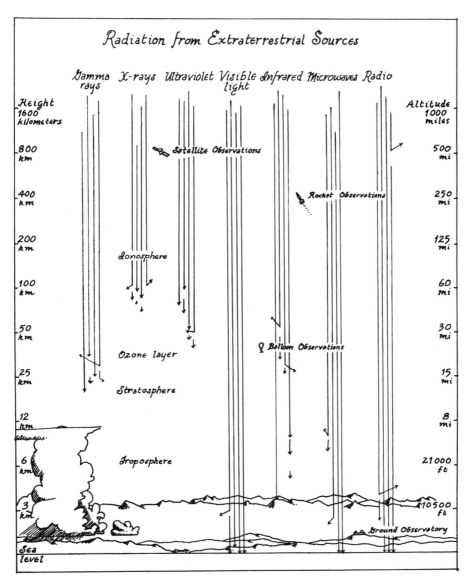

Figure 23.1. A schematic representation of incoming electromagnetic radiation with various wavelengths, and the success or failure of this radiation in penetrating the earth's atmosphere.

The radiation with the shortest wavelengths, gamma radiation, is absorbed in the stratosphere at altitudes of about 25 kilometers above the earth's surface. At still higher altitudes x rays and ultraviolet radiation are absorbed, while visible light radiation and certain wavelengths of infrared radiation come through the atmosphere to the earth's surface. The shorter-wavelength microwaves are absorbed by our atmosphere, but longer-wavelength microwaves, as well as radio waves (which have the longest wavelengths of the types of radiation shown in the diagram) penetrate to ground-based observatories.

does show a few "windows," through which we can observe celestial objects with earth-based telescopes. The partial absorption remains troublesome even for these windows, but we can leave most of the water vapor below if we observe from a high, isolated mountain like Mauna Kea (elevation 4 kilometers) on Hawaii.

Figures 23.1 and 23.2 show the transparency of the atmosphere to different types of electromagnetic radiation (photons). Photons of ultraviolet light, of x-radiation (wavelengths from 10^{-9} to 10^{-6} centimeter), and of gamma radiation (wavelengths less than 10^{-9} centimeter) are all removed at alti-

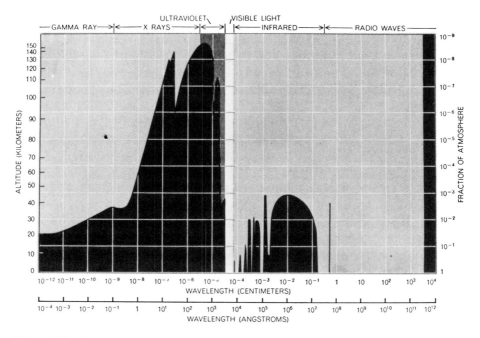

Figure 23.2. A more detailed outline of the penetration process shown in Figure 23.1.

The different types of electromagnetic radiation are indicated by their wavelengths, plotted along the horizontal axis. The top of the black area shows how far the types of radiation penetrate before being absorbed, measured either by altitude above the earth's surface (on the left vertical axis) or by the fraction of the atmosphere's mass above the level of absorption (shown on the right vertical axis). Diagram courtesy of *Scientific American* (W. H. Freeman and Co.).

tudes above 20 kilometers. Most of the ultraviolet photons (wavelengths 10^{-6} to 3×10^{-5} centimeter) are absorbed at higher altitudes, 50 to 150 kilometers. Infrared photons are absorbed at altitudes of 5 to 30 kilometers, while photons of visible light and of radio waves (wavelengths 2×10^{-1} to 5×10^3 centimeters) penetrate our atmosphere nearly unaffected. Radio waves with wavelengths longer than 5×10^3 centimeters (50 meters) are bent by the charged particles (free electrons and the ions from which they were separated) that form the "ionospheric" outer layers of the atmosphere. These long-wavelength radio waves therefore remain undetected at the earth's surface. Figure 23.2, which gives a detailed exposition of the summary given in Figure 23.1, shows for each wavelength the altitude at which significant absorption occurs and the fraction of the atmosphere (by mass) that lies above this altitude.

Even when the atmosphere does not absorb photons of a particular wavelength, it does produce a blurring of the images of objects seen through it. This effect, called "seeing," arises from local motions in the atmospheric gases. The "seeing" varies from place to place and from moment to moment in its effect, but as long as a sizable fraction of the atmosphere lies above us, even the largest telescope will show a blurry image of the finest points and lines. Except at the instants of best seeing, telescopes at ground-based observatories register a pointlike image as a blurred disk nearly a second of arc in diameter.

The atmosphere above us thus blurs our vision and removes many photons at wavelengths that might interest us. From the earth's surface, we can look through two openings in the wavelength spectrum—a narrow interval of visible light and a broader one of radio waves—plus a few partial clearings in the infrared regions. We can obtain improved observations at infrared wavelengths by placing telescopes in aircraft, in balloons, or on mountain peaks that can take the instruments 3 to 25 kilometers high. Such observations have revealed new "infrared sources" whose infrared radiation cannot be detected from the ground. However, to observe photons of ultraviolet or x-radiation, no balloon, no airplane, no mountain will suffice. For such observations we need a telescope outside our atmosphere, and so in the 1960s the National Aeronautics and Space Administration began a program of astronomical observations from earth-orbiting satellites.

Astronomy in Space

Why should we try to observe ultraviolet light and x rays from space? If we knew the answers to the fundamental questions of astronomy and cosmology, such as whether the universe will ever stop expanding, there might be no need for such observations. If we felt that we know enough about the basic processes of energy liberation that occur in the universe for us to construct a reasonable model of how things work, then we would probably be content to maintain only ground-based observatories. But in fact, we know only a disquietingly small amount about the strange things in space, and today with the advent of space observatories, we may just be starting to gather the data needed to understand astronomical events. Even so, many years of observation in the long-hidden wavelength intervals may be necessary to gain useful insights into the celestial activities around us, which today we can discuss and catalogue but not really comprehend. Twenty-five years of rocket research and fifteen years of satellite development have produced remarkable results, but ten or twenty more years of concentrated effort may elapse before the fundamental data have been acquired. Programs to observe from above the atmosphere have maintained a coherence and momentum that are impressive in view of the short time scale of popular interest and priority. It remains to be seen whether this steady pace will continue to the point of placing into earth orbit observatories that can study faint stars and distant galaxies.

What has been done so far? Ultraviolet astronomy started in 1946 with the firing from New Mexico of a German V-2 rocket that carried an instrument to make measurements of the flux from the sun, by far the most intense source of ultraviolet radiation in the sky, at wavelengths between 2×10^{-5} and 3×10^{-5} centimeter. Eventually, a series of rocket flights in the 1960s measured the solar spectrum at wavelengths from 3×10^{-6} to 3×10^{-5} centimeter. For solar observation, rocket flights are sufficient, because the great number of photons emitted by the sun allow observations to be made rapidly. But to observe fainter sources, it helps to have a satellite that can collect photons not just for seconds or minutes but for hours or even days.

With the beginning of the age of man-made satellites in 1957, astronomers and the National Aeronautics and Space Administration began to consider designs for unmanned space telescopes that could continuously study the sky in wavelength regions previously unobserved. A long history of technological development since then has allowed us to open our vision so long encumbered by our enfolding atmosphere. The problems of how to position a telescope accurately by remote control were solved during the "Stratoscope" series of balloon flights, with a system that could accurately track one position in the sky as the balloon gondola swung back and forth. These unmanned balloons still could not fly high enough to pierce the ozone layer, but they did provide photographs in visible light of the sun and the planets that had a clarity obtainable from the earth's surface only for fleeting instants. From a series of meetings held in the late 1950s and early 1960s came plans for a series of astronomical satellites such as the Orbiting Solar Observatories (OSO), the Small Astronomy Satellite (SAS), and the Orbiting Astronomical Observatories (OAO). The Copernicus satellite, last in the OAO series, is the heaviest and the most advanced of the unmanned scientific payloads to orbit around our planet.

The first of the OAO series to be launched successfully experienced an unfortunate power failure and did not transmit any astronomical data. However, the second of the series, OAO-2, was launched in 1968, and has compensated for that failure by producing data for more than four years, far longer than its designers had dared to hope. OAO-2 demonstrated that a space-bound telescope could track indefinitely, to a position that varied by less than 30 seconds of arc, an angle equal to the apparent width of a suspension-bridge cable seen from five miles away. Observations of the ultraviolet radiation from stars and other celestial objects made from OAO-2 produced new information concerning the nature of stellar atmospheres, the composition of comets, and the transparency of our atmosphere. Encouraged by this success, scientists and NASA designers and engineers produced OAO-C, which was designed to produce a pointing accuracy and spectroscopic resolution far greater than that of OAO-2.

The Copernicus Satellite

On August 21, 1972, an Atlas-Centaur rocket placed the 2200-kilogram OAO-C into an almost circular orbit 740 kilometers above the earth's surface. After launching, the satellite was named "Copernicus" in honor of the 500th anniversary of the astronomer's birthdate. Figures 23.3 and 23.4 show the spacecraft, which was built (as were the other OAO's) by the Grumman Aerospace Corporation. The primary instrument aboard the satellite is a reflecting telescope with a mirror 82 centimeters in diameter. In addition,

Figure 23.3. A photograph of the Copernicus satellite (OAO-C) in preparation for launching at Cape Kennedy, Florida. Note the large solar panels, which must of course be folded in during the launch operation.

Figure 23.4. An artist's rendition of how the Copernicus satellite looks in orbit.

The "light baffles" at one end of the instrument package keep stray light (such as sunlight) from entering the telescope optics and confusing the spectroscopic measurements of individual stars. Figures 23.3 and 23.4 are courtesy of the spacecraft manufacturers, Grumman Aerospace Corporation.

a smaller telescope designed by scientists at University College (London) can observe x rays with wavelengths from 10^{-8} to 7×10^{-7} centimeter. This x-ray experiment, though important, does not supplant the SAS satellite, which is entirely devoted to x-ray observations. At this writing (March 1973) SAS-1, also called "Uhuru," continues to survey x-ray sources from its equatorial orbit. (The satellite was launched from a platform near Kenya, and its name means "freedom" in Swahili.) The main telescope on board the Copernicus satellite can be held pointed toward a star with a tracking accuracy of 0.02 second of arc, equal to the angle subtended on the sky by the Apollo spacecraft when it was halfway to the moon! Such a fantastic accuracy would be useless for an earthbound telescope, because the motion of our atmosphere smears all images to a size at least ten times larger. Inside the satellite, light entering the telescope is reflected from the primary mirror to a secondary mirror that directs the beam onto a diffraction grating, which is a reflecting surface, ruled with thousands of parallel lines spaced regularly a few times 10^{-5} centimeter apart. Light waves with different wavelengths are reflected at different angles by such a grating, in accordance with the theories of optics calculated by Newton and his successors. In principle it is a simple matter to count the number of photons of various wavelengths by moving a detector from place to place. In practice the whole procedure requires some 50 million dollars and ten years of technological expertise.

Aboard the Copernicus satellite, the automated spectrometric system consists of a photomultiplier tube that moves from point to point along a circular track one meter in radius. At each point, this tube registers the photon counts for 14 seconds, thus determining the number of photons arriving from the source with wavelengths within a small interval. The counts in that wavelength interval are stored in the memory system, and the photomultiplier moves to the next position for 14 more seconds of counting. The workings of the satellite contain two pairs of photomultiplier tubes, set up so that one pair counts photons with wavelengths from 9.5×10^{-6} to 1.45×10^{-5} centimeter, while the other pair detects photons with wavelengths from 1.65×10^{-5} to 3×10^{-5} centimeter. The wavelength *resolution* (that is, the ability of the photomultipliers to discriminate between two nearby wavelengths) is 5×10^{-10} centimeter for the first pair of photomultiplier tubes and 10^{-9} centimeter for the second pair.

When the Copernicus satellite passes over one of its ground-based tracking stations (in North Carolina, Ecuador, and Chile), it relays to earth the spectral information it has obtained since the ground contact on the previous orbit. Scientists at Princeton University, who were responsible for the optical telescope in the Copernicus satellite, then perform complicated data reductions on the messages received by the ground station in order to obtain the results in a form useful for analysis of stars and interstellar matter. The satellite also receives instructions from the ground station that govern the following cycle of observation. Ten percent of the observing time is devoted to observations of the x-ray sources by the three small x-ray telescopes, which find x-radiation by the interactions that x-ray photons undergo in certain atoms that are placed inside detectors at the focus of the x-ray telescopes. The excitation of the gas atoms in the detector can be viewed by an automated system aboard the Copernicus satellite, and these results are beamed to the ground station along with the ultraviolet observations. All of the information storage and telemetry systems are powered by electricity generated in the satellite's solar panels.

What Copernicus Observes
The ultraviolet photons measured by the Copernicus satellite come from hot, bright, young stars that are located in the general vicinity of our own sun. Our Milky Way galaxy contains about 10^{11} stars, most of which have a smaller true brightness and a lower surface temperature than our sun. These average stars emit only a tiny fraction of their energies at ultraviolet wavelengths, and although these ultraviolet photons could tell us some of the finer points about how energy passes through the stars' atmospheres, detecting the photons would be too long a procedure to be a priority for the Copernicus satellite. Instead, the Copernicus observations focus on those stars with surface temperatures and true brightnesses much greater than the sun's. These stars are both rare and young by galactic standards. Objects with higher temperatures emit most of their photons at shorter wavelengths, and the stars which Copernicus observes emit the peak of their photon flux at wavelengths from 1.0×10^{-5} to 2.5×10^{-5} centimeter. Thus, ultraviolet observations of these stars focus on the bulk of their radiation, an impossible task from ground level.
Because these stars shine with true brightnesses one hundred to ten

thousand times the sun's, they are relatively easy to detect. However, their high energy output implies a rather short lifetime, since their energy supply is limited by the mass in the stars (at most ten or twenty times the sun's mass). Such bright, hot stars have lifetimes measured in tens of millions of years rather than the billions of years available to our sun as a liberator of energy. The Copernicus satellite, assigned a few days per star, can observe perhaps a hundred of the brightest young stars if it operates successfully for a year or two. These hundred or so stars lie within about 500 light-years of our sun. For comparison, the star nearest to our sun is four light-years away, and the diameter of the Milky Way galaxy is 100,000 light-years. But since we consider ourselves to occupy a nonunique position in space (astronomers call this the "Copernican principle"), we believe that the bright, young, hot stars in our vicinity are typical of their class. We can use the ultraviolet measurements from the Copernicus satellite to determine the structure and composition of the outer layers of these stars and to find their actual energy output (most of which occurs at ultraviolet wavelengths). Eventually we may be able to use this information to find how bright, young stars change their structure and energy output as they evolve.

In addition to improved observations of stellar atmospheres, the Copernicus satellite also provides much new information about the *interstellar medium,* the gas and dust spread among the stars. To study this tenuous medium, we must be able to separate its contribution to the observations from the stars' contribution, and the great spectral resolution of the Copernicus spectrometer allows us to do this. The interstellar gas and dust absorbs some of the photons emitted by bright stars. Such absorption is a highly variable function of the wavelengths of the photons. Since the outer layers of the stars' atmospheres produce a similar absorption, we must be able to discriminate between stellar and interstellar absorption effects. We can do this by using the superior ability of the Copernicus satellite to measure the absorption with extreme accuracy, carefully discriminating among the various wavelengths observed. The great spectral resolution of the photomultipliers produces highly accurate measurements of the amount of absorption produced at *each* wavelength.

Though astronomers have studied the interstellar medium for more than forty years, ground-based observations have given us scant information on the composition of interstellar gas and dust. This occurs because our atmospheric "windows" allow only a small fraction of the relevant information to reach us. It is true that some radio waves, which are produced by all hydrogen atoms even in the coldest regions of space, do pass through the atmosphere and provide valuable "21-centimeter" maps of the distribution of hydrogen atoms in our galaxy. But almost all of the other constituents that we think are important—helium, carbon, oxygen, nitrogen, neon, argon, sulfur, magnesium, silicon, manganese, phosphorus, and chlorine, to name a few—as well as the *molecules* of hydrogen and their compounds that become the dominant states for hydrogen in the denser interstellar clouds, all of these produce *no* detectable absorption in the wavelength intervals of visible light or radio waves. The constituents just named, however, can *all* produce absorption of photons with ultraviolet wavelengths, and the Copernicus satellite has observed absorption effects from all of these interstellar constituents, with the exception of helium and neon, whose absorption "lines" (that is, the absorption of photons of a very definite wavelength) occur at wavelengths shorter than 9.5×10^{-6} centimeter, the Copernicus satellite's lower limit. Until the era of satellite-borne telescopes, astronomers had detected only atomic hydrogen plus a few rare species such as calcium, sodium, titanium, and potassium in the general interstellar gas. (In addition, radio astronomers have recently detected relatively complicated molecules like formaldehyde, ammonia, and hydrogen cyanide in interstellar space, but these molecules occur only in special regions that have thousands or millions of times the density of the usual interstellar gas.) The new observations by the Copernicus satellite allow us to analyze systematically the composition of the material strewn through space. Most of this is gas—individual atoms, ions, or molecules—but a tiny fraction consists of much larger "dust" particles. These particles each contain perhaps a million atoms bound together and covered with an outer layer of unknown composition. Accurate measurements of the way the dust particles absorb photons from bright, hot stars may furnish the details of the composition of these interstellar dust grains, which remain unknown at the present time.

The connection between young stars and the interstellar medium goes beyond the fact that some photons from the stars are absorbed by the medium. The interstellar material forms about 1 percent of the mass in our galaxy, but what makes it important is that *all* stars must have condensed from interstellar gas and dust. Astronomers feel that they understand the broad

outlines of this condensation process, which produced stars that have a relatively narrow range of masses (all of the 10^{11} stars in our galaxy apparently
have masses ranging from one-tenth to one hundred times the sun's mass).
However, the details of the condensation depend on the relative abundances
of the various atoms, ions, and molecules within the interstellar gas. We believe that all of the elements except hydrogen and helium arose from nuclear fusion reactions inside stars which later exploded, spewing their contents throughout interstellar space for inclusion in a later generation of
stars. Recently, the x-ray detectors on the Copernicus satellite found that the
most recent stellar explosion ("supernova") in our galaxy left behind a large
region around the site of the explosion that is still emitting x rays, hundreds of years after the explosion occurred. Measurements from the Copernicus satellite of the abundance of the elements in the interstellar medium
should let us check the accuracy of the theory of element seeding by supernovae as we compare the composition of young stars with the composition
of the interstellar gas. In any event, the measurements of the composition,
density, temperature, and ionization of the medium between the stars that
should emerge from the Copernicus observations will tell us how the interstellar medium does behave. Preliminary results from the Copernicus satellite
show that the abundance ratios of the various elements in interstellar space
are approximately the same as the abundance ratios in stellar atmospheres,
but different lines of sight through the interstellar medium show abundance
variations that will require further study. In addition, the satellite observations show that certain suggestions for the heating and ionization processes
in interstellar space are not tenable, though we cannot yet be certain what
are the chief sources of energy input.

The Future of Space Astronomy
The Copernicus satellite provides us with unprecedented spectral measurements, ultraviolet wavelengths inaccessible to ground-based observatories.
Because astronomers judged this information most desirable, *no* pictures in
visible light come from the satellite, which instead analyzes the radiation
from celestial sources at each ultraviolet wavelength interval. Astronomers
hope that the next few years of spacebound observations will bring us closer
to the realization of two major goals.

First, the greater accuracy both in pointing and in wavelength resolution

attained by the Copernicus satellite should become available at shorter
ultraviolet and x-ray wavelengths. This will occur if the four High Energy
Astronomy Observatory (HEAO) satellites now in the design stage are in
fact completed and launched. The HEAO satellites will scan the sky to obtain more detailed information about the x-ray sources that the Uhuru satellite (SAS-1) is now discovering. Each HEAO will have more than a hundred
times the sensitivity of the Uhuru satellite, and will cover a wider range in
wavelength (from 1.2×10^{-10} to 10^{-6} centimeter, as compared to the range
from 2.5×10^{-7} to 10^{-7} centimeter for the Uhuru satellite).

Another planned improvement in space astronomy is the International Ultraviolet Explorer (IUE). Although the instruments aboard the IUE are to
cover the same wavelength range as those on the Copernicus satellite, the
IUE will have the advantage of being placed in a synchronous orbit, some
23,000 miles above the earth's surface. This means that the satellite will
circle the earth at the same rate as the earth rotates, so that the satellite will
in effect hover above a chosen point on the earth's surface. Such a stationary relationship to the ground station will allow the satellite to transmit
data with no delay for storage, so that scientists on the ground can analyze
the data immediately (in "real time," as they like to say) and redirect the
satellite's observations at once in response to what they learn from the first
observations.

In addition to the HEAO and IUE satellites, astronomers and space scientists have great hopes for the Large Space Telescope (LST) that is planned
for launching in 1980. The LST would finally use the freedom from atmospheric absorption and seeing to make observations with unparalleled
resolving power over the wavelength interval from 10^{-5} to 10^{-1} centimeter,
including part of the near ultraviolet, the visible, and the infrared spectral
regions. A Large Space Telescope would consist of a reflecting telescope
with a mirror 300 centimeters in diameter, placed into orbit with its auxil-
iary spectroscopic and telemetry systems, perhaps by a NASA space shuttle.
Such a telescope, ranking with the largest ground-based instruments, could,
once freed from the effects of seeing caused by the earth's atmosphere, observe sources one hundred times fainter than would be possible under the
atmospheric cover. From the earth's surface, we inevitably find images
smeared by seeing, so that almost pointlike objects have a larger size than
their true appearance would produce. This blurring prevents the detection

of faint pointlike objects, because any detection system has an inherent, random "noise" background, above which the signal from the object must rise before we can detect it. A spread-out image may fail to rise significantly above the noise background at any point, while the same object seen as a pointlike source would rise above the detection threshold at its true position. The LST, with its planned pointing accuracy of four one-thousandths (0.004) of a second of arc, offers the opportunity to lower the detection threshold by a factor of 100.

If galaxies all have the same true brightness, then since their apparent brightnesses decrease as the square of their distances from us, we could use this factor of 100 to discover and record galaxies ten times farther from us than the farthest known galaxies. With these observations, we might resolve the questions of whether all of space is finite or infinite and whether the expanding universe will continue to expand forever or else someday cease its expansion and start to contract. Furthermore, the LST could perform spectoscopic analysis on sources too faint to be analyzed in this way, wavelength by wavelength, from ground-based observatories. The LST should provide us with new observations about "quasi-stellar radio sources" that might tell us the true nature of these "quasars" and how their fantastic properties arise. Certainly we would find new information about the material spread among stars, and perhaps among galaxies, that could tell us how galaxies and stars ever condensed from a diffuse medium. The properties of the stars themselves would be revealed in a detail previously attainable only for the sun, so we could learn much more about the nature of peculiar and exploding stars.

Since the Second World War, astronomers have steadily benefited from the new technology that has allowed them to improve their capacity to detect radiation and to carry their detectors above the atmosphere. Each wavelength interval—radio, submillimeter, infrared, optical, ultraviolet, x ray, and gamma ray—has called for new detectors that are sensitive to photons of a given wavelength. Side by side with the space program, which has moved from balloons to rockets to satellites of increasing complexity and to planetary and deep-space probes, the development of detector systems has given us the capacity to use our ability to go outside our blurring and absorbing, life-giving ocean of air. This twin technological effort, to develop both photon detectors as well as the systems to deploy them, has been on the whole a highly organized process that brings us better and better astronomical information. If all goes well, the LST will reap the technological harvest of thirty years of space-oriented research. The LST's photon detection systems will improve on those aboard the OAO, HEAO, and IUE satellites to make full use of the large mirror's light-gathering power. The advent of the space shuttle, which could be used to visit the LST in orbit repeatedly, brings us the possibility of changing the LST's detection systems to allow for new developments in our technological capacity as well as for advances in our scientific understanding that may call for new approaches to the study of celestial objects.

The interdependence of spaceborne and ground-based observations will continue. Ground-based surveys of the sky will tell the satellites which objects are most interesting for further investigation at otherwise inaccessible wavelengths. Conversely, accurate positional measurements of the strange x-ray and ultraviolet emitters will allow us to mount ground-based observations that could reveal faint, peculiar objects associated with the emission at short wavelengths. It will always be cheaper to use ground-based equipment when possible; the satellite observations will permit us to determine the most efficient way to study different celestial phenomena.

Summary

Astronomy has entered a new era, in which the connections between the large and the small become ever more obvious, with high-energy astrophysics making daily use of the new discoveries of particle physics. The entire development of astronomy rests, of course, on the ability to make observations and to draw conclusions from them. In Copernicus's time, when the solar system marked the outer limits of the realm where humans could see changing phenomena, no telescopes existed to aid the human eye. The next four centuries saw the advent of optical telescopes, refined over three hundred years to the present giants of Palomar and Kitt Peak, and of radio antennas, now huge steerable dishes a hundred meters across.

To see farther, we must go outside our atomsphere. From our global fishbowl, which we daily pollute still more, even to the point where many ground-based observatories are threatened by the light from nearby urban development, humanity draws life but not vision. Only spaceborne telescopes can resolve many of the current mysteries that confront us. We have the chance to gain from observations of quasars, peculiar galaxies, newly formed stars, exploding stars, and the expansion of the universe some insight into the strange ways that matter can behave in states totally unknown to us on earth. At the same time, physicists probe deeper into the structure of the "elementary" particles, which may yet turn out to be complex agglomerations of still other particles. It may not be long before the poorly understood substructure of the world of particles and the poorly observed cosmos around us together provide us with an illumination of the violent behavior of the universe, a behavior not imagined until recent years on this quiet planet in a small corner of the Milky Way. Einstein once said that "the eternal mystery of the world is its comprehensibility." For the next steps toward astronomical understanding, we shall need a series of space observatories, working in interdependence with our ground-based observatories, so that we can pierce the atmospheric veil around us and gain an unhindered view of the universe.

Suggested Reading

1. *Astronomy and Astrophysics for the 1970's.* Vol. 1, National Academy of Sciences, Washington, D.C., 1972.

This report of the Astronomy Survey Committee of the National Academy of Sciences presents a thorough overview of the current status of astronomical research, together with plans for the next decade. The scientific discussions are nontechnical and understandable in comparison with most such reports.

2. *Frontiers in Astronomy.* Edited by O. Gingerich, W. H. Freeman and Co., San Francisco, 1970.

A compendium of the highly readable articles on the new astronomy from the last few years, first published in the *Scientific American*. The articles by H. Friedman on x-ray astronomy and by L. Goldberg on ultraviolet astronomy carry forward the discussions of this essay.

3. Colin Ronan, *Invisible Astronomy.* J. B. Lippincott, Philadelphia, 1972.

A description of the theoretical and observational problems of astronomy in the modern age, concentrating on the hidden portions of the spectrum of light. Quite entertaining and completely comprehensible.

Robert L. Carman

24 Lasers—Evolution and Technological Use

Many people view the invention of the laser (*l*ight *a*mplification by *s*timulated *e*mission of *r*adiation) as one of the most consequential technological breakthroughs of this century. It was a natural, though complicated, extension of the maser (*m*icrowave *a*mplification by *s*timulated *e*mission of *r*adiation) developed less than a decade before. Indeed some would say the laser was a natural outgrowth of the quantum theory formulation of matter which was developed in the late 1920s. Let us first examine the essential ingredients of a laser in order to explore the nature of the intellectual revolution involved in its invention. We shall find both Copernican and non-Copernican elements. Then we shall turn to the realized and potential impact of the laser on technology. The uses of a laser, as we shall see, are both highly diverse and quite amazing considering that the first laser was operated just twelve years ago!

What Is a Laser or Laser Light?

A laser is a device that converts various forms of energy, such as electric currents or light from the sun or flashlamps, into a special kind of light. Because of the close connection of a laser to an atom, it would be helpful first to characterize an atom. An atom is made up of a small central nucleus, surrounded by a cloud of electrons, where the shape and extent of this electron cloud is not arbitrary but can take on only certain forms or configurations, such as spherical, egg shaped, and so forth. Different configurations correspond to different amounts of energy being stored in the atom. The so-called ground state is the configuration with least energy. We must give the atom energy to excite it to a configuration of higher energy content. The atom acquires this energy in chunks (called quanta) of fixed size by collisions with another atom, a molecule, or an electron, resulting in a reduction in their velocities. Likewise, the atom can divest itself of this energy through collisions with a concurrent change in configuration. Most important, the atom can also emit or absorb energy in the form of light. All this in "chunks" of fixed size, the quanta, which in the case of light are called photons.

Light is a wave of electromagnetic energy whose color depends on its frequency, red light having a lower frequency than blue light, and both

University of California at Livermore, Livermore, California 94550.

lying at the extremes of the visible portion of the spectrum. Infrared light refers to light frequencies below red light, while ultraviolet light corresponds to light frequencies above blue light. Nearly three centuries ago, C. Huygens suggested that light might be a wave; and more recently it has become well accepted. Light is bent around a sharp corner (diffraction), and two intersecting light waves of the same color produce an alternating light and dark pattern in the overlap region (interference), demonstrating its wave properties. However, not until 1905 was light shown to be not only a wave but also a particle by M. Planck's explanation of the frequencies of light emitted by hot bodies and A. Einstein's explanation of a process known as the photoelectric effect. They reasoned that a large number of particles, called photons, carried the energy in an intense light beam, the energy of the photon being directly proportional to the frequency of the light wave. The light intensity depended only on the number of particles. Furthermore, these particles can carry energy to or from the atom. The burning of paper by focused sunlight proves that light carries energy. The dual wave-particle nature of light was later found to be a correct description for all matter, as completely explained by the quantum mechanical description of nature. (See V. F. Weisskopf's article, Chapter 14.)

Combining the particle nature of light with the quantum theory of the atom established the idea that emission or absorption of a photon involves a change of state (electron configuration) of the atom. This is true whether subtle changes in the electron configuration (involving small energy differences, hence small photon energies) or removal or recombination of an electron with the atom (involving large energy differences) is occurring. It also implies that in order for light to interact with an atom, the frequency of the light must be such that the corresponding photon energy closely matches the energy difference between two states of the atom or molecule.

Another important property of Nature is that an atom that finds itself in an excited state can emit a photon in two somewhat different ways called spontaneous and stimulated emission, thus relaxing the atom closer to the ground state. We illustrate the difference by an example. Consider two different neighborhoods, one a city block containing a large number of apartment houses, while the second is a group of barracks on an army base. Furthermore, we will assume that discipline is extreme on this army base, so that all of the soldiers must fall in each morning to a parade in which they march to work. We will now associate the houses with the atoms of the material and the people with the photons of light.

On the city block with apartment houses, the people come out at all different times, and usually travel in different directions. Even those who accidentally come out at the same time and are walking in the same direction usually are not walking in step as if they were part of a parade. This situation is like the random emission of photons by a group of atoms, and is called spontaneous emission. The atom emits its photon in a manner that is uncorrelated with all other atoms. The emitted light is referred to as being incoherent, as typified by an ordinary light bulb or flashlight.

Returning now to this regimented army base, as the parade of soldiers passes each barracks in the morning, the soldiers of the barracks fall into the ranks and get in step with all of the other marching soldiers and continue marching. This systematic arrangement is like the stimulated emission of photons by a group of atoms. However, this analogy is not complete because on our hypothetical army base, there is a certainty that the soldiers of a given barracks will fall into the ranks independent of the number of soldiers already marching. In the case of atoms, the probability that a photon will be emitted in this regimented way depends on the number of photons which are already regimented, and this is stimulated emission. In reality, there is a competition, since both spontaneous (haphazard) and stimulated (regimented) emission can occur simultaneously. Also the photons must be of the correct frequency for the particular excited state that the atom occupies in order for the emission of the photon to be stimulated. A. Einstein first proposed the concept of stimulated emission in 1916.

The soldiers (photons) all marching in one direction in unison, rather than randomly dispersing in all directions, is called spatial coherence in the case of photons, while all the photons being in step increases the intensity of light, much as the sound is louder of stepping soldiers in a drill unit who start by putting their foot down at the same time, then continuing at the same frequency. This latter effect is called temporal coherence in that all the waves start at the same time in their cycle and have the same frequency.

The systematic, regimented, stimulated emission of photons is one means by which a coherent light beam is constructed. However, coherent light can in principle be generated without a laser, but very inefficiently. One way is through the use of a frequency filter, lenses, and a small pinhole (see Figure 24.1), whereby the result is the same coherent light (both spatially and temporally) as produced by a laser. It is important to realize that laser light is different only in that it is coherent light generated by a laser through stimulated emission rather than by filtering. In practice, however, the laser is not only more efficient because of stimulated emission but is also capable of generating light beams that approach theoretical limits to the degree of smallness of the frequency or angular spread which is achievable.

How Does a Laser Work?

By the 1920s it was feasible for the laser to be invented. The concepts of distinct energy states of an atom (Weisskopf's article discusses this concept further), of the particlelike properties of light referred to as photons and of two kinds of photon emission processes as one of the several ways for de-exciting an atom were already established. There are two more practical ingredients that were required, however, before the first laser was constructed. One requirement is known as a population inversion, and the second is a photon storage technique, to be explained shortly. These are the two principal areas in which C. H. Townes and his co-workers made their breakthrough that first led to the maser (the microwave analogue of the laser); and later the theoretical work of A. L. Schawlow, C. H. Townes, A. M. Prokhorov, and N. G. Basov led to the development by T. Maiman, P. Sorokin, and A. Javan of the first operational lasers. For their contributions to the development of lasers and masers, C. H. Townes, A. M. Prokhorov, and N. G. Basov were recently awarded Nobel prizes.

In order to understand the idea of a population inversion, we should go back to the atom once again. Before, we noted that an atom could either emit or absorb a photon while changing its electron configuration or state. Let us pick a pair of states and call the state furthest from the ground state, state 2 (the upper state); and the one closest to the ground state or the ground state itself, state 1 (the lower state). Not only is the ground state of the atom the electron configuration possessing the least energy but also the state that is most preferred. However, through collisions with other atoms, molecules, electrons, and so on, some fraction of atoms become excited. By emitting a photon, let us assume that one excited atom changes state from state 2 (upper) to state 1 (lower). A second like atom in state 1 (lower) could now

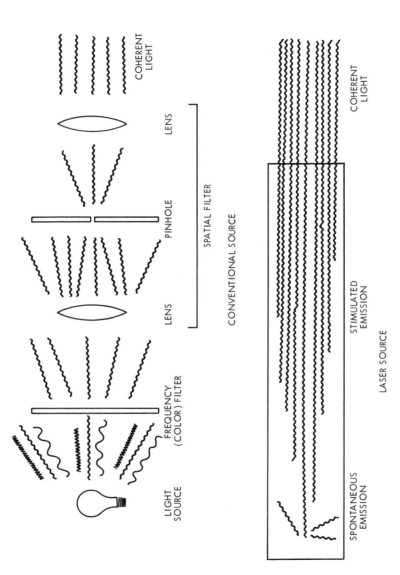

Figure 24.1. Two alternative techniques to obtain a coherent light beam (in principle).

The top system utilizes a frequency filter (absorbing or dielectrically coated glass) and a spatial filter in connection with an ordinary incoherent light source to produce very inefficiently the coherent light. The bottom system schematically illustrates the evolution of spontaneous through stimulated emission, characteristic of a laser, used for the efficient production of the same coherent light.

absorb this same photon arriving in a final state of state 2 (upper) and return to state 1 (lower), for example, by a collision process, resulting in no emitted photons. The energy due to motion of the collision partners can be absorbed or released during the collision leading to the change in the electron configuration of the atom. This illustrates that through absorption or emission, the atoms can either add photons or subtract photons from a light beam. The important question is when does a group of atoms add more photons than it subtracts from a light wave passing by, since both emission and absorption occur in the laser medium under almost all conditions? The answer is when more of the atoms are in the upper state than in the lower state. This is contrary to the situation usually found in nature where most atoms occupy the lower states, a condition referred to as a normal population distribution. For light beams passing through a medium with a normal population, only a net absorption of photons results. However, if we provide an energy source that is capable of pumping more atoms into the upper state from the lower state, then a net increase in photon numbers would result by passing a light beam through the material. Such an atomic population distribution is referred to as a population inversion, since we have interchanged the most occupied states. Thus, it is clear why the idea of a population inversion is important to the operation of a laser since it provides the opportunity for the light intensity to grow.

What kinds of energy sources can create population inversions? There are several different classes of such energy sources. The first ruby laser used an incoherent flashlamp, similar to a flashtube used by photographers. Subsequently, chemical reactions, high-energy electron beams that collide with atoms or molecules, high electrical currents in semiconductors, light from the sun, discharges to avalanche electrical breakdown in gases, have all been used as incoherent (chaotic) energy sources to achieve population inversions. Finally, coherent laser light has been used on some occasions to achieve population inversions. From this discussion it is clear why we have defined the laser as an energy converter.

Assuming it is possible to achieve a population inversion as indicated, and some spontaneously (randomly) emitted photons are being generated, it is now clear that the number of photons will tend to increase. As the number of photons that are traveling in a particular direction at the correct frequency becomes large, the possibility of stimulated emission then arises (see Figure 24.1). As indicated in the example with the soldiers, the stimulated photons will be emitted in step and travel in the same direction as the photons that are accomplishing the stimulating. For all the early lasers, it was necessary to force the spontaneously emitted photons to pass through the pumped medium many times in order to build up a large enough number of photons traveling in step in the same direction for stimulated (regimented) emission to occur. Among other important considerations, the means by which this directional photon storage is achieved was originally suggested in the classic paper by A. L. Schawlow and C. H. Townes in 1958. A pair of parallel mirrors are used, one on each side of the material that contains the atoms having a net population inversion. The photons now bounce back and forth between the mirrors building up first to the level where stimulated emission can occur, then continuing to build up until the number of photons that are lost from the beam is equal to the number of photons added to the beam on each pass. In order for events to occur as described, a significantly larger number of atoms must be in the upper than are in the lower state initially.

More recently, the requirement for mirrors has been eliminated completely in the case of a few types of lasers where such large population inversions are achieved that spontaneously (randomly) emitted photons at one end of a long tube, traveling down the axis, could be amplified to the stimulated emission level. Then, continuing down the tube, the number of photons grows so rapidly that, as the photons leave the other end of the tube, the atoms at the output end are of equal number in the upper and lower states. This equal population condition implies that as many photons would be absorbed from the beam as added to it via stimulated emission. This mirrorless type of laser is called a traveling-wave laser.

What Kinds of Lasers and Laser Light Exist?

We now turn to a brief description of various classes of lasers and some general properties of laser light which contribute to the large number of uses for a laser. First, there are two different kinds of lasers—those which have light outputs that are continuous in time, and those whose output exists for a short time and are called pulsed lasers. In the pulsed laser class, outputs of duration from greater than 1 hundredth (10^{-2}) of a second to 1 tentrillionth (10^{-13}) of a second have already been demonstrated. To understand how pulsed lasers operate, we return to the concept of a laser as an energy converter. We see that under circumstances where the upper state can store its energy for a long time (that is, the loss of population inversion due to spontaneous emission, collisions, and so on occurs very slowly compared with the gain of population inversion due to the pumping energy), it is possible to use stimulated emission as a trigger for the release of this stored energy, much as the breaking of a large dam triggers the release of stored water. These circumstances create the possibility of generating light pulses of short duration. A vivid demonstration of the potential impact of triggerable energy storage is that some currently operating lasers can generate more laser output power for a short time than all of the world's current electrical power generation capability combined, namely, greater than 1 trillion (10^{12}) watts.

Another property of laser light is that the angular spread of the beam can be reduced in practice to below that of any conventional light source. This is graphically demonstrated by the fact that if many currently operating lasers were pointed at the moon, using no elaborate equipment, they would illuminate a spot about 10 miles wide on the surface of the moon. This is to be contrasted with the best conventional searchlights that would illuminate a 25,000-mile-wide spot after traveling the 1-quarter-million-mile path. A further implication of the small angular spread is that the laser beam can be focused to very small spot sizes, namely, of the order of 10 one millionths (10^{-5}) of an inch for the very best current lasers.

There are a few lasers in the world today which combine the properties of the lowest angular spread and the highest output powers. Focusing these laser beams to their minimum possible spot size produces electric fields of the order of 5×10^{10} volts per centimeter, which is more than one hundred times the electric field that holds an atom together. Such a laser, if more gently focused in air, will cause a track of electrical breakdown over several feet, much like a bolt of lightning.

Lasers can also be classed by the state of material in which the population inversion exists. Solids, liquids, and gases or vapors have acted as a host for atoms that can be pumped to an excited state. Furthermore, lasers can be made in many shapes and sizes, some being as small as the head of a pin, others as large as 300 feet long and several feet in diameter. The output powers usually scale with size and vary from far less than from a small flashlight to more than from our sun. Finally, the cost and quality of materials required for constructing a laser range from materials frequently found around a home and a cost of about $100 to high-precision components and a cost approaching $100,000,000.

Another noteworthy property of some lasers is that the spread in output frequencies is so small and so constant that it is possible a laser will soon become the universal frequency standard. While the frequency of a laser is in the range of 10 million (10^7) to 10 billion (10^{10}) megacycles, some lasers have output frequency spreads of less than 1 kilocycle (1 thousandth megacycle) with the frequency content being constant to approximately one cycle. While not all laser outputs can boast such narrow frequency spreads, most laser outputs can have frequency spreads that are narrower than conventional narrow frequency width light sources.

What Are Lasers Used for?

To date, lasers have had a very significant impact on technology and research in a broad range of disciplines representing even more than the marriage of optics and electronics. These will be split up into six areas; namely, the measurement of distance and the definition of a straight line; materials (animate and inanimate) processing or treatment; laser communications and computers; pollution measurement and monitoring; photography; and basic research. There will be some overlap, and no attempt is made to be comprehensive, but the extraordinarily wide range of applications mentioned should be indicative of the possibilities.

One of the first applications was, not too astoundingly, a military use. Because of the good pointing ability or low angular spread, a laser beam could be easily aimed at a distant object. By measuring the round-trip transit time of a laser pulse, the distance or range to the object (that is, a tank or ship) could be determined (a range finder). These lasers produce relatively low energy outputs, so no damage to the object results. Since this early work similar lasers have been used extensively to measure accurately both distance and angle. Recently, the distance to the moon was measured to a precision of about 15 centimeters (about 1 foot) in this way, using the corner reflectors placed on the moon by some of our astronauts. The round-trip time for the light was about 2½ seconds. A study of how the earth-moon separation changes over several years is very significant to our understanding of gravity and our solar system. The field of surveying has been modified also by the accuracy that the laser can provide in measuring both distances and angles. More recently, this same capability has been exploited in measuring the motion of land across earthquake faults. In this same field, some potential use of lasers in the prediction of earthquakes could result from a new laser seismograph. This device measures the amplitude of vibrations that exist at the laser location. Vibrations are detected by mounting one of the mirrors of the laser cavity on a column buried deep in the ground. As the column vibrates, this mirror will twist away from parallelism with the second mirror that forms the laser cavity, resulting in a reduction or stopping of the laser output. In the field of civil engineering, lasers have been used extensively to define a straight line for grading roads and digging tunnels (for example, the tunnel under the San Francisco Bay). The Stanford linear accelerator and even other lasers are frequently aligned by using laser beams.

More sophisticated uses of the distance measuring capability of a low- to medium-power laser have been found; for example, a laser radar scheme has been developed. Operating in a similar way to conventional radar, laser radar can more accurately locate and track distant objects because of its precise aim. This capability would be very useful for example in tracking space satellites. A more subtle and possible future use of laser radar is to detect the reflected light created by clear air turbulence, which might some day allow an early warning of such navigational hazards. On the other hand, low-power lasers have been used to more accurately control machining tools, allowing the achievement of accuracies of .00001 inch to be obtained directly rather than .001 inch for manual and .0001 inch for the better nonlaser-controlled machines. In addition, parts can be inspected to even better accuracy. Of course, this inspection capability applies to both curved and flat surfaces, a fact that points up an important application, namely, the inspection of telescope mirrors and large radar or radio telescope antennas. A second of the many uses of this accuracy is in studying instabilities that occur in high-speed turbines or jet engines.

Turning to a second general area in which lasers are in wide use, we have the area of material processing, handling, or treatment. By focusing a medium- to high-output laser beam to a point, the resulting high energy density can be used to heat materials to high temperatures, much like the example of burning paper with focused sunlight. Because laser light can be focused to spots as small as .00001 inch, very small holes can be drilled in materials. Since all that is required is that the material being drilled absorb the light being used, the technique is applicable to a large range of materials that are, for example, brittle, soft, extraordinarily hard, or otherwise difficult to machine. In addition to drilling, welding and cutting can also be accomplished. Laser welding has been very important in areas where standard welding techniques do not work, such as welding dissimilar materials or some aircraft alloys. Furthermore, controlled atmospheres present no problem. Since no physical contact is required, contamination is also not a problem, while rapid and localized heating minimizes bulk changes in the materials being handled. Combining these with small welding sizes has caused a revolution in the field of microelectronics. Likewise, computer-controlled laser cutting, trimming, etching, and scribing has been revolutionary to integrated circuit fabrication. Finally, things like turbine blades can be balanced by computer-controlled laser machining (selected metal removal by the laser light) while the piece is spinning at high speeds. However, lasers can handle more kinds of materials than just metallic substances. Plastics, glasses, ceramics, rubber, gemstones, wood, even cloth have been cut, drilled, or otherwise treated by lasers; and lasers are becoming a part of all kinds of assembly lines. Even the area of quality control has been altered. By vaporizing small samples of material, the light emitted by the vaporized material indicates the quantity and kind of substances from which it was made. Thus, variations in composition can be detected readily, for example.

Similarly, low-power to high-power lasers have been used in medicine, as well as in biological research. The first medical use of low-power lasers was for the repair of detached retinas of the eye. Since then, many more surgical uses have been considered. While lasers are appearing in more hospitals, and are being used in many ways, they are not yet universally accepted as the best tool for various operations. One successful use of lasers has been their use as "cookie cutters" to remove corneas. Because a laser self-cauterizes a wound when used to cut tissue, operations where excessive bleeding is involved have seen some use of lasers. Because tattoos absorb more light than the skin in the immediate area, lasers have been successfully used to remove tattoos. As another example, a woman's Fallopian tubes have been burned away by a laser to achieve sterilization. Recently, large lasers have been used to vaporize and cauterize damaged tissue on patients who have been severely burned, thus avoiding infections associated with decaying tissue. On the horizon, the introduction of low-transmission-loss fiber optics (fine, long strands of plastic or glass that can each be less than .001 inch in diameter and carry light from one point to another irrespective of the shape to which the strands are bent) will possibly allow laser operations on the internal parts of the body without incisions. For example, preliminary experiments conducted on breaking up gallstones using this technique have had some success, where two sets of fibers were employed. One was used for the laser, while the other was used to observe what was happening.

In the area of biological research, lasers have been used to do microsurgery on individual cells; to inhibit the functions of enzymes; to determine the composition of restricted portions of cells by vaporizing those portions (10^{-14} to 10^{-16} gram); to activate a single nerve fiber allowing its function to be determined; to selectively damage single chromosomes (the portion of the cell which carries the information that makes people unique); and finally, to interrupt at various stages the process by which food is converted into living tissues, thus enhancing the understanding of the biochemistry of various living cells. Also, the compositional analysis technique referred to could potentially be used to diagnose diseases.

For many centuries, light has been used as a means of signaling or transmitting messages. With the advent of the laser, great predictions were made of its potential impact in the field of communications, being deemed the most important application by many. While the potential is truly great, as we shall see, progress has been slow. There are several reasons. Some of the problems require more technological breakthroughs, current communications systems meet most current demands, and the remaining problems that must be solved before the full potential of laser communications is realized are very hard. Nevertheless, some operational systems of a limited type do exist, for example, within some office buildings in Tokyo. Because laser light usually travels in one direction, point-to-point communications are the more important application. Furthermore, because light beams are greatly reduced in intensity by fog, clouds, and other adverse weather conditions, either outer space or controlled paths (vacuum pipes, fiber optics, and so on) on earth are envisioned.

Why were laser communications projected to be so important? Lasers can be used just as radio or microwave transmitters are used to transmit voice, television, codes, or other types of signals. Most radio and television systems broadcast at some frequency (the setting of the dial) and have a bandwidth (spread in frequencies) for various types of modulation (the voice or TV picture information) which is 10 percent of the broadcast frequency. Thus for AM radio, 1 megacycle (1000 kilocycles) is a typical broadcast frequency, and .1 megacycle is the bandwidth. This bandwidth is adequate to transmit ten telephone conversations, another important fact. Now suppose we consider a broadcast frequency of 1 billion (10^9) megacycles, which would correspond to blue light Also, assume that we maintain the tradition of 10 percent of the broadcast frequency being available for modulation. This would correspond to a 100-million (10^8) megacycle bandwidth. Thus this single beam could then carry 10 billion (10^{10}) telephone calls, a billion (10^9) AM radio stations, or 10 million (10^7) TV stations. Put another way, all of the world's current communications could be handled simultaneously by this one beam. Now, if we realize that there are lasers available from the infrared through the ultraviolet, while the same frequency could be used along different paths, we see that for the foreseeable future it is very unlikely that we could overfill such a potential. Even if 10 percent of this potential were realized, the capacity would be enormous.

The problem areas lie in developing efficient, very low loss ways to transmit the modulated laser light over long distances, in being able to amplify the modulated light beams at relay stations, and in developing efficient and fast enough modulators and demodulators to allow the full potentials of laser communications to be substantially realized. However, currently it appears feasible to replace intracity electronic telephone systems with an optical system employing fiber optic transmission lines.

One application of laser communications may be able to use much of this potential almost immediately, namely, a high-speed communications link between high-speed electronic computers. In fact, generalizing the concept of communications, we could include another similar potential use of tiny lasers, namely, as elements of a computer (an optical computer). Currently electronic elements (that is, transistors) are used in situations where they either conduct or do not conduct electrical currents to allow computers to perform mathematical operations. Similarly, lasers can be used either to emit or not to emit light, thus replacing the electronic elements. The advantages to be gained are that optical computers in principle could be made both larger and faster. The larger capacity comes from the sizes of achievable light beams compared to wires, and the small focal spot sizes. The potential faster speed stems from the fact that solid state electronics work reliably and flexibly down to about 1 billionth (10^{-9}) of a second, while light waves and integrated optics (a new field analogous to the field involved with integrated circuits in the electronics industry) appear to work at rates as fast as 1 trillionth (10^{-12}) of a second. Also lasers can be utilized in new techniques for high packing density information storage and retrieval, important to computers as well as elsewhere.

With ecology on the mind of much of our population, we find a fourth principal area of application of lasers, which comes from the study of the light emission and absorption spectra of atoms and molecules (spectroscopy). The principal application besides basic research is to the detection or monitoring of atmospheric pollution. The absorption or emission of different frequencies of light by atoms, molecules, or more complex structures is a type of "fingerprint" for that species, as mentioned previously. This fact can be used to detect the presence of a particular material in a laser beam path by monitoring the transmission of a laser beam, whose light frequency is absorbed by the material to be detected. By passing several laser beams over the same path, it is further possible to assess both the level and kinds of materials present. Various polluting gases from exhausts, smokestacks, and so forth, can readily be detected. On the horizon, however, is the possibility of extending this work to the identification and monitoring of such things as pollen, viruses, bacteria, and so on, and in media other than air.

Because of several unique properties that laser light combines, the field of photography is also being revolutionized. In the area of high-speed photography, the use of very short duration light pulses as "flash bulbs" for freezing fast action is obvious. However, new and improved methods of contrast enhancement, for example, have allowed low contrast, very grainy pictures, like those originally sent back from Mars by our satellite to be converted, through the use of lasers, into good photographs. Furthermore, photography using either infrared or ultraviolet light for some specialized applications has become far more feasible. Finally, the high coherence of laser light has allowed truly three-dimensional "photographs," indeed even "movies," to be made. The picture is recorded on film using a laser as in usual photography but in a very different form (a hologram). In order to see what has been photographed, the film must now be illuminated by a properly placed light source, which is usually also a laser. On the horizon is the commercialization of these techniques for home photography, TV, or cinema, for example. While one-color holography is far more developed at this time, it is now possible to obtain three-dimensional color pictures as well.

The final type of important "application" of lasers is to the field of fundamental research. We have already mentioned some uses of this type earlier, but there are many more, of which a few will be mentioned. The fact that chemical reactions proceed at a different rate when the reacting atoms or molecules are in excited rather than their ground state means that the course and nature of chemical reactions can be altered by laser light. Potentially, this could lead to numerous applications, one of which is the possibility of separating different isotopes of the same atoms. A second area is concerned with the possible extension of laser action to x rays or even gamma rays. In fact, it may be possible to stimulate the emission of the basic particles that constitute the nucleus (that is, mesons). A third area of research involves what could be an important intellectual revolution. Power has been transmitted in the form of electrical energy through wires for more than a century. With the recent development of very high power lasers that operate continuously, there is a possibility that someday power will be transmitted in the form of laser beams, thus eliminating the large losses in transmission lines. At this point, however, no high-efficiency converters from coherent light back to electricity exist, although the lasers can currently convert greater than 40 percent of the generated electrical power into coherent light. This potential application is not only important on earth but in outer space or for links between earth and outer space as well.

While on the subject of outer space, there is another potential application of lasers whose concept might go against one's intuition. Light beams not only carry energy but can exert a pressure upon impact. Most light sources with which we commonly come in contact are not large enough for this property to be conveniently manifested. However, by using this property, it may be possible to propel, for example, a space ship directly, although at this time the idea is very "farfetched." A second, more feasible way in which laser beams can propel things is to use the fact that intense laser light can cause materials to vaporize. An intense continuous type of laser that causes material to vaporize continuously in one direction can produce a jet engine type of action, forcing the unvaporized material forward at an ever-increasing speed. The rate of vaporization and the quantity of material available to be vaporized determine how high a speed can be achieved.

Potentially one of the most important areas of research which has recently come of age involves investigating the feasibility of initiating nuclear reactions by a laser beam, producing large amounts of thermonuclear energy in return (the laboratory-controlled form of our sun or of a hydrogen bomb). The implications of the success of this work are truly great indeed, since such a practical device could well represent the ultimate energy source needed by mankind. The concepts involved are quite simple, in principle. For thermonuclear reactions to take place in a controlled way, first a small amount of material (like solid hydrogen) must be made very hot (like our sun) and next the material must at the same time be very dense. The means by which a laser can hope to achieve this condition is first to use its ability both to heat and vaporize materials. Second, by employing light beams that are focused from all directions on a spherical target, the vaporizing material (which tends to propel the unvaporized material in the opposite direction) in this case can be used to compress the remaining material to very high densities. The required densities lie midway between that of an ordinary solid and of the most dense stars. Several groups throughout the world are pursuing this line of research at an ever-increasing rate. The minimum expected return is a better understanding of some of the physics of outer space, while achieving the ultimate of practical controlled thermonuclear fusion has consequences that could be so far-reaching as to be difficult to project.

Revolutions or Evolutions?

In order to arrive at this point in the scientific and technological development of lasers, we see that many intellectual revolutions as well as complex evolutions have occurred. Can we associate any of these developments with a Copernican revolution? Also, is it possible for truly Copernican-type revolutions to occur in a highly technological field? We answer both of these questions affirmatively but with qualifications. The abandonment of the concept that light was a wave only by M. Planck and A. Einstein, followed by the adoption that light was both a wave and a particle was truly a Copernican revolution. While Newton in 1704 was the first to consider light as possibly a stream of particles, his notion was not the same as Planck's or Einstein's. This early particle concept of light was abandoned for that of waves in the early nineteenth century because of the success of the compiled theory of electromagnetic waves by J. C. Maxwell in explaining a large body of experimental data.

The development of the concept of stimulated emission by A. Einstein is an important turning point, but a case where evolution rather than revolution is more appropriate, although for some people the change was more drastic than evolution suggests. A growing body of knowledge indicated that a process which was the inverse of absorption must exist, and spontaneous (random) emission was inadequate to explain these effects.

We see in the invention of the laser either a revolutionary or an evolutionary process depending on the perspective taken. Scientists at the time with strong interests in the study of the absorption and emission of radio waves and microwaves by atoms might view the invention of a laser as an excellent example of a scientific and technological breakthrough that was carefully built on a previous body of knowledge, thus involving an evolutionary process. On the other hand, many physicists too closely following the quantum mechanical description of light missed the possibility of coherence and its significance, as well as the consequence of stimulated emission. To this group, the more appropriate aspect was the reorienting of established ideas, therefore connecting the invention of the laser with an intellectual revolution. This raises a very important question, however: Are all intellectual revolutions Copernican? We have assumed in this discussion that a Copernican-type revolution is associated not only with the overturn of accepted ideas (an intellectual revolution) but also that these ideas were in the form of scientific theories that not only said what was correct but also what was incorrect. For Copernicus, either the earth or the sun was at the center of the solar system. From this point of view, an intellectual revolution associated with the invention of the laser could qualify as either

Copernican or non-Copernican depending on the extent to which laser or maser action was believed impossible prior to its invention. The author feels that more of a synthesis and less of a true overturning of ideas was involved in the invention of the laser.

There were several ideas mentioned in the section on the uses of lasers that would qualify as revolutionary, that is, the replacement of electrical transmission wires with laser beams. These intellectual revolutions are more appropriately associated with non-Copernican revolutions, since there was no theory which said that prior solutions to the problems of mankind were the only solutions.

Harrison Brown

25 Some Quasi-Copernican Revolutions in Man's Utilization of Energy

Introduction

Thus far in these discussions of "quasi-Copernican revolutions" emphasis has been placed almost entirely upon ways in which changed insights and attitudes of *individuals* concerning various aspects of nature have led to the overthrow of preconceived ideas and in turn to dramatically new levels of understanding. Copernicus, an individual thinking human being, conceived of the sun rather than earth as being the center of our planetary system. Another individual thinking human being conceived of our sun, and with it the millions of stars we see in the Milky Way, as component parts of a galaxy or "Island Universe" that he suggested might be but one of millions of similar galaxies which inhabit the vastness of space.

Recognizing the importance of the individual "flash of genius," we must also realize that over years and centuries an entire culture can develop collectively its own revolutionary insights concerning natural phenomena. More often than not the changing insights of a culture represent the summation of innumerable incremental changes in the attitudes of thousands of individual human beings over the course of many generations. Often the changed insights have enabled those who have subscribed to new ideas to survive while others have died.

Man's attitudes toward energy in its various forms provide dramatic illustrations of quasi-Copernican cultural revolutions that have so often in the stream of history enabled cultures which have acquired entirely new attitudes and insights to survive. These changing attitudes have taken many forms and collectively have made possible the emergence of the level of technology which is the mainspring of our present civilization.

Early Man

During the greater part of man's 2 million or so years of existence he has lived much as the other animals about him, at a level of energy consumption in the form of food or about 3000 calories per day. (Food calories are "large calories," each of which is equal to 1000 calories as normally defined.) To obtain this minimal quantity of food, people hunted and fished and gathered varieties of naturally occurring edible vegetable matter. The discovery of the controlled use of fire gave man protection from animals and made it possible for him to inhabit cooler climates. Cooking extended the range of

National Academy of Sciences, Washington, D.C. 20418.

food that could be eaten. While these new technologies gave rise to greatly increased numbers of human beings, they probably did not elevate per capita energy consumption more than a fewfold. And even with the controlled use of fire, the worldwide population of human beings living within the framework of a food-gathering culture could not have risen to a level much greater than about 10 million persons—about the population of the present city of New York.

Only a small proportion of the energy stored in a natural ecological system and which passes through it each year as the result of photosynthesis and metabolism is available for man's nourishment. As a result, a great deal of land area in the natural state is required to provide sufficient food to support an individual for a year. It has been estimated that a moderately fertile wild region 10,000 square miles in area offers sufficient animal and vegetable life to support about 5000 food gatherers.

Human beings are gregarious, but the very low density of available food places an upper limit on the size of a nomadic tribe or self-contained family group. The radius of operations of a hunting party from its base can be little more than about 15 miles, so under the circumstances no more than about 350 hunters and food gatherers could be supported by the food available within that distance. Thus, for the greater part of man's life on earth, humanity has been divided into some tens of thousands of small, primarily nomadic food-gathering tribes.

Domestication and Cultivation
Preconceived idea: Supplies of food are limited by the edible plants and animals as they coexist on earth in the natural state.
This idea, and the behavior associated with it, persisted for some 2 million years until new concepts began to emerge a few thousand years ago. First, man found that he could protect animals which were useful and destroy those which were not. Later he found that some useful animals could be bred in captivity. Still later he found that he could protect useful vegetation and encourage its growth. Later he learned to collect and plant seeds.

These new techniques increased enormously the number of persons who could be supported on a given area of land. Whereas two square miles of fertile land had been necessary to provide sustenance for an individual food gatherer, with the development of animal domestication and agriculture the same land could now provide nourishment for some 200 persons. This was indeed a major breakthrough.

When viewed on the time scale of human existence, the "agricultural revolution" came about explosively. Yet the new concepts do not appear to have been those of any one person or even many persons inhabiting any one location in time or space. Rather, the new concepts appear to have emerged as the result of small increments of experimentation undertaken by a large number of individuals during the course of many generations. The new technology was a collective development.

The changes in ways of life which were brought about by agriculture were enormous and came rapidly. By ensuring permanent food supply, the cultivation of plants made permanent settlement possible. Villages appeared, limited in size to some 200 to 300 persons by the food that could be produced within easy walking distance. As agriculture spread, population expanded rapidly.

In time, agricultural technology made it possible for a family to grow somewhat more food than it needed for itself, and this meant that some persons could engage in activities other than farming. This surplus, coupled with the emergence of transportation technology, made possible the emergence of the city, which in turn made possible the emergence of the great ancient civilizations.

Men and Animals as Prime Movers
Preconceived idea: per unit quantity of food eaten men have the same pulling power as horses.
Prior to Roman times, the primary motive power for tools and machinery was human muscle power. Although the ox had been used for many centuries for plowing and the horse had been used extensively for rapid transportation of people and materials, the concentration of energy for accomplishing specific objectives was generally obtained by the use of large gangs of men. Slavery was an accepted social notion in antiquity, and gangs of slaves were used to provide motive power for ships, for constructing roads and aqueducts, moving massive machines, and constructing great public works. Animals were seldom used for these purposes. Why?

In those days a horse when harnessed could pull four times as much as a man. But knowing that the animals each consume four times as much food as a man, the businessmen or project managers of the time could see no particular advantage to the use of animals, and they could see many disadvantages. Indeed the flour mills of ancient Rome were driven by two men or one donkey, and there appears to have been little economic incentive to prefer the latter.

We now know that the reason a horse could pull only four times as much weight as a man was the poor design of harness which resulted from ignorance of horse anatomy. The harnesses of the time prevented the horse from exerting its full available strength, which would have enabled it to pull fifteen times as much as a man. It was not until the Middle Ages that the breast-strap or postilion harness was acquired from the Mongol tribes that invaded Europe. The horse collar was also introduced and by the twelfth century was in general use throughout Europe. Following these developments, animals quickly replaced men as sources of power for the concentration of energy.

Power from Water and the Winds
Preconceived idea: It is better to invest capital in slaves and land than in equipment and machinery.

Although sailing vessels were used in antiquity and the waters of rivers were used to transport ships and materials downstream, wind and water were not used for motive power for tools and machinery until Roman times. The oldest known reference to a water mill was made by one Antipater of Thessalonica, who wrote in the first century B.C.:

Cease from grinding, ye women who toil at the mill; sleep late, even if the crowing cocks announce the dawn. For Demeter has ordered the nymphs to perform the work of your hands, and they, leaping down on the top of the wheel, turn its axle which, with its revolving spokes, turns the heavy concave Nisyran mill-stones.

These early mills were used primarily for grinding grain and had an output of about $\frac{1}{2}$ horsepower—which was about the same as mills operated by two slaves or a donkey. Such small mills gradually spread, reaching Ireland in the West and China in the East by the third and fourth centuries A.D.

A larger and considerably more efficient mill was constructed in the first century B.C. by a Roman engineer. At Venafro on the Tuliverno, near Naples, the remains of a mill have been found which had an output of about 3 horsepower and permitted the grinding of between 300 and 350 pounds of grain per hour. When we compare this with the output of but 15 pounds of grain per hour of a mill operated by two men or a donkey, we see that a truly major breakthrough had been achieved. The output of this early Roman mill was higher by far than that of any other power resource of antiquity.

In its time this early mill was more revolutionary and expanded industrial possibilities more than has nuclear energy expanded industrial possibilities today. Yet for reasons that appear to be very complex, the Roman (or Vitruvian) water mill spread very slowly. Many of the factors that contributed to this slow spread, as in the case of nuclear energy today, appear to be economic, social, and psychological in nature. Heavy investments were required in order to ensure a steady flow of water and to provide the machinery that could make use of a more concentrated output of energy. In an age when individual capital was invested primarily in slaves and land, there was little impulse toward industrialization that required appreciable capital investments which could depreciate with time.

The grinding of grain in Rome was centralized, and during the early Empire unemployment in the city was apparently a serious problem. Vespasian (A.D. 69–79) is said to have refused the building of a water-driven hoist "lest the poor have no work." Clearly, had the centralized grain grinders been converted to water mills, the situation would have been aggravated—a problem of a type that has occurred repeatedly in history.

There is no evidence of a rapid increase in the number of water mills until the fourth century A.D., some 400 years after their first appearance in Italy. There seems to have been an acute shortage of labor during that century, and this may have contributed to the spread.

The period in Europe following the disintegration of the Roman Empire was hardly favorable to technological progress. Indeed, the practical knowledge of machinery possessed by the Romans was not surpassed until the eighteenth century. Yet in spite of this, the number of water mills increased steadily between the tenth and twelfth centuries and appears to have doubled every 50 to 100 years. The mills came to be applied to mining and metallurgy as well as to the grinding of grain.

The windmill was apparently unknown to both the Greeks and the Romans. It was of eastern origin and spread slowly in Europe in the thirteenth century. This source of concentrated power, like the water mill, was limited by nature, a factor that undoubtedly contributed to its slow spread. Also, laborers objected to their installation. In the sixteenth century, for example, craftsmen protested against them, claiming that they would throw many persons out of work. A mechanical saw, driven by a windmill, was destroyed near London by a riotous mob in 1768. Nevertheless, by the mid-eighteenth century the waters and winds together were providing the greater part of the mechanical power available in Europe.

The Linking of Coal to Iron
Preconceived idea: Metallic iron of high quality can only be obtained from its ores on a large scale by using charcoal.

Furnaces that could achieve sufficiently high temperatures to permit reduction of iron ore to the metal did not appear on the world scene until about 1100 B.C. in the Middle East. From that point on in history the use of metals became truly widespread. The availability of relatively inexpensive iron tools greatly accelerated the rate of spread of human population as huge forested areas were transformed into agricultural land.

England entered the business of iron production richly endowed with iron ore. Forests were also abundant, and the trees were used to produce charcoal, which in turn was used to reduce the iron oxide to the metal. These resources enabled England to become a major producer of iron for the world.

By the sixteenth century the island, which had once been covered with trees, was confronted by a shortage of timber. Deforestation had taken place rapidly, in part to increase the area of agricultural land, to provide wood for building, fuel for heating, and charcoal for the iron industry. Faced by the shortage of wood, increased use was made of coal, and by the seventeenth century coal was being substituted for wood in a number of localities as a fuel.

Numerous efforts were made to use coal as a substitute for charcoal in the manufacture of iron, but although coal will reduce iron ore to the metal, the impurities in it render the properties of the metal quite unsatisfactory. By 1709 Abraham Darby, after much effort, learned to drive off the volatile materials in coal by preheating, and succeeded in using the product, called "coke," to smelt iron ore. Although the product was reasonably satisfactory, in part because the coal he used had a low proportion of sulfur to begin with, coke-produced pig iron did not rival charcoal iron in quality until about 1750. From that time on, however, England's vast coal resources were available for iron manufacture and the iron industry grew very rapidly. As the practice of smelting with coke spread, the quality of the iron improved and the costs dropped, thus making the metal more generally available and encouraging its more widespread use.

The linking of coal to iron has been second only to the development of agriculture in its impact upon the course of human history.

The Steam Engine

Preconceived idea: Energy can be concentrated on a large scale only through the use of water, wind, or animals.

The linking of coal to iron greatly increased the demand for coal. Miners had to dig ever deeper shafts, and underground water became a major limiting factor in coal mining. In 1705 Thomas Newcomen and John Calley designed an engine which was moved by steam and which could furnish the power to pump water from the mines. By the mid-eighteenth century more than 100 Newcomen engines were in use. In 1765 James Watt constructed a steam engine that required only one-third the coal consumed by a Newcomen engine per unit of power output, and the following year a steam engine was used for the first time for purposes other than pumping water— it was used to blow air through a blast furnace. By 1785 the steam engine had been harnessed to power looms and spinning machines, in addition to pumps. These developments quickly led to the consolidation of small industries into large factories. It was only a matter of time before the steam engine was applied to both land and sea transport and eventually to the farm.

Following the invention of the steam engine, technological developments followed each other rapidly. By the beginning of the nineteenth century men had conceived of the screw-propeller steamboat, the threshing machine, and the sewing machine. The same century saw the appearance of the water turbine, the internal combustion engine, and the steam turbine. In the early twentieth century the gas turbine and the jet appeared. The electrical generator came into existence and revolutionized communications and the transmission of power. Two new energy sources, petroleum and natural gas, came into widespread use early in the twentieth century and were soon seriously competing with coal.

Following the advent of the steam engine, per capita demands for energy increased rapidly. In part this increase resulted from the replacement of human labor by machine labor. In part it resulted from the fact that people want luxuries, and the new sources of energy, properly harnessed, made those luxuries available. But in large measure the increase stemmed from the fact that the ready availability of large quantities of energy created a new environment for the human species within which a new culture emerged. Machine culture rapidly replaced agriculture. What were formerly luxuries became necessities.

The Nuclear Revolution
Preconceived idea: Although mass and energy are theoretically convertible to each other, practical mechanisms for releasing energy from mass on a large scale are beyond man's reach.

Early in the twentieth century it became clear that mass and energy can be converted to each other and it was repeatedly emphasized that our largest ocean liners could cross the Atlantic powered only by a teaspoonful or so of water properly processed and harnessed. Unfortunately mechanisms for accomplishing this were unknown. In the early 1930s Leo Szilard seriously studied the possibility of generating a chain reaction in beryllium, but this turned out not to be possible.

The eventual interpretation of experiments that involved the exposure of uranium to slowly moving neutrons was in a very real sense "Copernican" and demanded entirely new approaches in our thinking. Once it was really recognized that uranium nuclei were dividing into two fragments, releasing neutrons in the process, it was but a matter of time before it was recognized that a chain reaction might be possible and that nuclear energy might be liberated on a large scale. Through the work of Fermi, Seaborg, and many others this possibility was reduced to practicality in 1942 when the first nuclear reactor was activated and again in 1945 when the first nuclear explosive was detonated.

The first nuclear power plants appeared in the 1950s. By 1970 it was recognized that in spite of numerous problems, such as the safe disposal of radioactive wastes and the design of efficient "breeder" reactors that could utilize the available uranium efficiently, nuclear power plants were probably here to stay.

While nuclear physics was blossoming in the 1930s it came to be recognized that stars are powered by thermonuclear reactions. Hans Bethe and others developed theories that explained reasonably well many aspects of energy generation within stars. Elaborations of these theories led eventually to the development of the thermonuclear bomb, first successfully exploded in 1954. Developments in recent years make it seem likely that thermonuclear reactions may eventually be harnessed for the practical generation of power.

Nuclear Fuels
Preconceived idea: Uranium and thorium, like coal, are fossil fuels of limited availability.
Uranium and thorium are not very abundant in nature, and even with the successful development of "breeder" reactors deposits of ores of uranium and thorium that are presently exploitable could not possibly satisfy the energy needs of a highly industrialized world for a very long period of time. The history of copper mining has shown, however, that if there is a real need for a metal, very lean ores can be mined economically. Only a few decades ago, for example, the copper ores that were processed contained on the average several percent copper. As these ores were consumed, leaner ones were used. Today copper is successfully being extracted from ores that contain as little as 0.5 percent of the element, and in the years ahead we will undoubtedly process ores that contain even smaller concentrations.

Uranium and thorium are far less abundant in nature than is copper. How lean can such ores be and still be useful to man? In 1955, the author and his colleague Leon Silver demonstrated that uranium and thorium can be readily extracted from ordinary granite yielding an energy profit equivalent to about 15 tons of coal per ton of granite processed. Although the need for processing such extremely dilute ore for uranium and thorium is far removed in time, the principle that emerges from this finding is in its own way revolutionary. Basically, the leanest of earth substances, ordinary rocks, are at mankind's disposal to provide a substantial proportion of the basic raw materials for the maintenance of a high level of civilization, world-wide, for a very long time.

Using the rocks of the earth's crust, a world population of some 10 billion persons could be supported for literally millions of years at a per capita level of energy consumption equal to that in the United States today. If thermonuclear power is made practical, as seems likely, additional millions of years of fuel supplies will be added to our inventory of energy resources.

Problems and Prospects

In the future, can we expect that there will be additional "quantum jumps" in our attitudes toward and in our exploitation of energy? Two developments make it appear that the answer might be negative. First, our newly found ability to release energy on a large scale suddenly through the use of thermonuclear explosives threatens to bring our entire civilization to extinction. Indeed, it is conceivable that all mankind might one day be destroyed in a nuclear holocaust. Second, we are now suffering from a number of serious environmental difficulties that have been precipitated by a combination of industrialization, urbanization, transportation, and population.

There are those who suggest that we must search for a new style of life—one that is less oriented toward affluence. They correctly point out that our present society is extremely wasteful. We could recycle materials more effectively; we could walk more and drive less; we could consume less and still live comfortably. We could control our environment. We could limit population growth.

At the same time here we are: the first creature to emerge along the tortuous path of evolution endowed with the power of conceptional thought, capable of wondering about its past and expressing concern about its future. We study the stars and wonder about the past and future of our universe. We have now harnessed power that enables us to leave the earth. We have sent men to the moon and instruments to the planets. We wonder about life elsewhere in our galaxy. Would it not be the greatest of tragedies were we at this stage of our development to go the way of previous life forms—the way of extinction?

If we manage not to let civilization die, if instead we convert it into the beautiful and effective mechanism that it can be for enabling humanity to achieve its full potential, I suspect we will experience new quantum jumps in the utilization of energy. For one thing, those areas of the world where poverty and hunger are widespread and that embrace some two-thirds of mankind have rates of energy consumption per capita more than an order of magnitude less than the per capita consumption in the rich countries. The elimination of poverty would necessitate a two- to threefold increase in energy consumption worldwide. Increasing population will necessitate yet another doubling.

Beyond these obvious factors, as man's understanding of his planet grows, he will be able to make use of its resources more effectively. One can imagine a fantastic array of engineering efforts that are based upon far better understanding of nature than we now possess which will enable mankind to achieve levels of development far removed from the status quo. As the food gatherer could not imagine the world of agriculture and the world of steam-powered industry was alien to the peasant-village farmer, perhaps we too are unable to comprehend the next age of man.

Name Index